Biodiversity Loss Assessment for Ecosystem Protection

Ashok Kumar Rathoure
Saraca Research Inc., India

IGI Global
PUBLISHER of TIMELY KNOWLEDGE

A volume in the Advances in Environmental Engineering and Green Technologies (AEEGT) Book Series

Published in the United States of America by
IGI Global
Engineering Science Reference (an imprint of IGI Global)
701 E. Chocolate Avenue
Hershey PA, USA 17033
Tel: 717-533-8845
Fax: 717-533-8661
E-mail: cust@igi-global.com
Web site: http://www.igi-global.com

Library of Congress Cataloging-in-Publication Data

Names: Rathoure, Ashok K., 1983- editor.
Title: Biodiversity loss assessment for ecosystem protection / edited by Ashok Rathoure.
Description: Hershey PA : Engineering Science Reference, 2024. | Includes bibliographical references. | Summary: "This book provides relevant theoretical frameworks and the latest empirical research findings in the biodiversity loss and conservation area. The major objective of current edited book is to highlight the challenges of biodiversity conservation and current scenario of existing biodiversity uses along losses due to anthropogenic activities"-- Provided by publisher.
Identifiers: LCCN 2024010156 (print) | LCCN 2024010157 (ebook) | ISBN 9798369333303 (hardcover) | ISBN 9798369345108 (paperback) | ISBN 9798369333310 (ebook)
Subjects: LCSH: Biodiversity conservation.
Classification: LCC QH75 .B585 2024 (print) | LCC QH75 (ebook) | DDC 333.95/16--dc23/eng/20240422
LC record available at https://lccn.loc.gov/2024010156
LC ebook record available at https://lccn.loc.gov/2024010157

This book is published in the IGI Global book series Advances in Environmental Engineering and Green Technologies (AEEGT) (ISSN: 2326-9162; eISSN: 2326-9170)

British Cataloguing in Publication Data
A Cataloguing in Publication record for this book is available from the British Library.

Advances in Environmental Engineering and Green Technologies (AEEGT) Book Series

Sang-Bing (Jason) Tsai
Zhongshan Institute, University of Electronic Science and Technology of China, China & Wuyi University, China
Ming-Lang Tseng
Lunghwa University of Science and Technology, Taiwan
Yuchi Wang
University of Electronic Science and Technology of China Zhongshan Institute, China

ISSN:2326-9162
EISSN:2326-9170

Mission

Growing awareness and an increased focus on environmental issues such as climate change, energy use, and loss of non-renewable resources have brought about a greater need for research that provides potential solutions to these problems. Research in environmental science and engineering continues to play a vital role in uncovering new opportunities for a "green" future.

The **Advances in Environmental Engineering and Green Technologies (AEEGT)** book series is a mouthpiece for research in all aspects of environmental science, earth science, and green initiatives. This series supports the ongoing research in this field through publishing books that discuss topics within environmental engineering or that deal with the interdisciplinary field of green technologies.

Coverage

- Air Quality
- Alternative Power Sources
- Green Transportation
- Electric Vehicles
- Waste Management
- Radioactive Waste Treatment
- Green Technology
- Water Supply and Treatment
- Cleantech
- Policies Involving Green Technologies and Environmental Engineering

IGI Global is currently accepting manuscripts for publication within this series. To submit a proposal for a volume in this series, please contact our Acquisition Editors at Acquisitions@igi-global.com or visit: http://www.igi-global.com/publish/.

The Advances in Environmental Engineering and Green Technologies (AEEGT) Book Series (ISSN 2326-9162) is published by IGI Global, 701 E. Chocolate Avenue, Hershey, PA 17033-1240, USA, www.igi-global.com. This series is composed of titles available for purchase individually; each title is edited to be contextually exclusive from any other title within the series. For pricing and ordering information please visit http://www.igi-global.com/book-series/advances-environmental-engineering-green-technologies/73679. Postmaster: Send all address changes to above address.

Titles in this Series

For a list of additional titles in this series, please visit: http://www.igi-global.com/book-series/advances-environmental-engineering-green-technologies/73679

Reshaping Environmental Science Through Machine Learning and IoT
Rajeev Kumar Gupta (Pandit Deendayal Energy University, India) Arti Jain (Jaypee Institute of Information Technology, India) John Wang (Montclair State University, USA) and Rajesh Kumar Pateriya (Maulana Azad National Institute of Technology, India)
Engineering Science Reference • copyright 2024 • 438pp • H/C (ISBN: 9798369323519) • US $295.00 (our price)

Advancements in Climate and Smart Environment Technology
Jamal Mabrouki (Faculty of Science, Mohammed V University in Rabat, Morocco) and Mourade Azrour (Faculty of Sciences and Techniques, Moulay Ismail University of Meknes, Errachidia, Morocco)
Engineering Science Reference • copyright 2024 • 292pp • H/C (ISBN: 9798369338070) • US $295.00 (our price)

Harnessing NanoOmics and Nanozymes for Sustainable Agriculture
Vishnu D. Rajput (Academy of Biology and Biotechnology, Southern Federal University, Rostov-on-Don, Russia) Abhishek Singh (Faculty of Biology, Yerevan State University, Armenia) Karen Ghazaryan (Faculty of Biology, Yerevan State University, Armenia) Athanasios Alexiou (Department of Science and Engineering, Novel Global Community Educational Foundation, Hebersham, Australia) and Abdel Rahman Mohammad Said Al-Tawaha (Department of Biological Sciences, Al Hussein bin Talal University, Jordan)
Engineering Science Reference • copyright 2024 • 531pp • H/C (ISBN: 9798369318904) • US $345.00 (our price)

Machine Learning and Computer Vision for Renewable Energy
Pinaki Pratim Acharjya (Haldia Institute of Technology, India) Santanu Koley (Haldia Institute of Technology, India) and Subhabrata Barman (Haldia Institute of Technology, India)
Engineering Science Reference • copyright 2024 • 333pp • H/C (ISBN: 9798369323557) • US $295.00 (our price)

Convergence Strategies for Green Computing and Sustainable Development
Vishal Jain (Sharda University, India) Murali Raman (Asia Pacific University of Technology, Malaysia) Akshat Agrawal (Amity University, Guragon, India) Meenu Hans (K.R. Mangalam University, India) and Swati Gupta (Center of Excellence, K.R. Mangalam University, India)
Engineering Science Reference • copyright 2024 • 332pp • H/C (ISBN: 9798369303382) • US $270.00 (our price)

Biosorption Processes for Heavy Metal Removal
Pinki Saini (University of Allahabad, India)
Engineering Science Reference • copyright 2024 • 371pp • H/C (ISBN: 9798369316184) • US $275.00 (our price)

IGI Global
PUBLISHER of TIMELY KNOWLEDGE

701 East Chocolate Avenue, Hershey, PA 17033, USA
Tel: 717-533-8845 x100 • Fax: 717-533-8661
E-Mail: cust@igi-global.com • www.igi-global.com

The book is dedicated to
Master Chaitanya A. Rathoure "Dev"

Editorial Advisory Board

Table of Contents

Detailed Table of Contents

This study aims to analyze some of the theoretical, conceptual, and methodological implications between the anthropology and the environment and sustainability. Departing from the assumption that anthropological and ethnographic analysis and interpretive research on organizational anthropology has an emergent and symbolic role rooted in social processes and with no a priori assumptions, with explanations and categories defined by the organizational actors, which had an impact on sustainability and the environment through the green marketing practices. The method employes is based on an analytical and descriptive issues from the theoretical and empirical literature review leading to reflective analysis. It is concluded that ethnographic and anthropological methodology is pertinent to be used in the analysis of its interactions between anthropology, the environment and sustainability as a new specialty in organizational studies. Also, the study proposes the implementation of green marketing habits and practices on sustainability, drawing from the organizational anthropology.

The chapter delves into the imperative task of detecting ecological changes to ensure the sustainability of our planet's ecosystems. It emphasizes the multidisciplinary approach required, combining field observations, remote sensing technologies, data analytics, and community engagement. Through tracking key indicators such as biodiversity, species abundance, and environmental parameters, scientists unveil subtle shifts and overarching trends in ecosystems worldwide. From forests to oceans, diverse habitats undergo scrutiny, aided by sophisticated analytical techniques that sift through vast datasets to identify patterns and predict future trajectories. Understanding the causes and consequences of ecological changes is essential for effective management and conservation, with human activities and natural factors often playing significant roles. The chapter concludes with a call to action for collective efforts towards sustainability, highlighting the role of individuals, communities, governments, and organizations in safeguarding the health and integrity of Earth's ecosystems.

Assessing ecological risk and evaluating biodiversity loss are essential components of conservation planning and management. This chapter examines methodologies and approaches used in these assessments to understand the complex interactions between human activities and ecosystems. The chapter explores various techniques, including measuring species richness and abundance, assessing habitat fragmentation and connectivity, evaluating threats to biodiversity, and modeling ecosystem dynamics under different scenarios. Additionally, the importance of stakeholder engagement and participatory approaches in conservation decision-making is highlighted. By integrating these methodologies and approaches, researchers and policymakers can gain valuable insights into ecological risk and develop effective strategies to mitigate biodiversity loss and safeguard ecosystems for future generations.

The chapter explores the various stressors impacting biodiversity and emphasizes the pivotal role of environmental impact assessment (EIA) in mitigating these threats. Biodiversity faces challenges such as habitat loss, climate change, pollution, invasive species, overexploitation, and disease outbreaks. These stressors, driven by human activities, endanger ecosystems and species worldwide. The chapter underscores how EIA promotes sustainable practices by evaluating the environmental implications of development projects. It encourages the adoption of eco-friendly technologies, conservation-oriented designs, and compliance with environmental regulations. Public participation in the EIA process ensures that local communities and stakeholders contribute to decision-making, fostering awareness about biodiversity.

Human activities, including pollution, invasive species, habitat degradation, overfishing, and climate change, are causing a decline in marine biodiversity and ecosystems. These issues disrupt food webs, habitat loss, and interfere with marine organisms' health and reproduction. Climate change also strains marine life. Conservation efforts and sustainable management techniques are needed to protect marine biodiversity, including pollution management, habitat protection, sustainable fishing methods, and climate change mitigation. Integrated strategies addressing pollution, habitat protection, sustainable fishing, and climate change mitigation are crucial for preserving marine ecosystems.

The variability of life on Earth is called biodiversity. Biodiversity takes into account all the living organisms present on Earth. Utilitarian values include the many basic needs humans obtain from biodiversity such as food, fuel, shelter, and medicine. Further, ecosystems provide crucial services such as pollination, seed dispersal, climate regulation, water purification, nutrient cycling, and control of agricultural pests. Biodiversity has cultural value to humans as well, for spiritual or religious reasons for instance. The intrinsic value of biodiversity refers to its inherent worth, which is independent of its value to anyone or anything else. Biodiversity loss refers to the decline or disappearance of biological diversity, understood as the variety of living things that inhabit the planet, its different levels of biological organization and their respective genetic variability, as well as the natural patterns present in ecosystems. The present case study is taken from mining project for extraction of coal by underground semi-mechanized mining bearing area of 659 ha.

As societies strive for progress and economic growth, the imperative to balance development with the preservation of natural ecosystems becomes increasingly pressing. The concept of harmonizing development with nature emphasizes the need to ensure that human activities are conducted in a manner that sustains biodiversity and ecological integrity. Central to achieving this balance is the implementation of comprehensive environmental impact assessments (EIAs) that incorporate considerations for sustainable biodiversity conservation. By employing innovative tools and methodologies and engaging stakeholders at all levels, EIAs can inform decision-makers about the potential impacts of development projects on biodiversity and ecosystems, thus facilitating informed decision-making and promoting sustainable development practices. Through collaborative efforts and a commitment to environmental stewardship, the harmonization of development with nature can be realized, ensuring the long-term health and resilience of our planet's ecosystems for generations to come.

This study provides a comprehensive analysis of green spaces in Punjab Province, Pakistan, from 2019 to 2023. The OpenStreetMap (OSM) data has been employed to gain insight into green spaces focusing on leisure, natural, and land use tags. These findings indicate a consistent decline in green spaces across all key categories over the five-year period. This decline raises concerns about habitat loss, biodiversity depletion, and the potential impact on ecosystem services. Recommendations for policymakers include strengthening habitat protection, promoting sustainable land use planning, investing in restoration, enhancing urban green infrastructure, and raising public awareness. Collaboration among stakeholders is essential to effectively address these challenges and safeguard biodiversity for future generations.

Biodiversity conservation is imperative for the well-being of our planet, impacting ecosystem stability, productivity, and resilience against environmental stressors. This chapter emphasizes the significance of biodiversity in maintaining ecological balance and outlines key conservation strategies. Pollution and climate change pose significant threats to biodiversity, leading to disruptions in ecosystems and loss of species. Moreover, climate change-induced shifts in species abundance and distribution necessitate holistic conservation approaches. Plant ecology plays a pivotal role in conservation by studying species interactions and ecosystem dynamics. Conservation strategies must consider ecological tolerances and promote sustainable resource use in agriculture, forestry, and ecosystem restoration. Protected areas serve as vital biodiversity reserves, offering essential ecosystem services and supporting cultural and aesthetic values. Overall, effective conservation strategies are essential for preserving biodiversity and ensuring the long-term sustainability of our planet.

Chapter 11

Asad Azeem, Department of Plant Breeding and Genetics, University of Layyah, Layyah, Paksiatan
Sami Ul-Allah, Department of Plant Breeding and Genetics, Bahauddin Zakariya University, Multan, Pakistan
Sana Khan, Department of Plant Breeding and Genetics, University of Agriculture, Faisalabad, Pakistan
Sonia Azeem, Department of Zoology, Government College University, Faisalabad, Layyah Campus, Layyah, Pakistan
Huma Shahid, Department of Environmental Science, Government College University, Faisalabad, Pakistan
Muhammad Zia Ul Haq, Department of Agronomy, University of Agriculture, Faisalabad, Pakistan
Waqas Ahmad, Department of Horticulture, University of Layyah, Layyah, Pakistan

In this chapter, the authors have analyzed the important role of different strategies such as satellite-based monitoring, remote sensing approach, as well as geographic information systems (GIS) in biodiversity conservation. Through the exploration of the multi-dimensional role of satellites, the research papers emphasize LiDAR and multispectral imaging as well as the effective integration of GIS. These techniques demonstrate how they can be used to tackle today's biodiversity problems. Spatial mapping and informed decision-making can be used to solve the complex problems created by the loss of biodiversity at the core level. And lastly, this chapter stresses the importance of all the techniques in protecting ecosystems, by using integrated approaches which combine the advantages of all the techniques and preserve biodiversity for a long time.

Chapter 12

José G. Vargas-Hernandez, Tecnológico Nacional de México, Mexico
Omar C. Vargas-González, Tecnológico Nacional de Mexico, Ciudad Guzmán, Mexico

This study aims to analyze the relationships between policymaking and strategies for socio-ecological, environmental, and green socio-technological innovation. The analysis departs from the assumption that the environment of a socio-ecological and socio-technological innovation can be fostered by policymaking and strategies aimed to provide the transformation of pure research on scientific knowledge into innovative commercial products and services. The method employed is analytic-reflective based on the theoretical and empirical literature review. It is concluded that policymaking and strategies accelerate the creation and development processes of socio-ecological, environmental, and green socio-technological innovation.

In light of the problems posed by fast global development and its effects on the environment, environmental sustainability has become a crucial concern. A possible route for sustainable development lies in the integration of environmental genomics, as standard methodologies might not be sufficient to solve these complex difficulties. Advanced genomic technologies are used in the interdisciplinary field of environmental genomics to research the biodiversity, relationships, and functions of species in their natural settings. The purpose of this study is to provide novel approaches and exploratory concepts for sustainable development using environmental genomics.

High economic development is caused to increase ecological footprint (EF). Increasing EF creates an impediment for sustainable development. Previous studies applied several methods and indicators to examine EF in different countries. However, very few studies could examine the originating factors of EF using empirical models. Hence, this chapter measures EF for SAARC countries during 2000 – 2022. It explains the comparative performance of SAARC countries in EF using a time trend analysis. Thereupon, it also examines the impact of certain indicators on EF in SAARC countries using robust empirical analysis. There seems to be significant diversity in EF due to disparity in economic development and its enhancing factors in SAARC countries. EF increases as industrialization, greenhouse gas, inflation, GDP per person employed and age dependency ratio increase. Digitalization, renewable energy consumption, technological change and life expectancy rate showed a negative impact on EF.

Foreword

The intricate tapestry of life on Earth faces an unprecedented threat: biodiversity loss. We are witnessing a decline in the number and variety of species, as well as the disruption of the vital roles they play within ecosystems. This alarming trend jeopardizes the very foundation of our planet's health and the services nature provides.

Biodiversity Loss Assessment for Ecosystem Protection arrives at a critical juncture. It tackles this crucial challenge head-on by providing a comprehensive exploration of methods and approaches for assessing biodiversity loss. By equipping readers with the necessary knowledge and tools, this book empowers them to become active participants in safeguarding our natural heritage.

This volume dives into the practical aspects of inventorying, assessing, and monitoring biodiversity. It delves into established methods for analysing species richness and community composition, as well as techniques for assessing habitat quality. The book also explores cutting-edge approaches like genetic analysis, which can reveal valuable information about population health and adaptation potential. Additionally, it examines functional traits that can indicate an ecosystem's resilience in the face of environmental change.

Biodiversity Loss Assessment for Ecosystem Protection recognizes that effective conservation goes beyond just technical expertise. It acknowledges the critical role of cost-benefit analysis in developing strategies that are both effective and economically viable.

Ultimately, the book's emphasis remains firmly on solutions. By providing a clear understanding of the drivers of biodiversity loss, the contributors pave the way for the development and implementation of targeted conservation actions. They champion in situ conservation, aiming to retain and protect biodiversity within its natural habitats.

This book is a valuable resource for a broad audience. Ecologists, conservation biologists, policymakers, environmental scientists, and anyone concerned about the future of our planet will find its insights invaluable. *Biodiversity Loss Assessment for Ecosystem Protection* not only provides a stark assessment of the challenges we face, but also offers the promise of a path forward. This path is paved with robust scientific understanding informing effective action to safeguard Earth's irreplaceable biodiversity.

Dr. Ashok K. Rathoure's contribution in this critical area is truly monumental. His work in the area of biodiversity loss assessment will serve as a milestone, guiding conservation efforts as we grapple with this major global problem.

Vishnu D Rajput
Southern Federal University, Rostov-on-Don, Russia

Preface

The book *Biodiversity Loss Assessment for Ecosystem Protection* provides a comprehensive understanding of the various factors contributing to biodiversity loss and the importance of ecosystem protection. The book offers insights into the current state of global biodiversity, the threats posed by human activities, and the potential consequences of continued loss of biodiversity.

The current book delves into the pressing challenges of biodiversity conservation and the current state of existing biodiversity. It emphasizes the importance of inventory, assessment, and monitoring as vital components of effective biodiversity conservation and management. The conservation of biodiversity is crucial, as it plays a key role in maintaining the health and resilience of ecosystems, as well as providing numerous benefits to human societies. The book offers a comprehensive understanding of the various factors contributing to biodiversity loss, the threats posed by human activities, and the potential consequences of continued loss of biodiversity. It also presents a range of assessment methods and tools that can be used by policymakers, researchers, and practitioners to monitor and protect ecosystems. The book is an essential resource for anyone interested in understanding and addressing the complex issue of biodiversity loss and its impact on ecosystems around the world.

By providing a comprehensive overview of the challenges and consequences of biodiversity loss, as well as offering practical assessment methods and tools, the book serves as a valuable guide for anyone seeking to contribute to the protection and preservation of ecosystems.

INSIDE THIS ISSUE

In this regard, the first chapter is devoted background on biodiversity conservation and its importance. This chapter delves into biodiversity, unveiling life intricate tapestry sustaining our planet. It defines biodiversity, explores genetic, species, and ecosystem complexity. Methods to measure biodiversity and challenges in comprehension are elucidated. Chapter highlights biodiversity ecological, economic, and cultural significance, from ecosystem stability to cultural identity. Human-induced threats like habitat destruction and climate change are discussed, alongside global conservation efforts, success stories, and challenges. The symbiotic relationship between biodiversity and climate change is explored, emphasizing diverse ecosystem's role in mitigating environmental challenges. Biodiversity impact on human health, medicinal value, and consequences of its loss are uncovered. Emerging trends in biodiversity research, technological innovations, and global policies shaping conservation future. It's concluding stressing collective action to preserve life intricate web for current and future generations' benefit, aiming to inspire responsibility and appreciation for biodiversity.

The second chapter is based on anthropological implications between the environment and sustainability. This study aims to analyze some of the theoretical, conceptual, and methodological implications between the anthropology and the environment and sustainability. Departing from the assumption that anthropological and ethnographic analysis and interpretive research on organizational anthropology has an emergent and symbolic role rooted in social processes and with no a priori assumptions, with explanations and categories defined by the organizational actors, which had an impact on sustainability and the environment through the green marketing practices. The method employes is based on an analytical and descriptive issues from the theoretical and empirical literature review leading to reflective analysis. It is concluded that the ethnographic and anthropological methodology is pertinent to be used in the analysis of its interactions between anthropology, the environment and sustainability as a new specialty in organizational studies. Also, the study proposes the implementation of green marketing habits and practices on the sustainability drawing from the organizational anthropology.

The third chapter delves into the imperative task of detecting ecological changes to ensure the sustainability of our planet's ecosystems. It emphasizes the multidisciplinary approach required, combining field observations, remote sensing technologies, data analytics, and community engagement. Through tracking key indicators such as biodiversity, species abundance, and environmental parameters, scientists unveil subtle shifts and overarching trends in ecosystems worldwide. From forests to oceans, diverse habitats undergo scrutiny, aided by sophisticated analytical techniques that sift through vast datasets to identify patterns and predict future trajectories. Understanding the causes and consequences of ecological changes is essential for effective management and conservation, with human activities and natural factors often playing significant roles. The chapter concludes with a call to action for collective efforts towards sustainability, highlighting the role of individuals, communities, governments, and organizations in safeguarding the health and integrity of Earth's ecosystems.

The fourth chapter, emphasizing the ecological risk and evaluating biodiversity loss are essential components of conservation planning and management. This chapter examines methodologies and approaches used in these assessments to understand the complex interactions between human activities and ecosystems. The chapter explores various techniques, including measuring species richness and abundance, assessing habitat fragmentation and connectivity, evaluating threats to biodiversity, and modeling ecosystem dynamics under different scenarios. Additionally, the importance of stakeholder engagement and participatory approaches in conservation decision-making is highlighted. By integrating these methodologies and approaches, researchers and policymakers can gain valuable insights into ecological risk and develop effective strategies to mitigate biodiversity loss and safeguard ecosystems for future generations.

The fifth chapter explores the various stressors impacting biodiversity and emphasizes the pivotal role of Environmental Impact Assessment (EIA) in mitigating these threats. Biodiversity faces challenges such as habitat loss, climate change, pollution, invasive species, overexploitation, and disease outbreaks. These stressors, driven by human activities, endanger ecosystems and species worldwide. The chapter underscores how EIA promotes sustainable practices by evaluating the environmental implications of development projects. It encourages the adoption of eco-friendly technologies, conservation-oriented designs, and compliance with environmental regulations. Public participation in the EIA process ensures that local communities and stakeholders contribute to decision-making, fostering awareness about biodiversity.

The sixth chapter signifies the marine impact assessment. Human activities, including pollution, invasive species, habitat degradation, overfishing, and climate change, are causing a decline in marine biodiversity and ecosystems. These issues disrupt food webs, habitat loss, and interfere with marine organisms' health and reproduction. Climate change also strains marine life. Conservation efforts and sustainable management techniques are needed to protect marine biodiversity, including pollution management, habitat protection, sustainable fishing methods, and climate change mitigation. Integrated strategies addressing pollution, habitat protection, sustainable fishing, and climate change mitigation are crucial for preserving marine ecosystems.

The seventh chapter emphasizes the biodiversity loss estimation on stream diversion. The variability of life on Earth is called Biodiversity. Biodiversity takes into account all the living organisms present on Earth. Utilitarian values include the many basic needs humans obtain from biodiversity such as food, fuel, shelter, and medicine. Further, ecosystems provide crucial services such as pollination, seed dispersal, climate regulation, water purification, nutrient cycling, and control of agricultural pests. Biodiversity has cultural value to humans as well, for spiritual or religious reasons for instance. The intrinsic value of biodiversity refers to its inherent worth, which is independent of its value to anyone or anything else. Biodiversity loss refers to the decline or disappearance of biological diversity, understood as the variety of living things that inhabit the planet, its different levels of biological organization and their respective genetic variability, as well as the natural patterns present in ecosystems. The case study presented in this chapter has been taken from mining project which is for extraction of coal by underground semi-mechanized mining bearing area of 659 ha.

The eighth chapter signifies the Harmonizing Development with Nature. As societies strive for progress and economic growth, the imperative to balance development with the preservation of natural ecosystems becomes increasingly pressing. The concept of harmonizing development with nature emphasizes the need to ensure that human activities are conducted in a manner that sustains biodiversity and ecological integrity. Central to achieving this balance is the implementation of comprehensive Environmental Impact Assessments (EIAs) that incorporate considerations for sustainable biodiversity conservation. By employing innovative tools and methodologies and engaging stakeholders at all levels, EIAs can inform decision-makers about the potential impacts of development projects on biodiversity and ecosystems, thus facilitating informed decision-making and promoting sustainable development practices. Through collaborative efforts and a commitment to environmental stewardship, the harmonization of development with nature can be realized, ensuring the long-term health and resilience of our planet's ecosystems for generations to come.

The ninth chapter focuses on the comprehensive analysis of green spaces in Punjab Province, Pakistan, from 2019 to 2023. The OpenStreetMap (OSM) data has been employed to gain insight into green spaces focusing on leisure, natural, and land use tags. Authors findings indicate a consistent decline in green spaces across all key categories over the five-year period. This decline raises concerns about habitat loss, biodiversity depletion, and the potential impact on ecosystem services. Recommendations for policymakers include strengthening habitat protection, promoting sustainable land use planning, investing in restoration, enhancing urban green infrastructure, and raising public awareness. Collaboration among stakeholders is essential to effectively address these challenges and safeguard biodiversity for future generations.

The tenth chapter signifies the biodiversity conservation. It is imperative for the well-being of our planet, impacting ecosystem stability, productivity, and resilience against environmental stressors. This chapter emphasizes the significance of biodiversity in maintaining ecological balance and outlines key conservation strategies. Pollution and climate change pose significant threats to biodiversity, leading to disruptions in ecosystems and loss of species. Moreover, climate change-induced shifts in species abundance and distribution necessitate holistic conservation approaches. Plant ecology plays a pivotal role in conservation by studying species interactions and ecosystem dynamics. Conservation strategies must consider ecological tolerances and promote sustainable resource use in agriculture, forestry, and ecosystem restoration. Protected areas serve as vital biodiversity reserves, offering essential ecosystem services and supporting cultural and aesthetic values. Overall, effective conservation strategies are essential for preserving biodiversity and ensuring the long-term sustainability of our planet.

The eleventh chapter describes the important role of different strategies such as satellite-based monitoring, remote sensing approach, as well as Geographic Information Systems (GIS) in biodiversity conservation. Through the exploration of the multi-dimensional role of satellites, the research papers emphasize LiDAR and multispectral imaging as well as the effective integration of GIS: these techniques demonstrate how they can be used to tackle today's biodiversity problems. Spatial mapping and informed decision-making can be used to solve the complex problems created by the loss of biodiversity at the core level. And lastly, this chapter stresses the importance of all the techniques in protecting ecosystems, by using integrated approaches which combine the advantages of all the techniques and preserve biodiversity for a long time.

The twelfth chapter emphasizes the policymaking and strategies for socio-ecological, environmental, and green socio-technological innovation. This study has the aim to analyze the relationships between policymaking and strategies for socio-ecological, environmental, and green socio-technological innovation. The analysis departs from the assumption that the environment of a socio-ecological and socio-technological innovation can be fostered by policymaking and strategies aimed to provide the transformation of pure research on scientific knowledge into innovative commercial products and services. The method employed is the analytic-reflective based on the theoretical and empirical literature review. It is concluded that policymaking and strategies accelerates the creation and development processes of socio-ecological, environmental, and green socio-technological innovation.

The thirteenth chapter clarifies the problems posed by fast global development and its effects on the environment, environmental sustainability has become a crucial concern. A possible route for sustainable development lies in the integration of environmental genomics, as standard methodologies might not be sufficient to solve these complex difficulties. Advanced genomic technologies are used in the interdisciplinary field of environmental genomics to research the biodiversity, relationships, and functions of species in their natural settings. The purpose of this study is to provide novel approaches and exploratory concepts for sustainable development using environmental genomics.

The fourteenth chapter signifies the Ecological Footprint and its Enhancing Factors in SAARC Countries. High economic development is caused to increase ecological footprint (EF). Increasing EF creates an impediment for sustainable development. Previous studies applied several methods and indicators to examine EF in different countries. However, very few studies could examine the originating factors of EF using empirical models. Hence, this chapter measures EF for SAARC countries during 2000 – 2022. It explains the comparative performance of SAARC countries in EF using a time trend analysis. Thereupon, it also examines the impact of certain indicators on EF in SAARC countries using robust empirical analysis. There seems to be significant diversity in EF due to disparity in economic development and its enhancing factors in SAARC countries. EF increases as industrialization, greenhouse gas, inflation, GDP per person employed and age dependency ratio increase. Digitalization, renewable energy consumption, technological change and life expectancy rate showed a negative impact on EF.

CONCLUSION

The compilation of chapters presents a comprehensive overview of biodiversity conservation, anthropological implications on sustainability, ecological assessments, and various strategies for mitigating environmental threats. Each chapter underscores the intricate interplay between human activities and ecological systems, emphasizing the urgency of collective action to preserve biodiversity and ensure the sustainability of our planet. Here is key conclusions include:

1. *Biodiversity Conservation:* Biodiversity is fundamental to ecosystem stability, human well-being, and cultural identity. Effective conservation strategies must address human-induced threats like habitat destruction and climate change while promoting collective responsibility and appreciation for biodiversity.
2. *Anthropological Implications:* Anthropological and ethnographic methodologies play a crucial role in understanding the interactions between human societies, the environment, and sustainability. Green marketing practices and organizational anthropology offer avenues for promoting sustainable behaviors and harmonizing development with nature.
3. *Ecological Assessments:* Detecting and understanding ecological changes are essential for sustainability. Multidisciplinary approaches, including field observations, remote sensing, and data analytics, are necessary for tracking biodiversity trends, identifying threats, and informing conservation efforts.
4. *Conservation Planning and Management:* Methodologies for assessing ecological risk and biodiversity loss are critical for conservation planning. Stakeholder engagement and participatory approaches enhance decision-making processes, leading to more effective conservation strategies.
5. *Environmental Impact Assessment (EIA):* EIA serves as a tool for mitigating threats to biodiversity by evaluating the environmental implications of development projects. Public participation in the EIA process ensures that local communities contribute to decision-making, fostering awareness and promoting sustainable practices.
6. *Marine Impact Assessment:* Human activities pose significant threats to marine biodiversity, necessitating conservation efforts and sustainable management practices to protect marine ecosystems from pollution, habitat degradation, overfishing, and climate change.
7. *Policy Making and Innovation:* Policymaking and strategies are crucial for fostering socio-ecological, environmental, and green socio-technological innovation. Integrating advanced technologies, such as environmental genomics, can provide novel approaches for sustainable development.
8. *Ecological Footprint and Enhancing Factors:* High economic development increases ecological footprint, posing challenges for sustainable development. Understanding the factors influencing ecological footprint, such as industrialization and renewable energy consumption, is essential for crafting effective policies and strategies.

In conclusion, the synthesis of research presented in the book entitled *Biodiversity Loss Assessment for Ecosystem Protection* underscores the complex challenges facing biodiversity conservation and sustainability. It emphasizes the need for holistic approaches, innovative technologies, and collaborative efforts across various sectors to safeguard the health and integrity of Earth's ecosystems for current and future generations.

Ashok K. Rathoure
Saraca Research Inc. India

Acknowledgment

This book is the result of the dedicated effort of numerous individuals. We would like to acknowledge the help of all the people involved in this project and, more specifically, to the authors and reviewers that took part in the review process. Without their support, this book would not have become a reality.

First, we would like to thank each one of the authors for their contributions. Our sincere gratitude goes to the chapter's authors who contributed their time and expertise to this book.

Second, we wish to acknowledge the valuable contributions of the reviewers regarding the improvement of quality, coherence and content presentation of chapters. Most of the authors also served as referees; we highly appreciate their double task.

Additionally, we would like to thank our Editorial Advisory Board.

Our Production Coordinator skillfully managed the production of the textbook. We wish to thank the publisher for the sustained interest shown by him during the entire work. A sincere expression of the thanks goes to the entire staff of IGI Global.

Generous participation of faculties and students to eliminate errors in the text and to eliminate errors in the text and to refine the presentation will be greatly acknowledged.

Last and not least, we beg forgiveness of all those who have been with us over the course of the years and whose names we have failed to mention.

Ashok K. Rathoure

Introduction

Our planet teems with life, a vibrant tapestry woven from an extraordinary diversity of species interacting within complex ecosystems. This biodiversity is not merely a beautiful spectacle; it is the very foundation of a healthy planet, providing essential services that sustain human well-being. From the clean air we breathe to the fertile soil that nourishes our crops, from the natural water filtration systems to the regulation of climate, biodiversity underpins our very existence.

However, this intricate web of life faces a grave threat: biodiversity loss. Driven by human activities such as habitat destruction, climate change, and overexploitation, the number and variety of species are declining at an alarming rate. This decline disrupts the delicate balance of nature, jeopardizing the vital services ecosystems provide and ultimately, the future of our planet.

Biodiversity loss is a pressing global issue that poses significant threats to the health and functioning of ecosystems, as well as the well-being of human societies. The book *Biodiversity Loss Assessment for Ecosystem Protection* aims to provide a comprehensive understanding of the various factors contributing to biodiversity loss, the threats posed by human activities, and the potential consequences of continued loss of biodiversity.

The book offers insights into the current state of global biodiversity, the challenges faced by ecosystems worldwide, and the importance of effective conservation and management strategies. It also presents a range of assessment methods and tools that can be used by policymakers, researchers, and practitioners to monitor and protect ecosystems.

This book is an essential resource for anyone interested in understanding and addressing the complex issue of biodiversity loss and its impact on ecosystems around the world. By providing a thorough examination of the factors contributing to biodiversity loss, the threats posed by human activities, and the potential consequences of continued loss, the book equips readers with the necessary knowledge and tools to contribute to the protection and preservation of ecosystems.

THE CRITICAL ROLE OF ASSESSMENT

In the face of this crisis, effective conservation strategies are paramount. But how can we effectively protect what we cannot measure? This is where biodiversity loss assessment emerges as a crucial tool. By accurately assessing the extent and rate of biodiversity loss, we gain a clear understanding of the problem. This knowledge is essential for:

- **Prioritizing Conservation Efforts:** Identifying areas with the most significant declines allows us to focus limited resources on the most critical needs.
- **Identifying Drivers of Loss:** Through assessment, we can pinpoint the root causes of biodiversity loss, enabling us to address them at the source.
- **Monitoring Progress:** Biodiversity assessment allows us to track the effectiveness of conservation actions over time, ensuring we are moving in the right direction.

A Comprehensive Approach

Biodiversity Loss Assessment for Ecosystem Protection takes a comprehensive approach to this critical subject. We delve into a range of methodologies and approaches for assessing biodiversity loss, exploring established techniques like analysing species richness and community composition, alongside cutting-edge tools like genetic analysis and functional trait studies.

This book recognizes that effective conservation goes beyond the purely technical. Cost-benefit analysis is a key element, ensuring that conservation strategies are both practical and economically viable. The ultimate goal, however, remains firmly focused on solutions.

By providing a clear understanding of the drivers of biodiversity loss, the contributors to this volume pave the way for the development and implementation of targeted conservation actions. The emphasis rests on "in situ" conservation, aiming to retain and protect irreplaceable biodiversity within its natural habitats.

This book is a valuable resource for a broad audience. Ecologists, conservation biologists, policymakers, environmental scientists, and anyone concerned about the future of our planet will find its insights invaluable. *Biodiversity Loss Assessment for Ecosystem Protection* is not just a stark assessment of the challenges we face; it also offers the promise of a path forward – a path paved with robust scientific understanding that will inform effective action to safeguard Earth's irreplaceable biodiversity.

Chapter 1
Biodiversity and Its Importance: An Introduction

Prashantkumar B. Sathvara
https://orcid.org/0000-0002-3548-5584
Nims Institute of Allied Medical Science and Technology, Nims University Rajasthan, Jaipur, India

J. Anuradha
Nims Institute of Allied Medical Science and Technology, Nims University Rajasthan, Jaipur, India

Sandeep Tripathi
Nims Institute of Allied Medical Science and Technology, Nims University Rajasthan, Jaipur, India

Sanjeevi R.
Nims Institute of Allied Medical Science and Technology, India

ABSTRACT

This chapter delves into biodiversity, unveiling life intricate tapestry sustaining our planet. It defines biodiversity, explores genetic, species, and ecosystem complexity. Methods to measure biodiversity and challenges in comprehension are elucidated. The chapter highlights biodiversity ecological, economic, and cultural significance, from ecosystem stability to cultural identity. Human-induced threats like habitat destruction and climate change are discussed, alongside global conservation efforts, success stories, and challenges. The symbiotic relationship between biodiversity and climate change is explored, emphasizing diverse ecosystems' role in mitigating environmental challenges. Biodiversity's impact on human health, medicinal value, and consequences of its loss are uncovered. Emerging trends in biodiversity research, technological innovations, and global policies shaping conservation future are discussed. The chapter concludes by stressing collective action to preserve life intricate web for current and future generations' benefit, aiming to inspire responsibility and appreciation for biodiversity.

DOI: 10.4018/979-8-3693-3330-3.ch001

INTRODUCTION

The Living Planet Report is a comprehensive assessment conducted by the World Wide Fund for Nature (WWF) that evaluates the health of the planet and the state of global biodiversity. The report uses data from various sources to track trends in wildlife populations, ecosystems, and human impacts on the environment. The finding that wildlife populations have declined by an average of 69% over the past 50 years is a stark indicator of the magnitude of biodiversity loss occurring globally. This decline isn't uniform across all regions or species, as certain areas may experience more severe declines due to specific environmental pressures or human activities (Prashantkumar B. Sathvara *et al.*, 2023). Human activities such as deforestation, urbanization, and conversion of natural habitats for agriculture or infrastructure development lead to the loss and fragmentation of ecosystems, reducing available habitat for wildlife and contributing to species decline. Rising temperatures, altered precipitation patterns, and more frequent extreme weather events disrupt ecosystems and alter the distribution and abundance of species, leading to declines in biodiversity. For instance, populations in Latin America and the Caribbean have faced an average decline of 94%, while global freshwater species have declined by 83% on average (Singh, 2012).

Biodiversity is declining rapidly due to several factors: The primary direct cause of biodiversity loss is land use change, mainly for large-scale food production, which drives an estimated 30% of biodiversity decline globally (Griffiths *et al.*, 2004). Overfishing, overhunting, and overharvesting for food, medicines, and timber drive around 20% of biodiversity loss. Climate change is the third most significant direct driver of biodiversity loss, accounting for 14% of the decline. Non-native species introduced to new environments can outcompete native species for resources, prey on them, or transmit diseases. Pollution, including that associated with widespread pesticide use and overuse of fertilizer, also contributes to biodiversity loss.

Human activities have significantly altered the natural processes of extinction and evolution, leading to a dramatic decrease in biodiversity over the past hundred years. This includes deforestation, pollution, urbanization, and overfishing, which lead to habitat destruction and degradation. There are ongoing efforts to reach a "nature-positive" target, which involves not just halting the loss of biodiversity but actively restoring the planet's living species. The goal is to halt and reverse nature loss by 2030 on a 2020 baseline and achieve full recovery by 2050. This involves increasing the health, abundance, diversity, and resilience of species, populations, and ecosystems (Bull & Stach, 2014). The IUCN is developing a technical approach to support the delivery and measurement of Nature-Positive outcomes.

The Nature of Biodiversity

Biodiversity refers to the variety of life on Earth at all its levels, from genes to ecosystems, and the ecological and evolutionary processes that sustain it. Biodiversity includes not only the world's species but also the genetic differences within species. This can refer to genetic variation within a population of a species or between different species. It also includes a variety of ecosystems such as those that occur in deserts, forests, wetlands, mountains, lakes, rivers, and agricultural landscapes. In each ecosystem, living creatures, including humans, form a community, interacting with one another and with the air, water, and soil around them. The concept of biodiversity also encompasses the idea of species richness, which refers to the number of different species in a particular area or ecosystem. Species richness is an important element of biodiversity, but biodiversity includes much more than just species richness. Understanding the nature of biodiversity is crucial because each species can contribute to the way ecosystems function and to the services they provide to humanity (Mulongoy & Chape, 2004). Biodiversity, therefore, touches every aspect of our lives – from our health to our economy, to our well-being. It's a complex and fascinating subject that is fundamental to life on Earth.

Biodiversity, short for biological diversity, refers to the variety of life forms on Earth, encompassing all levels of biological organization, from genes and species to ecosystems. It is a measure of the richness and complexity of living organisms and the ecological systems they inhabit. It includes the diversity of genes, alleles, and genetic traits present in individuals of the same species (Kevin J. Gaston & John I. Spicer, n.d.). Genetic diversity is crucial for the adaptation and resilience of populations to environmental changes, as it provides the raw material for evolution through natural selection.

Species diversity refers to the variety and abundance of different species within a given area or ecosystem. It includes the total number of species present and the relative abundance of each species. High species diversity is indicative of a healthy and balanced ecosystem, as each species plays a unique role in ecosystem functioning and stability (Macdonald, 2023).

Ecosystem diversity refers to the variety of ecosystems or habitats present in a region. It includes different types of biomes, such as forests, grasslands, wetlands, deserts, and aquatic ecosystems like rivers, lakes, and oceans. Ecosystem diversity is important for supporting a wide range of species and ecological processes, as each ecosystem provides unique resources and environmental conditions (Gaston & Spicer, 2013). Biodiversity exists at multiple spatial scales, ranging from local habitats and ecosystems to global biogeographic regions. It is influenced by various factors, including climate, topography, geology, and human activities. Biodiversity is not evenly distributed across the planet, with some regions being hotspots of species richness and endemism, while others have lower levels of diversity. Biodiversity is essential for the health and functioning of ecosystems and the services they provide to humanity. It supports ecosystem stability, resilience to environmental changes, nutrient cycling, pollination, pest control, and the provision of food, medicines, and other resources (*Global Biodiversity*, n.d.). Biodiversity also has intrinsic value, contributing to the beauty, cultural heritage, and spiritual significance of the natural world.

Biodiversity encompasses the variety of life forms and ecological systems on Earth, including genetic diversity within species, species diversity within ecosystems, and the diversity of ecosystems themselves. Protecting and conserving biodiversity is essential for sustaining life on Earth and ensuring the well-being of present and future generations.

The History of Biodiversity

The history of biodiversity is a testament to the resilience of life on Earth, and its ability to adapt and evolve in response to changing environments. However, the current rapid decline in biodiversity due to human activities is a cause for concern. Life on Earth began around 3.5 to 4 billion years ago with simple, single-celled organisms known as prokaryotes. These organisms lacked a nucleus and other complex cell structures. Over time, these organisms evolved and diversified into a wide range of forms, including the development of more complex cells known as eukaryotes, which have a nucleus and other specialized structures (Doran & Richardson, 2010).

Around 1 billion years ago, the first multicellular organisms appeared. This was a significant milestone in the history of biodiversity as it paved the way for the development of complex life forms. These organisms had cells that specialized to perform different functions, leading to the development of tissues and organs. Approximately 540 million years ago, during a period known as the Cambrian explosion, there was a rapid increase in the diversity of life forms. This period saw the emergence of most of the major groups of animals that exist today, including the first vertebrates (Purvis & Hector, 2000). Throughout Earth's history, there have been five major mass extinctions, each of which had a significant impact on biodiversity. These events led to a drastic decrease in biodiversity, followed by a slow process of recovery and diversification. The most recent of these, the Cretaceous-Tertiary extinction event, led to the extinction of the dinosaurs and many other species. In recent history, human activities have had a significant impact on biodiversity. Habitat destruction, overexploitation, pollution, and climate change are causing an unprecedented loss of biodiversity (Sheehan, 2001). The current rate of species extinction is estimated to be hundreds to thousands of times higher than the natural background rate. Today, despite the ongoing loss of species, Earth is still home to a vast array of life forms. However, the rate of species extinction is much higher than the natural background rate, indicating a significant biodiversity crisis. This is often referred to as the sixth mass extinction (Vitt & Pianka, 2005). Understanding the history of biodiversity helps us appreciate the richness and complexity of life on Earth. It also underscores the importance of conserving biodiversity for future generations.

The Spatial Distribution of Biodiversity

The spatial distribution of biodiversity is not uniform across the globe. It is influenced by a variety of factors, including environmental conditions, geographical features, and human activities. The latitudinal gradient, also known as the latitudinal diversity gradient, is a well-established pattern in ecology that describes the variation in species richness as one moves from the equator toward the poles (Wilson *et al.*, 2004). This gradient is characterized by higher species richness in tropical regions near the equator, gradually decreasing towards the poles.

One of the primary drivers of the latitudinal diversity gradient is the amount of solar energy received by different regions of the Earth. Tropical regions near the equator receive more solar energy compared to Polar Regions. This higher solar energy input leads to warmer temperatures and increased primary productivity in tropical ecosystems. The availability of abundant energy supports diverse habitats and provides a wide range of resources for species, leading to higher species richness (Sanjeevi *et al.*, 2017; Taheri *et al.*, 2012). Solar energy drives photosynthesis, the process by which plants and other photosynthetic organisms convert sunlight into energy. In tropical regions, where solar radiation is high and temperatures are warm throughout the year, primary productivity is generally higher compared to polar regions with lower solar energy input and colder temperatures. Higher primary productivity in tropical ecosystems results in more food resources available to support diverse species assemblages.

Tropical regions tend to experience more stable climates compared to temperate and polar regions, which may be subject to more extreme fluctuations in temperature and precipitation. The stable climate of tropical regions provides a favorable environment for species persistence and speciation, allowing diverse communities of plants and animals to thrive over long periods (Navarro *et al.*, 2015). Tropical regions often exhibit greater habitat heterogeneity, with diverse ecosystems such as rainforests, wetlands, mangroves, and coral reefs coexisting nearby. This diverse array of habitats provides a wide range of niches and resources that can support a greater variety of species. In contrast, temperate and polar regions may have fewer habitat types and less habitat heterogeneity, leading to lower species richness (Murguía *et al.*, 2016a).

Altitudinal gradients, also known as elevation gradients, represent another important aspect of biodiversity distribution, particularly in mountainous regions. Mountainous regions typically exhibit distinct altitudinal belts, each characterized by unique environmental conditions and ecosystems (Kusumoto *et al.*, 2020; Sathvara *et al.*, 2023). These belts often include the montane, subalpine, alpine, and nival zones, each with its own set of plant and animal communities adapted to the specific conditions of that elevation range. As altitude increases, environmental conditions become more extreme. Temperatures generally decrease with altitude, leading to colder climates at higher elevations. Additionally, oxygen levels decrease, atmospheric pressure decreases, and exposure to harsh weather conditions such as wind, snow, and ice increases. Species inhabiting mountainous regions have evolved various adaptations to cope with the challenges of high-altitude environments. These adaptations may include physiological adjustments to low oxygen levels, changes in body size and shape to minimize heat loss, specialized behaviors for foraging and reproduction, and tolerance to extreme weather conditions (Kusumoto *et al.*, 2020).

Biodiversity along elevation gradients often follows a non-linear pattern, with peak species richness occurring at mid-elevations, typically in the montane zone. This is often referred to as the mid-domain effect or the mid-elevation peak hypothesis. Species richness may decline at higher elevations due to the harsh environmental conditions and limited availability of suitable habitats and resources. Along elevation gradients, there is also a phenomenon known as species turnover or beta diversity, where species composition changes as one moves from lower to higher elevations. This turnover is influenced by factors such as changes in temperature, precipitation, habitat structure, and interspecies interactions. Mountain ecosystems are hotspots of biodiversity and provide essential ecosystem services such as water regulation, carbon storage, and habitat for a variety of species. However, they are also highly vulnerable to environmental changes, including climate change, habitat fragmentation, and land use pressures (Abbasi *et al.*, 2013; Gizachew, 2021). Conservation efforts in mountainous regions aim to protect these unique ecosystems and the species they support. Altitudinal gradients in mountainous regions play a crucial role in shaping biodiversity patterns, with distinct ecosystems and species adaptations occurring at different elevations. Understanding these gradients is essential for conservation planning and management, particularly in the face of ongoing environmental changes.

Indeed, the distribution of biodiversity in aquatic environments, such as oceans, seas, and freshwater bodies, is influenced by various factors, including depth. Light availability decreases rapidly with increasing depth in aquatic environments, as sunlight is absorbed and scattered by water molecules and particles. This reduction in light penetration limits the photosynthetic activity of plants and algae, which form the base of many aquatic food webs. As a result, primary productivity decreases with depth, leading to lower overall species richness in deeper waters (Ferrier & Guisan, 2006).

Along with decreasing light availability, other environmental conditions change with depth in aquatic ecosystems. These include variations in temperature, pressure, dissolved oxygen levels, and nutrient concentrations. These changing environmental factors can affect the distribution and abundance of species, with some organisms adapted to specific depth ranges and environmental conditions. Despite the general trend of decreasing species richness with depth, there may be pockets of increased species richness near the seabed, particularly in areas with complex habitat structures such as coral reefs, rocky outcrops, or underwater ridges. These habitats provide shelter, food, and breeding sites for a diverse array of species, including fish, invertebrates, and algae (Jetz *et al.*, 2012). Additionally, nutrient-rich sediments and upwelling currents can enhance productivity and support diverse communities near the seabed.

The type of substrate or bottom composition also influences species distribution and diversity in aquatic ecosystems. Different substrates, such as sand, mud, gravel, or rocky substrates, provide unique habitats for various organisms, and species composition may vary depending on substrate type. Water temperature gradients and nutrient availability can also influence species distribution patterns in aquatic environments (Heino, 2009). Some species may prefer warmer or cooler temperatures, while others may thrive in nutrient-rich or nutrient-poor waters. These factors, along with depth-related changes in environmental conditions, contribute to the overall diversity of aquatic communities. The factors that influence species richness and distribution in aquatic ecosystems are essential for effective conservation and management strategies. Protecting critical habitats, maintaining water quality, and mitigating human impacts such as pollution, overfishing, and habitat destruction are key steps in preserving the biodiversity of aquatic environments for future generations (Ankitkumar B Rathod *et al.*, 2022; Jo *et al.*, 2019).

Human activities have become a dominant force shaping biodiversity patterns and ecosystems worldwide. Urban areas often experience extensive habitat loss and fragmentation as natural landscapes are converted into built environments, such as cities and suburbs. This conversion of natural habitats into urban landscapes disrupts ecosystems, displaces native species, and reduces overall biodiversity. Urbanization also introduces pollution, noise, light pollution, and other stressors that can further degrade habitats and impact wildlife populations (Sponsel, 2013).

Agriculture is a major driver of biodiversity loss globally. The expansion of agricultural land for crop cultivation and livestock grazing leads to the conversion of natural habitats, such as forests, grasslands, and wetlands, into agricultural landscapes. This habitat conversion results in the loss of biodiversity, as native species are displaced or unable to survive in the altered environment (Mossman *et al.*, 2015). Agricultural practices such as monoculture farming, pesticide use, and irrigation can also degrade soil quality, water resources, and ecosystems. Logging and deforestation for timber extraction, agriculture, and infrastructure development are significant contributors to habitat loss and biodiversity decline, particularly in tropical rainforests and other forested regions. Deforestation not only destroys habitat for countless plant and animal species but also releases carbon stored in trees into the atmosphere, contributing to climate change. Logging operations can also fragment forests, disrupt ecological processes, and increase the risk of wildfires (Mishra *et al.*, 2016; UN, 1992).

Pollution from various sources, including industrial activities, transportation, agriculture, and waste disposal, poses a significant threat to biodiversity. Pollution can contaminate air, water, and soil, harming plants, animals, and ecosystems. Chemical pollutants, such as heavy metals, pesticides, and toxic substances, can accumulate in the environment and bioaccumulate in food chains, posing risks to human health and wildlife populations. Pollution can also lead to eutrophication, algal blooms, habitat degradation, and biodiversity loss in aquatic ecosystems. Overexploitation of natural resources, including overfishing, hunting, and poaching, can deplete populations of wildlife and fish to unsustainable levels, leading to declines in biodiversity and ecosystem health (Venter *et al.*, 2018). Unsustainable harvesting practices, such as illegal logging, bush meat hunting, and the trade of endangered species, threaten the survival of many species and contribute to ecosystem degradation and species extinctions.

Conservation efforts in areas impacted by human activities often focus on restoring degraded habitats, conserving remaining biodiversity hotspots, implementing sustainable land use practices, and mitigating threats such as pollution and habitat fragmentation. These efforts may involve establishing protected areas, implementing habitat restoration projects, promoting sustainable agriculture and forestry practices, regulating pollution and resource extraction, and raising awareness about the importance of biodiversity conservation. By addressing the root causes of biodiversity loss and promoting sustainable development practices, we can work towards safeguarding biodiversity for future generations (UN, 1992). The spatial distribution of species within a habitat is a crucial aspect of biodiversity assessment and conservation planning. Species distribution patterns are influenced by the suitability of habitats for different species. Each species has specific habitat requirements, including factors such as temperature, humidity, substrate type, vegetation cover, and availability of food and shelter. Suitable habitats provide the necessary resources and environmental conditions for species to survive and reproduce, leading to the establishment of stable populations (Soberón & Peterson, 2009). The availability of resources such as food, water, and nesting sites plays a significant role in shaping species distribution patterns. Species tend to concentrate in areas where resources are abundant and easily accessible. Resource-rich habitats support higher population densities and greater species diversity compared to resource-poor habitats. Competition for resources, such as food, territory, and mates, can influence species distribution patterns by shaping the spatial arrangement of individuals within a habitat. Competitive interactions between species may result in the exclusion of certain species from particular areas or the partitioning of resources among coexisting species (Ankitkumar B Rathod *et al.*, 2023; Scherer *et al.*, 2017). Understanding the dynamics of interspecific competition can help explain observed patterns of species distribution and coexistence.

Predation and herbivores can also influence species distribution patterns by affecting the behavior and distribution of prey and plant species. Predators may concentrate in areas with abundant prey populations, leading to spatial clustering of predator-prey interactions. Similarly, herbivores may preferentially feed on certain plant species or habitats, influencing plant distribution patterns and community composition. The ability of species to disperse and colonize new habitats plays a crucial role in determining their spatial distribution patterns. Species with high dispersal abilities may be more widely distributed and occur in a variety of habitats, while species with limited dispersal abilities may be restricted to specific habitats or geographic regions (Abbasi *et al.*, 2012; Murguía *et al.*, 2016b). Understanding the dispersal mechanisms of species can help predict their potential range expansions or contractions in response to environmental changes.

By studying species distribution patterns and the factors influencing them, scientists can gain valuable insights into the ecological processes shaping biodiversity. This information can be used to identify key habitats for conservation, prioritize areas for protected area designation, and develop strategies to mitigate the impacts of habitat loss, fragmentation, and environmental change. Additionally, understanding how species may respond to future environmental changes can help inform proactive conservation measures aimed at safeguarding biodiversity and ecosystem integrity (Kullberg & Moilanen, 2014). Conservation planning at the national scale is a complex and multi-faceted process that aims to identify and prioritize areas for biodiversity conservation to maximize conservation benefits while considering various ecological, social, and economic factors. Conservation planners use a variety of data and analytical tools to identify priority areas for biodiversity conservation. This may include mapping species distributions, habitat types, ecosystem services, and areas of high ecological significance (Joppa *et al.*, 2013). By assessing the distribution of biodiversity and evaluating the importance of different habitats and ecosystems, planners can identify areas with high conservation value that warrant protection.

In addition to identifying areas of high conservation value, conservation planners also assess the threats and pressures facing biodiversity within these areas. This may include human activities such as habitat destruction, pollution, overexploitation of natural resources, and invasive species. By understanding the primary threats to biodiversity, planners can prioritize conservation actions and allocate resources to address these threats effectively. Conservation planning must take into account human activities and land use practices that may impact biodiversity. This may involve analyzing land tenure, zoning regulations, development plans, and socioeconomic factors that influence land use decisions (Hominick *et al.*, 1996). By understanding the interactions between human activities and biodiversity, planners can develop strategies to minimize negative impacts and promote sustainable land management practices.

Conservation planners use spatial optimization techniques to identify the most efficient allocation of limited resources for conservation. This involves balancing competing priorities, such as maximizing species representation, minimizing threats, and considering connectivity between protected areas. By optimizing spatial priorities, planners can maximize the effectiveness of conservation efforts and achieve the greatest conservation benefits with available resources. Conservation planners apply principles such as complementarity and representativeness to guide decision-making and prioritize areas for conservation. Complementarity refers to selecting areas that maximize the representation of unique habitats, species, and ecosystems, ensuring that conservation efforts capture the full range of biodiversity. Representativeness involves selecting areas that are typical or characteristic of different ecosystems and habitats, providing a comprehensive representation of regional biodiversity (Dunn *et al.*, 2014; Sanjeevi, 2011). Conservation planning often involves engaging with stakeholders, including government agencies, local communities, indigenous peoples, non-governmental organizations, and other relevant actors. Stakeholder engagement helps ensure that conservation plans are inclusive, transparent, and aligned with the interests and priorities of local communities and stakeholders.

Conservation planning at the national scale is a collaborative and interdisciplinary process that aims to balance conservation goals with socioeconomic priorities and land use considerations. By identifying priority areas for conservation, assessing threats, and optimizing spatial priorities, conservation planners can guide effective conservation actions that protect biodiversity, preserve ecosystem services, and promote sustainable development for future generations. Understanding these factors is essential for developing effective conservation strategies and managing biodiversity in the face of ongoing threats such as climate change and habitat destruction (Mossman *et al.*, 2015). By considering the complex interactions between environmental, ecological, and human factors, conservation efforts can be targeted toward protecting and restoring ecosystems and species diversity for the benefit of present and future generations.

The Importance of Biodiversity

Biodiversity is of paramount importance for ecosystems and human societies due to its fundamental role in supporting life, maintaining ecosystem function, and providing numerous benefits to humanity. Biodiversity is essential for the stability and resilience of ecosystems. Diverse ecosystems are better able to withstand environmental disturbances, such as climate change, natural disasters, and invasive species, due to the presence of a variety of species with different ecological roles and functions (Velado-Alonso *et al.*, 2024). Biodiversity enhances ecosystem resilience by providing redundancy, adaptability, and ecosystem services that contribute to ecosystem health and functionality.

Biodiversity underpins a wide range of ecosystem services that are vital for human well-being. These ecosystem services include provisioning services (e.g., food, water, timber, medicine), regulating services (e.g., climate regulation, pollination, water purification, flood control), cultural services (e.g., recreation, spiritual and cultural values, aesthetic enjoyment), and supporting services (e.g., nutrient cycling, soil formation, photosynthesis). Biodiversity loss can compromise the delivery of these ecosystem services, affecting human health, livelihoods, and quality of life. Biodiversity represents the genetic diversity within and between species, providing a rich reservoir of genetic resources that can be utilized for agriculture, medicine, biotechnology, and other applications. Genetic diversity within species ensures the resilience and adaptability of populations to changing environmental conditions, enabling species to evolve and survive in dynamic ecosystems (Plumptre *et al.*, 2024). Biodiversity loss threatens the loss of valuable genetic resources and the potential for future innovation and discovery.

Biodiversity is essential for food security and sustainable agriculture. Diverse ecosystems support a wide variety of crops, livestock breeds, and wild food sources that contribute to human nutrition and livelihoods. Genetic diversity within crop and livestock species provides the foundation for breeding programs to develop resilient and productive varieties that can adapt to changing environmental conditions, pests, and diseases. Biodiversity loss in agriculture can lead to reduced resilience, increased vulnerability to pests and diseases, and diminished agricultural productivity. Biodiversity holds significant cultural, spiritual, and aesthetic value for human societies (Nicholson *et al.*, 2024; Ramakrishnan & Jayaraman, 2019). Many cultures and indigenous peoples have deep spiritual and cultural connections to the natural world and rely on biodiversity for traditional knowledge, cultural practices, and spiritual beliefs. Biodiversity also contributes to the aesthetic beauty of landscapes, providing opportunities for recreation, tourism, and cultural enrichment.

Biodiversity plays a crucial role in mitigating and adapting to climate change. Forests, wetlands, and other ecosystems act as carbon sinks, sequestering carbon dioxide from the atmosphere and helping to regulate the global climate. Biodiversity ecosystems are more resilient to climate change impacts, such as extreme weather events, sea-level rise, and habitat loss, and can provide natural solutions for climate change adaptation, such as coastal protection, carbon storage, and water regulation. Biodiversity is indispensable for the functioning of ecosystems, the provision of ecosystem services, and the well-being of human societies (Cerri *et al.*, 2024). Recognizing the importance of biodiversity and taking proactive measures to conserve and sustainably manage biodiversity is essential for ensuring the resilience and sustainability of ecosystems and the prosperity of future generations.

The Human Impact on Biodiversity

Human activities have a profound impact on biodiversity, leading to widespread biodiversity loss and ecosystem degradation worldwide. One of the primary drivers of biodiversity loss is habitat destruction and fragmentation caused by human activities such as deforestation, urbanization, agriculture, and infrastructure development. Large-scale conversion of natural habitats into agricultural land, urban areas, roads, and industrial sites leads to the loss of habitat for countless plant and animal species. Habitat fragmentation further exacerbates the problem by dividing remaining habitats into smaller, isolated

Table 1. Importance of biodiversity

Importance of Biodiversity	Description	Reference
Ecosystem Stability	Biodiversity contributes to the resilience and stability of ecosystems, making them better able to withstand environmental changes, such as climate change and disease outbreaks.	(Smith *et al.*, 1999)
Economic Value	Biodiversity provides numerous economic benefits, including food, medicine, raw materials, and ecotourism opportunities.	(Maffi, 2005)
Ecological Services	Biodiversity supports essential ecological services such as pollination, nutrient cycling, soil formation, water purification, and climate regulation.	(Tetior, 2015)
Genetic Resources	Biodiversity encompasses a vast array of genetic resources, which are crucial for breeding programs, agriculture, and the development of new medicines and technologies.	(Takov & Djia, 2020)
Cultural Importance	Biodiversity holds cultural significance, inspiring art, literature, spirituality, and traditional knowledge systems.	(Sandemose, 2016)
Ethical Considerations	Preserving biodiversity is an ethical imperative, as all species have intrinsic value and the right to exist.	(Palese, 2013)

patches, which can disrupt ecological processes, increase edge effects, and reduce habitat quality and connectivity (Xue *et al.*, 2024).

Overexploitation of natural resources, including overfishing, hunting, poaching, and logging, threatens the survival of many species and ecosystems. Unsustainable harvesting practices deplete populations of wildlife, fish, and timber to unsustainable levels, leading to declines in biodiversity, species extinctions, and disruption of food webs and ecosystem functioning. Illegal wildlife trade and poaching also contribute to the decline of many endangered species, including iconic megafauna such as elephants, rhinos, and tigers. Pollution from various sources, including industrial activities, agriculture, transportation, and waste disposal, poses a significant threat to biodiversity and ecosystem health. Chemical pollutants such as heavy metals, pesticides, fertilizers, and plastics can contaminate air, water, and soil, harming plants, animals, and microorganisms. Pollution can disrupt ecological processes, degrade habitats, impair reproductive success, and contribute to species extinctions (Robinson *et al.*, 2024). Pollution in aquatic ecosystems, such as marine plastic pollution and oil spills, threatens marine biodiversity and ecosystem health.

Climate change is a major driver of biodiversity loss, exacerbating existing threats and impacting ecosystems and species worldwide. Rising temperatures, changing precipitation patterns, extreme weather events, and sea-level rise are altering habitats, disrupting ecological processes, and causing shifts in species distributions and phenology. Climate change threatens the survival of many species, particularly those with specialized habitat requirements or limited dispersal abilities, and can lead to widespread extinctions and ecosystem transformations. The introduction of invasive alien species, facilitated by human activities such as international trade, travel, and agriculture, poses a significant threat to native biodiversity. Invasive species can outcompete native species, prey on native fauna and flora, introduce diseases, alter habitat structure, and disrupt ecosystem functioning (Mondière *et al.*, 2024). Invasive species can cause declines or extinctions of native species, degrade ecosystems, and lead to loss of biodiversity and ecosystem services.

Biodiversity loss has far-reaching implications for ecosystem functioning, human well-being, and the stability of the planet. Loss of biodiversity can lead to reduced ecosystem resilience, loss of ecosystem services, diminished food security, increased vulnerability to natural disasters, and loss of cultural and aesthetic values. Biodiversity loss also undermines the ability of ecosystems to provide essential services such as climate regulation, water purification, pollination, and nutrient cycling, with profound implications for human health, livelihoods, and quality of life. Human activities are driving widespread biodiversity loss and ecosystem degradation, with far-reaching consequences for ecosystems, species, and human societies (Kosnik *et al.*, 2024). Addressing the root causes of biodiversity loss, promoting

Table 2. Human impact on biodiversity

Impact	Description
Decline in Wildlife Populations	Wildlife populations have declined by an average of 69% in the past 50 years.
Overconsumption of Natural Resources	Humanity's consumption of the Earth is fast outpacing the Earth's biocapacity.
Exploitation and Land Degradation	Exploitation of the land's resources, particularly for energy and food, negatively impacts biodiversity.
Deforestation	Deforestation is another major human impact on the environment.
Threatened Species	It is estimated that one million animal and plant species are threatened with extinction.

sustainable land use and resource management practices, conserving critical habitats, and mitigating climate change is essential for safeguarding biodiversity and ensuring the long-term health and sustainability of ecosystems and the planet.

The Maintenance of Biodiversity

Maintaining biodiversity requires a concerted effort and the implementation of various strategies and measures aimed at conserving species, habitats, and ecosystems while promoting sustainable practices. Establishing and effectively managing protected areas, such as national parks, wildlife reserves, and marine protected areas, is critical for conserving biodiversity. Protected areas provide refuge for threatened species, preserve intact ecosystems, and safeguard important habitats from habitat destruction, overexploitation, and other threats. Proper management of protected areas involves monitoring biodiversity, enforcing regulations, and engaging local communities in conservation efforts. Restoring degraded habitats and ecosystems is essential for reversing biodiversity loss and enhancing ecosystem resilience (Zoffoli *et al.*, 2024). Habitat restoration projects involve reforestation, wetland restoration, coral reef rehabilitation, and other measures to restore natural habitats and improve ecosystem function. Restoring habitat connectivity and creating green corridors can help facilitate species movement and dispersal, supporting biodiversity conservation in fragmented landscapes. Promoting sustainable land use practices, such as sustainable agriculture, forestry, and urban planning, is crucial for maintaining biodiversity while meeting human needs. Sustainable agriculture practices include agroforestry, organic farming, integrated pest management, and crop diversification, which minimize the use of chemical inputs, conserve soil and water resources, and protect biodiversity. Sustainable forestry practices involve selective logging, forest certification, and forest restoration to ensure the long-term health and productivity of forests while conserving biodiversity.

Implementing measures to protect and manage endangered species is essential for preventing species extinctions and conserving biodiversity. This may include habitat protection, captive breeding programs, reintroduction efforts, and measures to combat poaching, illegal wildlife trade, and habitat destruction. Conservation efforts should also address the underlying threats to species survival, such as habitat loss, pollution, and climate change. Engaging local communities and indigenous peoples in conservation efforts is essential for achieving sustainable biodiversity conservation. Community-based conservation approaches involve empowering local communities to participate in decision-making, natural resource management, and conservation activities (Ribeiro-Souza *et al.*, 2024). This may include providing incentives for conservation, promoting traditional knowledge and practices, and supporting sustainable livelihoods that are compatible with biodiversity conservation.

Raising awareness about the importance of biodiversity conservation and the threats facing ecosystems and species is essential for fostering public support and engagement in conservation efforts. Education programs, outreach initiatives, and public campaigns can help increase awareness about biodiversity, promote sustainable lifestyles, and mobilize action for conservation. Promoting environmental education in schools, universities, and communities can empower individuals to become stewards of biodiversity and advocates for sustainable development. Collaboration and cooperation at the international level are essential for addressing global challenges and achieving biodiversity conservation goals. International agreements and conventions, such as the Convention on Biological Diversity (CBD), provide frameworks for countries to work together to conserve biodiversity, share knowledge and resources, and implement measures to address common threats (Pesce *et al.*, 2024). International cooperation also involves supporting developing countries in their efforts to conserve biodiversity and build capacity for sustainable development.

Maintaining biodiversity requires a holistic approach that integrates conservation, sustainable development, and community engagement (Otake *et al.*, 2024). By implementing strategies such as protected area management, habitat restoration, sustainable land use practices, community-based conservation, education, and international cooperation, we can work towards conserving biodiversity for future generations while promoting human well-being and sustainable development.

CONCLUSION

This chapter offers a comprehensive exploration of the multifaceted concept of biodiversity. It delves into various aspects of biodiversity, ranging from its fundamental nature to its historical origins, spatial distribution, importance for ecosystems and societies, human impacts, and strategies for conservation and maintenance. The concept of biodiversity, elucidates its diverse forms and the intricate web of life it encompasses. It then traces the historical evolution of biodiversity, shedding light on its origins and the changes it has undergone over time. Exploring the spatial distribution of biodiversity, the book elucidates why certain regions boast higher biodiversity levels than others, unraveling the geographical patterns that underpin biodiversity hotspots and their significance for global conservation efforts. Highlighting the critical importance of biodiversity, the book underscores its indispensable role in sustaining ecosystems and supporting human societies. It emphasizes the myriad benefits that biodiversity provides, from ecosystem services to cultural and aesthetic values, underscoring the urgency of its conservation. Examining the human impact on biodiversity, the book unveils the detrimental effects of human activities on the planet's rich tapestry of life, from habitat destruction to overexploitation and pollution. It underscores the far-reaching implications of biodiversity loss for ecosystems, species, and human well-being. Lastly, this chapter delves into strategies and measures for conserving and maintaining biodiversity, emphasizing the imperative of adopting sustainable practices to safeguard the planet's biological heritage for future generations.

REFERENCES

Abbasi, T., Sanjeevi, R., Anuradha, J., & Abbasi, S. A. (2013). Impact of Al 3+ on sludge granulation in UASB reactor. *Indian Journal of Biotechnology*, *12*(2), 254–259.

Abbasi, T., Sanjeevi, R., Makhija, M., & Abbasi, S. A. (2012). Role of Vitamins B-3 and C in the Fashioning of Granules in UASB Reactor Sludge. *Applied Biochemistry and Biotechnology*, *167*(2), 348–357. doi:10.1007/s12010-012-9691-y PMID:22549583

Ankitkumar, B. Rathod, Prashantkumar B Sathvara, Aviral Tripathi, J. Anuradha, Sandeep Tripathi, & R. Sanjeevi. (2022). Vegetational Cartography Analysis Utilizing Multi-temporal Ndvi Data Series: A Case Study from Rajkot District (Gujarat), India. *Tianjin Daxue Xuebao, 55*(04), 490–497. https://doi.org/ doi:10.17605/OSF.IO/UGJYM

Ankitkumar, B Rathod, Anuradha, J., Prashantkumar B Sathvara, Tripathi, S., & Sanjeevi, R. (2023). Vegetational Change Detection Using Machine Learning in GIS Technique: A Case Study from Jamnagar (Gujarat). *Journal of Data Acquisition and Processing*, *38*(1), 1046–1061. doi:10.5281/zenodo.7700655

Bull, A. T., & Stach, J. E. M. (2014). An Overview of Biodiversity-Estimating the Scale. In A. T. Bull (Ed.), *Microbial Diversity and Bioprospecting* (pp. 15–28). ASM Press. doi:10.1128/9781555817770.ch2

Cerri, F., Louis, Y. D., Fallati, L., Siena, F., Mazumdar, A., Nicolai, R., Zitouni, M. S., Adam, A. S., Mohamed, S., Lavorano, S., & Galli, P. (2024). Mangroves of the Maldives: A review of their distribution, diversity, ecological importance and biodiversity of associated flora and fauna. *Aquatic Sciences*, *86*(2), 44. doi:10.1007/s00027-024-01061-2

Doran, N. E., & Richardson, A. M. (2010). History of biodiversity conservation, protected areas and the conservation movement. *History and Philosophy of Science and Technology*, *2*, 55–88.

Dunn, D. C., Ardron, J., Bax, N., Bernal, P., Cleary, J., Cresswell, I., Donnelly, B., Dunstan, P., Gjerde, K., Johnson, D., Kaschner, K., Lascelles, B., Rice, J., von Nordheim, H., Wood, L., & Halpin, P. N. (2014). The convention on biological diversity's ecologically or biologically significant areas: Origins, development, and current status. *Marine Policy*, *49*, 137–145. doi:10.1016/j.marpol.2013.12.002

Ferrier, S., & Guisan, A. (2006). Spatial modelling of biodiversity at the community level. *Journal of Applied Ecology*, *43*(3), 393–404. doi:10.1111/j.1365-2664.2006.01149.x

Gaston, K. J., & Spicer, J. I. (2013). *Biodiversity: An introduction*. John Wiley & Sons. https://books.google.com/books?hl=en&lr=&id=0Bjp2o5CVnQC&oi=fnd&pg=PT6&dq=Biodiversity:+An+overview&ots=Y2X6YSmcEx&sig=KHEre9jRGafSgjzx7GfRWb1tK-w

Gizachew, G. T. (2021). Spatial-temporal and factors influencing the distribution of biodiversity: A Review. *Scientific Reports in Life Sciences*, *2*(4), 1–19.

Global Biodiversity. (n.d.). National Geographic. https://education.nationalgeographic.org/resource/global-biodiversity

Griffiths, H. I., Kryštufek, B., & Reed, J. M. (Eds.). (2004). *Balkan Biodiversity*. Springer Netherlands., doi:10.1007/978-1-4020-2854-0

Heino, J. (2009). Biodiversity of aquatic insects: Spatial gradients and environmental correlates of assemblage-level measures at large scales. *Freshwater Reviews*, *2*(1), 1–29. doi:10.1608/FRJ-2.1.1

Hominick, W. M., Reid, A. P., Bohan, D. A., & Briscoe, B. R. (1996). Entomopathogenic Nematodes: Biodiversity, Geographical Distribution and the Convention on Biological Diversity. *Biocontrol Science and Technology*, *6*(3), 317–332. doi:10.1080/09583159631307

Jetz, W., McPherson, J. M., & Guralnick, R. P. (2012). Integrating biodiversity distribution knowledge: Toward a global map of life. *Trends in Ecology & Evolution*, *27*(3), 151–159. doi:10.1016/j.tree.2011.09.007 PMID:22019413

Jo, H., Kim, D.-K., Park, K., & Kwak, I.-S. (2019). Discrimination of spatial distribution of aquatic organisms in a coastal ecosystem using eDNA. *Applied Sciences (Basel, Switzerland)*, *9*(17), 3450. doi:10.3390/app9173450

Joppa, L. N., Visconti, P., Jenkins, C. N., & Pimm, S. L. (2013). Achieving the Convention on Biological Diversity's Goals for Plant Conservation. *Science*, *341*(6150), 1100–1103. doi:10.1126/science.1241706 PMID:24009391

Kevin, J. Gaston & John I. Spicer. (n.d.). *Biodiversity: An Introduction, 2nd Edition | Wiley* (2nd ed.). Wiley. https://www.wiley.com/en-in/Biodiversity:+An+Introduction,+2nd+Edition-p-9781405118576

Kosnik, M. B., Schuwirth, N., & Rico, A. (2024). Harnessing Computational Methods to Characterize Chemical Impacts on Biodiversity. *Environmental Science & Technology Letters*, *11*(3), 185–194. doi:10.1021/acs.estlett.3c00865

Kullberg, P., & Moilanen, A. (2014). How do recent spatial biodiversity analyses support the convention on biological diversity in the expansion of the global conservation area network? *Natureza & Conservação*, *12*(1), 3–10. doi:10.4322/natcon.2014.002

Kusumoto, B., Costello, M. J., Kubota, Y., Shiono, T., Wei, C., Yasuhara, M., & Chao, A. (2020). Global distribution of coral diversity: Biodiversity knowledge gradients related to spatial resolution. *Ecological Research*, *35*(2), 315–326. doi:10.1111/1440-1703.12096

Macdonald, D. W. (2023). *Biodiversity Conservation: A Very Short Introduction*. Oxford University Press., doi:10.1093/actrade/9780199592272.001.0001

Maffi, L. (2005). LINGUISTIC, CULTURAL, AND BIOLOGICAL DIVERSITY. *Annual Review of Anthropology, 34*, 599–617. doi:10.1146/annurev.anthro.34.081804.120437

Mishra, S., Anuradha, J., Tripathi, S., & Kumar, S. (2016). In vitro antioxidant and antimicrobial efficacy of Triphala constituents: Emblica officinalis, Terminalia belerica and Terminalia chebula. *Journal of Pharmacognosy and Phytochemistry*, *5*(6), 273–277.

Mondière, A., Corson, M. S., Auberger, J., Durant, D., Foray, S., Glinec, J.-F., Green, P., Novak, S., Signoret, F., & van der Werf, H. M. (2024). Trade-offs between higher productivity and lower environmental impacts for biodiversity-friendly and conventional cattle-oriented systems. *Agricultural Systems*, *213*, 103798. doi:10.1016/j.agsy.2023.103798

Mossman, H. L., Panter, C. J., & Dolman, P. M. (2015). Modelling biodiversity distribution in agricultural landscapes to support ecological network planning. *Landscape and Urban Planning*, *141*, 59–67. doi:10.1016/j.landurbplan.2015.04.010

Mulongoy, K., & Chape, S. (2004). *Protected areas and biodiversity: An overview of key issues*. Policy Commons. https://policycommons.net/artifacts/1376714/protected-areas-and-biodiversity/1990978/

Murguía, D. I., Bringezu, S., & Schaldach, R. (2016a). Global direct pressures on biodiversity by large-scale metal mining: Spatial distribution and implications for conservation. *Journal of Environmental Management*, *180*, 409–420. doi:10.1016/j.jenvman.2016.05.040 PMID:27262340

Murguía, D. I., Bringezu, S., & Schaldach, R. (2016b). Global direct pressures on biodiversity by large-scale metal mining: Spatial distribution and implications for conservation. *Journal of Environmental Management*, *180*, 409–420. doi:10.1016/j.jenvman.2016.05.040 PMID:27262340

Navarro, J., Coll, M., Cardador, L., Fernández, Á. M., & Bellido, J. M. (2015). The relative roles of the environment, human activities and spatial factors in the spatial distribution of marine biodiversity in the Western Mediterranean Sea. *Progress in Oceanography*, *131*, 126–137. doi:10.1016/j.pocean.2014.12.004

Nicholson, E., Andrade, A., Brooks, T. M., Driver, A., Ferrer-Paris, J. R., Grantham, H., Gudka, M., Keith, D. A., Kontula, T., Lindgaard, A., Londono-Murcia, M. C., Murray, N., Raunio, A., Rowland, J. A., Sievers, M., Skowno, A. L., Stevenson, S. L., Valderrabano, M., Vernon, C. M., & Obura, D. (2024). Roles of the Red List of Ecosystems in the Kunming-Montreal Global Biodiversity Framework. *Nature Ecology & Evolution*, *8*(4), 1–8. doi:10.1038/s41559-023-02320-5 PMID:38332025

Otake, Y., Yamamichi, M., Hirata, Y., Odagiri, H., & Yoshida, T. (2024). Different photoperiodic responses in diapause induction can promote the maintenance of genetic diversity via the storage effect in *Daphnia pulex*. *Proceedings of the Royal Society B: Biological Sciences, 291*(2016), 20231860. 10.1098/rspb.2023.1860

Palese, E. (2013). Ethics without Morality, Morality without Ethics—Politics, Identity, Responsibility in Our Contemporary World. *Open Journal of Philosophy*, *03*(03), 366–371. doi:10.4236/ojpp.2013.33055

Pesce, S., Mamy, L., Sanchez, W., Artigas, J., Bérard, A., Betoulle, S., Chaumot, A., Coutellec, M.-A., Crouzet, O., Faburé, J., Hedde, M., Leboulanger, C., Margoum, C., Martin-Laurent, F., Morin, S., Mougin, C., Munaron, D., Nélieu, S., Pelosi, C., & Leenhardt, S. (2024). The use of copper as plant protection product contributes to environmental contamination and resulting impacts on terrestrial and aquatic biodiversity and ecosystem functions. *Environmental Science and Pollution Research International*. Advance online publication. doi:10.1007/s11356-024-32145-z PMID:38324154

Plumptre, A. J., Baisero, D., Brooks, T. M., Buchanan, G., Butchart, S. H., Bowser, A., Boyd, C., Carneiro, A. P., Davies, T., Elliot, W., Foster, M., Langhammer, P. F., Marnewick, D., Matiku, P., McCreless, E., Raudsepp-Hearne, C., Tordoff, A. W., Azpiroz, A. B., Trisurat, Y., & Upgren, A. (2024). Targeting site conservation to increase the effectiveness of new global biodiversity targets. *One Earth*, *7*(1), 11–17. doi:10.1016/j.oneear.2023.12.007

Prashantkumar, B. Sathvara, J. Anuradha, Sandeep Tripathi, & R. Sanjeevi. (2023). Impact of climate change and its importance on human performance. In Insights on Impact of Climate Change and Adaptation of Biodiversity (1st ed., pp. 1–9). KD Publication.

Purvis, A., & Hector, A. (2000). Getting the measure of biodiversity. *Nature*, *405*(6783), 212–219. doi:10.1038/35012221 PMID:10821281

Ramakrishnan, S., & Jayaraman, A. (2019). Global Warming and Pesticides in Water Bodies. In *Handbook of Research on the Adverse Effects of Pesticide Pollution in Aquatic Ecosystems* (pp. 421–436). IGI Global. https://www.igi-global.com/chapter/global-warming-and-pesticides-in-water-bodies/213519 doi:10.4018/978-1-5225-6111-8.ch023

Ribeiro-Souza, P., Haji, J., Oshima, J., Lima, F., Lima-Silva, B., Pires, J., Ribeiro, M., & Graipel, M. (2024). Under pressure: Suitable areas for neotropical cats within an under protected biodiversity hotspot. *Remote Sensing Applications: Society and Environment*, *34*, 101155. doi:10.1016/j.rsase.2024.101155

Robinson, J. M., Breed, A. C., Camargo, A., Redvers, N., & Breed, M. F. (2024). Biodiversity and human health: A scoping review and examples of underrepresented linkages. *Environmental Research, 246*, 118115. doi:10.1016/j.envres.2024.118115 PMID:38199470

Sandemose, J. (2016). The Motley World of "International Values": Modes of Production on the World Market. *Theoretical Economics Letters, 06*(03), 507–533. doi:10.4236/tel.2016.63058

Sanjeevi, R. (2011). *Studies on the treatment of low-strength wastewaters with upflow anaerobic sludge blanket (UASB) reactor: With emphasis on granulation studies*. Centre for Pollution Control and Environmental Engineering.

Sanjeevi, R., Haruna, M., Tripathi, S., Singh, B., & Jayaraman, A. (2017). Impacts of Global Carbon Foot Print on the Marine Environment. *International Journal of Engineering Research & Technology (Ahmedabad), 5*, 51–54.

Sathvara, P. B., Anuradha, J., Sanjeevi, R., Tripathi, S., & Rathod, A. B. (2023). Spatial Analysis of Carbon Sequestration Mapping Using Remote Sensing and Satellite Image Processing. In *Multimodal Biometric and Machine Learning Technologies* (pp. 71–83). John Wiley & Sons, Ltd. doi:10.1002/9781119785491.ch4

Scherer, L., Curran, M., & Alvarez, M. (2017). Expanding Kenya's protected areas under the Convention on Biological Diversity to maximize coverage of plant diversity. *Conservation Biology, 31*(2), 302–310. doi:10.1111/cobi.12792 PMID:27346759

Sheehan, P. M. (2001). History of marine biodiversity. *Geological Journal, 36*(3–4), 231–249. doi:10.1002/gj.890

Singh, J. S. (2012). Biodiversity: An Overview. *Proceedings of the National Academy of Sciences, India Section B: Biological Sciences*. 10.1007/s40011-012-0112-3

Smith, V. H., Tilman, G. D., & Nekola, J. C. (1999). Eutrophication: Impacts of excess nutrient inputs on freshwater, marine, and terrestrial ecosystems. *Environmental Pollution (Barking, Essex: 1987), 100*(1–3), 179–196. doi:10.1016/S0269-7491(99)00091-3

Soberón, J., & Peterson, A. T. (2009). Monitoring biodiversity loss with primary species-occurrence data: Toward national-level indicators for the 2010 target of the convention on biological diversity. *Ambio, 38*(1), 29–34. doi:10.1579/0044-7447-38.1.29 PMID:19260344

Sponsel, L. E. (2013). Human impact on biodiversity, overview. Encyclopedia of Biodiversity, 4, 137–152.

Taheri, M., Foshtomi, M. Y., Noranian, M., & Mira, S. S. (2012). Spatial distribution and biodiversity of macrofauna in the southeast of the Caspian Sea, Gorgan Bay in relation to environmental conditions. *Ocean Science Journal, 47*(2), 113–122. doi:10.1007/s12601-012-0012-8

Takov, P., & Djia, V. (2020). Wittgenstein's Ethical Individualism as a Foundation for Environmental Ethics. *Open Journal of Philosophy, 10*(04), 427–442. doi:10.4236/ojpp.2020.104030

Tetior, A. (2015). New Philosophy of Plural World, New Ethics of Empathy. *OAlib, 02*(01), 1–6. doi:10.4236/oalib.1101193

UN. I. (1992). Convention on biological diversity. *Treaty Collection.* https://www.ipcc.ch/site/assets/uploads/2001/09/doc4f.pdf

Velado-Alonso, E., Kleijn, D., & Bartomeus, I. (2024). Reassessing science communication for effective farmland biodiversity conservation. *Trends in Ecology & Evolution.* https://www.cell.com/trends/ecology-evolution/abstract/S0169-5347(24)00032-6

Venter, O., Magrach, A., Outram, N., Klein, C. J., Possingham, H. P., Di Marco, M., & Watson, J. E. M. (2018). Bias in protected-area location and its effects on long-term aspirations of biodiversity conventions. *Conservation Biology*, *32*(1), 127–134. doi:10.1111/cobi.12970 PMID:28639356

Vitt, L. J., & Pianka, E. R. (2005). Deep history impacts present-day ecology and biodiversity. *Proceedings of the National Academy of Sciences of the United States of America*, *102*(22), 7877–7881. doi:10.1073/pnas.0501104102 PMID:15867150

Wilson, R. J., Thomas, C. D., Fox, R., Roy, D. B., & Kunin, W. E. (2004). Spatial patterns in species distributions reveal biodiversity change. *Nature*, *432*(7015), 393–396. doi:10.1038/nature03031 PMID:15549106

Xue, P., Minasny, B., Wadoux, A. M. J.-C., Dobarco, M. R., McBratney, A., Bissett, A., & De Caritat, P. (2024). Drivers and human impacts on topsoil bacterial and fungal community biogeography across Australia. *Global Change Biology*, *30*(3), e17216. doi:10.1111/gcb.17216 PMID:38429628

Zoffoli, M. L., Topouzelis, K., Traganos, D., Change, G., Traganos, D., Lee, C. B., Blume, A., Poursanidis, D., Čižmek, H., & Deter, J. (2024). The seagrass Posidonia oceanica is the main habitat-forming species of the coastal Mediterranean, providing millennial-scale ecosystem services including habitat provisioning, biodiversity maintenance, food security, coastal protection, and carbon sequestration. *Blue Carbon: Beyond the Inventory*, 186.

Chapter 2
Anthropological Implications Between the Environment and Sustainability

José G. Vargas-Hernandez
Tecnológico Nacional de México, Mexico

Elizabeth García-Domínguez
Tecnológico Superior de Jalisco, Zapopan, Mexico

Omar C. Vargas-González
https://orcid.org/0000-0002-6089-956X
Tecnológico Nacional de Mexico,Ciudad Guzmán, Mexico

ABSTRACT

This study aims to analyze some of the theoretical, conceptual, and methodological implications between the anthropology and the environment and sustainability. Departing from the assumption that anthropological and ethnographic analysis and interpretive research on organizational anthropology has an emergent and symbolic role rooted in social processes and with no a priori assumptions, with explanations and categories defined by the organizational actors, which had an impact on sustainability and the environment through the green marketing practices. The method employes is based on an analytical and descriptive issues from the theoretical and empirical literature review leading to reflective analysis. It is concluded that ethnographic and anthropological methodology is pertinent to be used in the analysis of its interactions between anthropology, the environment and sustainability as a new specialty in organizational studies. Also, the study proposes the implementation of green marketing habits and practices on sustainability, drawing from the organizational anthropology.

INTRODUCTION

The earth has reached the limits of its sustainability. Anthropological ecology has theoretical and practical implications for organizational sustainability through the integration of social and environmental dimen-

DOI: 10.4018/979-8-3693-3330-3.ch002

sions. Sustainability ecology is a resource available to manage environmental changes in areas such as market forces, technology, geographical locations, natural resources, etc. The ecology of organizational sustainability shaped by environmental factors and industrial organizational structures.

Anthropology has made relevant contributions to the sustainable empirical knowledge available to humanity (Brightman & Lewis, 2017). Different theoretical perspectives of anthropology, such as essentialism and romanticism, need dialogue and learning from indigenous groups living in ecological sustainability, even though most people live in urban areas. Anthropology has a relationship with sustainability in the steady-state equilibrium models of anthropology considered by sustainability. The sustainable system can reproduce itself over a period without undermining its own conditions and absorb its incremental changes without collapsing. The Cretaceous is a short-term sustainable biosphere that ends abruptly and catastrophically. The contributions of scientific anthropology can lead to changes in sustainability that preserve the Anthropocene.

Theorizing and research into the perspectives of the humanities since the Anthropocene are developing rapidly and challenging indigenous cosmologies of environmental sustainability (Sörlin, 2007; Vetlesen, 2020; Rivers, 1922). Anthropocene anthropology considers an interconnected society and a culture limitless to sustainability that are conceptualized as ecological devastation and climate change. Sociocultural anthropology studies sustainable societies that are supposed to be able to reproduce changes indefinitely and incrementally. An anthropocentric framing of the environment wrapped in the government's decontamination policy.

The anthropocentric bias of policies opens the way to profit-oriented green capitalist policies based on the growth of the market economy by acknowledging environmental damage (Argyrou, 2005; Fox and Alldred, 2020). An economically and socially sustainable society in capitalism is committed to growth, expansion, and technological progress in contradiction to the ecological sustainability pointed out by anthropology based on the Gaia hypothesis of the destructive fossil economy (Latour, 2018; Hornborg, 2019). Sources of historical ecology analysis used in human interactions and integrations with socio-ecosystem dynamics to foster sustainable environmental relationships. Human sustainability is threatened by global nature. Human activities shape the environment before the advent of agriculture, resulting in anthropogenic landscapes.

Historical ecology fosters sustainable relationships and interactions between human communities and ecosystems to maintain the environment with the goal of human well-being focused on international policy to improve socio-ecosystems and develop sustainable economies (European Commission, 2017; Food and Agriculture Organization of the United Nations, 2017; United Nations Environment Programme, 2012; United Nations General Assembly, 2017). Socio-ecosystem environmental history aims to cross the economic, social, humanities and natural sciences such as history, anthropology, archaeology, paleoecology, etc.

The globalization of modernity has increased discontent and its unintended consequences presenting new theoretical and empirical challenges for anthropology (Livi Bacci, 2017) such as, for example, the shift towards concern for climate change in the context of environmental sustainability identified as the result of facts and weak progress. The anthropological framework has a significant difference with social sustainability threatened with the exogenous of modernity. Sustainable societies were irreparably invaded by colonialism. Political ecology incorporates ecosocialist governmentality into forest conservation and is a powerful tool in collective efforts toward social sustainability. State-owned forest enterprises led to sustainable activities (Zhang et al., 2014).

The renewal of anthropological theory recognizes that the planet is unsustainable, leading to anthropogenic climate change, overexploitation of natural resources, environmental destruction (McNeill and Peter, 2016). Recent hermeneutical readings of Noah draw attention to interventions that leave God aside that have proven neither sustainable nor effective due to rejection, bias, and plurality of climate reception coexisting within a society (Nunn, 2017). Sustainable society makes obsolete the older functionalist and functionalist-structural anthropology that requires reevaluation.

Only a few anthropological studies are identified as reception studies and the most recent texts treat climate change as a discursive reality with observations to the environment (Fiske et al., 2014). Contemporary anthropology is concerned with the life of a damaged planet, limited by the concept of Anthropocene to be replaced by the concept of capitalocene that accelerates the cause of environmental destruction is the economic system and not humanity. The concept of homocene emphasizes similarities, standardization, and simplification. Another proposal is the concept of plantationocene considering relevant the standardized characteristics of the plantation rather than the plantations. (Moore, 2016; Wolford, 2021; Mann, 2011; Crutzen and Eugene, 2000).

From anthropology, psychology, sociology, and management sciences arise diverse perspectives of anthropological culture in common attributes of shared meaning, social construction affected by history and environment and cognitive and symbolic-cultural layers. Social ecology incorporates the biological orientation of human and anthropological ecologies to integrate socially embedded human agency with natural environmental contexts. Anthropology and geography generate sociocultural conditions and relationships different from natural environments. On cultural translation reflections implied by anthropological processes and power dynamics, including environmental transformations and climate change (Asad, 1986; Callison, 2014; Di Giminiani and Haines, 2020; Mathur, 2017).

The reception and realization of the treatment of climate change are part of the risks of environmental anthropology with implications with crucial issues according to anthropological communication, prophecy, translations, trust, guilt, experience, ideology, historical narrative, religion, homeland (Rudiak-Gould, 2011). Anthropological studies of climate reception concerned with climate change involve historically constructed space and time and not environmental and biophysical dynamics, leading to unequal scales, places, and ecosystems (Orlove et al., 2014).

Multidisciplinary research analyzes the complexity of contextual and interactive economic, social, political and environmental relationships with the health system. Theories on environmental and sustainability philosophy and other disciplines focus on the human-environmental relationship as a promising model of sustainability. A way of life is elucidated by a theory to appropriate sustainable relationships in natural environments (Weston, 1995). The adaptive Anthropocene supports the critique of radical ecology that combines neo-Marxist analysis of environmental change and critical study of race.

Meta-analyses predict a weak relationship between pro-environmental attitudes and behaviors, which are mediated by other variables (Bhuian et al., 2014; Tarfaoui and Salah Zkim, 2017; Zeng et al., 2017). This study aims to analyze the environment that leads to sustainability by focusing on its implications between these two concepts. Finally, some conclusions are presented.

LITERATURE REVIEW

Biocultural in anthropology has been used in biocultural studies dealing with biological and cultural environments that influence human biology and well-being prior to the use of biocultural diversity.

Biocultural studies within anthropology are a term used before biocultural diversity to broaden scope, including reciprocity in human and environmental relations.

Biocultural studies in anthropology refer to the reciprocal relationship between humans and nature that moves from environmental determinism to environmental probabilism (Hoke & Schell, 2020; Lewthwaite, 1966). Biculturality refers to the methodological integration of cultural, environmental, and biological data. Biocultural studies methodically integrate cultural environmental and biological data (McElroy, 1990). Biocultural studies in anthropology influence the biological, environmental, and cultural factors of human biology. Biocultural anthropology deals with the physical and social environment as deterministic influences on human biology and well-being.

The rationality model proposed by Nozawa (2015) is different from the techno-scientific policy of decontamination and radioecology through the spatial and temporal links between humans, non-humans, and the environment (Creager, 2013). Competing relationality models distribute harms that transcend individual suffering to the environment. Ecoculturalism in cultural, environmental communication and psychology link ecology and culture and advocate sociocultural identity inseparable from ecology and recognize local knowledge and memories in a dynamic link between landscapes and ecocultural identities that promote resilience (Pretty, 2011; Berry, 2004; Samnang, 2015; *Chen, et al.* 2012; Daryl Slack, 2008; Milstein and Castro-Sotomayor, 2020; Hoffmann, 2020; Dahake, 2020; Zedler and Stevens, 2018).

Ethnoecology makes relevant the type of knowledge in which humanity and its natural environments are integrated into sustainability. Ecology and ethnoecology study the relationships between organisms and the physical, biological and social factors with which they come into contact, based on a principle of political ecology that contemporary human society and the natural environment are not under the control of processes of exploitation and deterioration (Toledo, 1997). Environmental ethics, cultural identities and land-based populations are implicit in sustainable practices. Environmental ethics contributes to sustainability through the incorporation of viable ways of life to value the natural environment as the component of human-environmental communities.

The ecology of liberation is based on the idea that the time has come for the liberation of humanity and the natural environment (Marcuse, 1972; Gómez Hinojosa, 1990). Liberation ecology movements are more concerned with environmental integrity for livelihoods, rights, and social justice (Peet & Watts, 1996, p. 39). Local politicians and government officials are concerned about challenges to sustainability in local communities as the sustainable goals of urban public administration. The state's efforts to remove and contain pollutants transform the local ecology through decontamination by presupposing residents connected to the land, ancestors, nonhumans, and the environment. Government institutions promote a suitable business environment for start-ups.

The theorization of sustainability issues studies institutional intertwining to produce a narrow social structure aimed at self-correcting systems to restore equilibrium (Radcliffe-Brown, 1940; Bateson, 1972). University-industry interactions should focus on attracting companies that seek an environment that offers physical and intellectual infrastructure and represents a diversity of advanced technologies leading to product and service research and development, assembly, manufacturing, service support, logistics, etc. University-based local innovation systems are environments that facilitate the achievement of shared goals and conditions through the collaboration of start-ups from various technological sectors that interact with market and non-market segments that lead to the formation of networks, clusters and agglomerations that enable cooperation for technological innovation through technologies.

Companies that are leaders in sustainability address environmental and social issues and costs that seek to eliminate waste in the production and consumption of products. Companies must increase their commitments to be more effective and credible in environmental sustainability (Borel-Saladin & Turok, 2013). The foundations of the conceptual framework of university systems of local innovation based on the consolidation of structures must be able to cope with the environmental uncertainty of crucial components.

According to Pimenta-Bueno (2002), these components consist of their technological affinities between companies, startups tend to form agglomerations and clusters related to their technological innovation and commercial dimensions formed from local innovation systems derived from similarities of technological innovation in their research and development group and the local innovation system university affinities reinforced by interpersonal physical proximity coexistence in environments, incubators, technology park and non-professional.

The objective model of local ecology must make visible spiritual connections to place, land, and environment to generate meaning. Links can be broken, and pollution rejected without regard to what exists between humans, nonhumans, and the environment. Live Eco semiotics in the post-fallout environment and attend multi-scalarity to illuminate the tensions between competing relationality models to articulate consequences. Ecological semiotics builds on Perisean's semiotics, expanding the linguistic study of sign-mediated relationships and processes between humans and the natural environment.

Urban political ecology takes a guide from history to distinguish between different forms of nature agency in relation to anthropogenic environmental disturbance. The conjunctions of poverty, racism, and inadequate housing are associated with an unhealthy built environment and dangerous indoor organisms that lead to pests (Biehler 2013). Emissions and pollution are local public health concerns that tend to refer to the concepts of economic, social and environmental factors.

The theoretical approach to radioactive fallout links human behaviors and actions, technological capabilities, atmospheres, environments, and ecologies in a new formation of contamination (Masco, 2015). The theory provides a fallout framework for articulating specificities and making sense of connectivity between humans, nonhumans that cannot be captured by a scientific model. Scientific problematization of the post-fallout environment and state use of an objective and factual model of local ecology, ruptures and repair.

Post-fallout ecology tends to threaten the local model of relationality between animate and inanimate animals, animals and plants connected to the environment (Goldfarb, 2016; Nozawa, 2015). The anthropocentric assumption about human separability from the environment justifies damaging lives, values, historicity, relationships, things, and futurity (Morimoto, 2022).

Sustainability

The concept of sustainability was used in forestry in the eighteenth century and has since been used to describe economic policies sustained over a long period of time (Raworth, 2017). The term sustainability was used in Saxony as Nachhaltigkeit in the early eighteenth century, according to anthropologists (Brightman and Lewis, 2017). The concept of sustainability incorporates changes from the perspective of evolution. The concept of sustainability is a common denominator for different disciplines that question the long-term survival on the planet of humans and non-humans with the side effects on imperialist and capitalist efforts.

The concept of sustainability interrelates the economic, social, and environmental and has shifted towards the environmental dimension with an anthropocentric climate change impact. The definition of sustainability is related to the economic, social, and environmental dimensions dominated in academic and political circles. Sustainable ways of life are determined by physical and socio-economic and personal commitment to ethical principles respectful of nature.

When social and environmental factors intersect, the scenario becomes bearable, which guarantees good quality in goods and services, but does not ensure economic stability or an equitable perspective. On the other hand, if environmental and economic factors are sustained, the scenario is viable, since there is the availability of resources for the desired economic activities, however, it does not ensure that it is equitable or bearable over time. Finally, in terms of economic and social, equity is ensured that is transformed into access to employment without discrimination based on gender, race or culture, but should not be done at the expense of environmental problems (Lira, 2018).

The resilience discourse is related to continued sustainability for life security that brings relevance to communities and organizations (Petterson, Kensek, & Noble, 2017). Resilience is linked to sustainable systems, but it is not equivalent. Resilience contributes to sustainable energy efficiency and sustainable building practices and materials, supporting sustainable communities and buildings, supporting strategies for performance. Energy slave maps are visual rationalizations of industrialized energy provision, socio-political, environmental sustainability, aesthetic and rational expressions of human life that lead to political environmental agency, which has an impact on global ideology and politics (Rawes, 2019).

There is a consensus among social scientists in cultural systems that community sustainability is intentionally developed in structured resource management and conservation strategies (Trosper, 2009; Campbell and Butler 2010; Moss, 2011; Mathews and Turner 2017). Public goods, assets and resources are intertwined with private ones to advance the philosophy of sustainability in community benefits, environmental and social responsibility, and improving the well-being of citizens (Mog, 2004). A sustainable project should be supported by analysis to emphasize community economic growth towards organizational interventions (Sisaye, 2012). There are some contradictions and tensions between sustainability and growth (Brightman & Lewis, (Eds.) (2017). Long-term sustainable community projects and programs are used in emerging economies.

Sustainable human land-use policies and ecosystem management increase resilience and reduce harmful effects. However, land uses have become monocultures and pastures following the generalized model of modernization of sustainability (Toledo, 1994). Humans with different knowledge, skills, and abilities inhabit their niches to inform strategies for healthy, safe, and sustainable urban green areas through urban design, urban planning, and land use and mobility management (Guxens, et al., 2018; Buffel and Phillipson, 2016; Hamraie, 2017). The capacity of urban green land accumulates energy to supply sustainable natural resources, sunlight and air for leisure and exercise activities and ecological functions for urban economic activities (Pfaffenbichler, Emberger, Shepherd, 2010).

Environmentally sustainable policy for water and land, for example, poses new challenges for anthropological participation. For example, cotton is a variety originally grown by indigenous people in Mexico that had better quality, it is a good example as the resurgence of research in anthropology of plantations at the time such as the 'Plantationocene' Work on these plantations emphasizes the exploitation of people and plants, the destruction of livelihoods and local and community environmental environments (Haraway, 2016 ; Li, 2007). Indigenous tea cultivation in the Zomia highlands stands as a non-coercive and sustainable alternative to colonial plantations. Wilderness areas have been inhabited where Indigenous and extractive inhabitants practice sustainable activities.

The principles that lead to sustainability in local communities and their environments threatened by unsustainable practices, according to Toledo (1996) include self-sufficiency and autonomy that lead to a strategy of resistance to external domination of both humans and the environment. Exogamous changes affect sustainability in a socio-ecological sense. Risk assessments in the sustainability analyst focus on risk-sharing strategies, investment and acquisition strategies, financing, etc. with the aim of formulating and implementing strategies for lean operations, sustainable growth strategies, survival, etc. (Bowden et al., 2010).

Global ecological sustainability is possible if it resembles that of the remaining Indigenous peoples (Escobar, 2020). Socio-ecological sustainability is linked to ideology, modernity, capitalism, industrialism, and economic growth (Hornborg, 2019). Social ecology, natural ecosystems, cultural and anthropological ecology are relevant topics to address issues of organization, management, and ecology of sustainable networks.

Socioecology is relevant to the public discourse of sustainability with an impact on local communities determined to mitigate negative effects. Local socioecology and sustainability are reserved for the well-being and health of the community with an impact on prosperity, employment, civility, and security. Social ecology embodies certain social movements with practical consequences for human well-being in natural sustainability. Sustainable enabling practices and beliefs are a type of cultural matrix. The practices of populations are the crucial characteristic to achieve sustainability in a cultural matrix. Multidimensional aspects of site conservation and enhancement of cultural heritage and biodiversity management, land use and public uses of sustainable territorial management, cultural and tourism management, etc. (Zan and Lusiani, 2011).

Anthropological and sociological ecologies provide the theoretical framework for the integration of sustainability ecology into accounting reporting (Hopwood, 2009, 2010) in specific conceptual areas such as financial accounting, managerial accounting, and auditing. Existing models of sustainability based on ethnoecology are already threatened to maintain the long-term sustainability of natural environments as livelihoods of communities (Toledo, 1997).

Social capital plays a significant role in the vital component of cooperation in innovative process environments. In networked environments, cooperative relationships between members can take a long time to develop, but once established. The term network is widely used as an environment characterized by independent actors and developing relationships aligned with complementary objectives of collaboration and cooperation (OECD, 2001). Cooperative relationships and ties need time to allow convergence of members' objectives in a networked environment.

Innovation environments with different benchmarks indicate the possibility of implementing local innovation systems, even in regions with adverse conditions. Business innovation focuses on new methods of creating a new culture of sustainability to promote innovation and respond to environmental uncertainties. An innovation incubator is conceived with the aim of developing an environment to incubate innovation and not incubate companies, developing innovative products and services based on institutional arrangements to assume the role of crossover agents (de Melo, 2002).

Sustainability activities address responsibility practices, life cycle analysis, material and resource flows, eco-efficient use of renewable energies and adoption of environmental and social objectives (Kempera & Ballantine, 2019; McDonagh y Prothero, 2014). Ecotourism is an approach to sustainably commercializing common natural resources and mobilizing ecosystem and environmental services to benefit local communities (Briassoulis, 2002). General statements for sustainable marketing are culturally oriented behaviors and cannot be translated into paradigms of consumer behavior (Holt, 2012). There are several approaches to sustainable marketing and consumer research that contrast with unsustainable marketing (Kempera and Ballantine, 2019).

The nature of data infrastructure has implications for value related to environmental sustainability and biodiversity conservation. The infrastructure data gap remains a discrepancy with sustainable governance and environmental justice (Edelman, 2013; Zoomers et al., 2016). The biodiversity and sustainability of the natural environment of rural areas in Mexico depends on the degree of autonomy of the communities, so it is necessary to promote new forms of local autonomy and social structures that facilitate the reconstruction of sustainable communities and the production of goods and services that lead to expanding the provision of environmental services (Barkin, 1998).

The macromarketing framework recognizes culturally instantiated eco-economies that lead to sustainable consumer market economics challenges (Polanyi, 1947/2001; Sahlins, 1976, 1996[REMOVED HYPERLINK FIELD]; Slater and Tonkiss, 2001). Anthropology has several alternatives to sustainability in contrast to neoliberal capitalism, represented ethnographically (Kulick, 2019). The sustainability of transformative marketing transcends the neoliberal theory of social exchange and naturalistic ontology to frame the marketing system.

Systemic assessments of the needs of stakeholders, shareholders, customers and consumers in sustainable marketing should offer alternatives to social sharing (Kempera and Ballantine, 2019). The holistic systems theory of marketing does not support sustainable eco economics, but is the precondition for the social exchange approach towards sustainable and green marketing (Thomas, 2018).

Atheoretical approach to the practice of sustainable consumption behaviors involving heterogeneity within or between cultures (Bartiaux, 2008). The critical perspective of macromarketing addresses the challenges posed by sustainable marketing, consumption, economic, social and environmental that leads to the adoption of systems thinking (Mittelstaedt et al., 2014). The market economy approach to sustainable marketing and consumption does not address toxic relationships (Cahen-Fourot, 2020; Kramer, 2021; Stoner, 2021; Trujillo, 2021) and the environment rooted in the DSP, our modern social imaginary (Taylor, 2002).

Transformational sustainable eco economics replaces the construction of nature with rights in the diversity of living beings involved in economic processes that lead to circulation systems to meet the environmental costs of producing, distributing and consuming a sustainable economy (Peattie, 2001). Unsustainable production and sustainable consumption require the transition to a sustainable society (Kempera & Ballantine, 2019). Sustainable consumption and consumption practices are among the impossible and voluntary anti-consumers at the social level as alienated subjects of the eco-economic process (McDonagh and Prothero, 2014).

Eco economics is a post-humanist and neo-animist circulation of resources that addresses sustainable human livelihoods in biotic communities identified as the modern social imaginary in macro-marketing (Taylor, 2002). Animistic ontology is critical of marketing and consumer research, arguing with latest ideas that offer guidance for diagnosing a more sustainable circulation and consumption of natural resources. The transitive and complementary principles of circulation help natural resources circulate more sustainably between plants, animals and humans.

Neo animism is an eco-economic system that challenges the dominant sustainable paradigm by taking advantage of behavioral gaps in sustainable consumption. The ecosystem costs of economic relations include exchanges, gifts, predation, foreground, etc. between human and non-human communities. The neo-animist perspective from the approaches of ontology and epistemology is oriented towards a more sustainable ecocentric economy of people that brings sustainable marketing and consumption to critics (Descola, 2012; Kohn, 2013; Tsing, 2015; Viveiros de Castro, 2007).

Ontological comparisons of different approaches to sustainable and green marketing and consumption involve the Kuhnian critique (Kuhn, 1962) of the dominant forms of thought identified in the ontological and epistemological contrasts between naturalism and animism. Theoretical frameworks of sustainable marketing and consumption at the organizational and individual levels of behavior and attitude models are underpinned by nurturing relationships for sustainable practices identified in naturalism and animism.

Managerial green marketing focuses on consumer decision factors in firm-level strategy (Hult & Tomas, 2011). Sustainable marketing relies on ontological and epistemological foundations to address marketing relationships with non-human environments (McDonagh & Prothero, 2014; Thomas, 2018). Green consumers can pay a premium for sustainable products (Macdonald & Uncles, 2007; Kempera and Ballantine, 2019). Green consumer demand for sustainable products and services supports sustainable marketing (Wymer and Polonsky, 2015; Peattie and Crane, 2005; Kempera and Ballantine, 2019).

Sustainable green marketing and consumerism have institutional and environmental limitations and regulations. Green buying intent and consumer environmental attitude (Malik et al., 2017) lead to mass green consumerism, although the effects on global consumerism are dubious due to continued increases in global energy demand and carbon emissions. The critical sociological perspective accepts the autonomy of consumers to identify sustainable behaviors, cognitive ability, and knowledge. Consumers find the tools to reassess behaviors that challenge unsustainable consumption and exert regressive normative influences (Cherrier et al., 2012; Kemper and Ballantine, 2019).

Marketing is being challenged by sustainability marketing, social marketing and macromarketing that need deeper changes to meet the challenges of sustainability (Belz & Peattie, 2010). The dominant social perspective demonstrates the nature of micromarketing and micro-managerialist approaches by recognizing that sustainable green marketing and consumption (Taylor, 2002; Bradshaw and Zwick, 2016; McDonagh & Prothero, 2014, p. 1199) cannot be addressed through consumer choice embedded in regulatory and material constraints (Cherrier et al., 2012[REMOVED HYPERLINK FIELD]; Shove, 2003a, 2003b[REMOVED HYPERLINK FIELD]; Slater and Tonkiss, 2001).

Responsible consumers in the process of self-realization and improvement are the guarantees for the perpetuation of an unsustainable marketing process (Lambert, 2019; Stoner, 2021, Yngfalk, 2019[REMOVED HYPERLINK FIELD]). Green marketing and consumerism could be closer to sustainable marketing focused on consumer actions and business choices within the market economy. Sustainability marketing is a strategic and practical approach to changing marketing mixes that incorporates stakeholder perspectives (Belz and Peattie, 2010; Kempera and Ballantine, 2019).

Sociology and anthropology are integrated into organizational management oriented to sustainability co-evolution. The creation, conservation, maintenance, protection and sustainable management of sustainable environmental sites are more linked to private financial interests and political logic. The alternative incorporation of corporate ethics, corporate social responsibility and sustainability in organizations is not limited to available expertise and changes in the organizational environment. Sustainable topics in organizational development should include financial and managerial reporting. Economic and financial reporting is mandatory and regularly disseminated closely associated with external sustainable reporting systems.

Organizational sustainability is evolving in the management of environmental natural resources and the sustainability of corporate finance and accounting reporting in organizations. Integration of organizational sustainability into organizations' green resource endowments with the goal of achieving performance improvements. The organizational sociological and anthropological approach studies the relationship between ecology and sustainability accounting reporting. Sensitivity to social soundness has implications for sustainability accounting reporting to promote conservation of environmental resources (Sisaye et al. 2004). Sustainability accounting reports can be used strategically to develop alternative plans to identify and analyze environmental risks, mitigate market volatility, manage the environmental crisis, and create opportunities to maintain market share. Sustainability accounting reports support the development, growth, and expansion of the organization.

It is also used to evaluate the cost-benefit analysis, return on investment, and potential risks of environmental projects and programs. External sustainability accounting reports reveal on the balance sheet economic, social and environmental performance to develop sustainable metrics and organizational operation processes (Savitz and Weber, 2006; Sustainability Index, 2010; Koehler et al., 2005). The role of sustainability accounting reporting in advancing environmental issues has been criticized (Hopwood and Unerman 2010). Integrated sustainability accounting on environmental and ecological issues is relevant to the management of the long-term livelihood of natural resources to promote economic profitability and the social and environmental well-being of individuals, organizations, and societies.

The following are some of the indicators of sustainability and environmental financial evaluation as shown in table 1.

Ecology supported by sociological and organizational anthropological perspectives is integrated into sustainability financial and accounting reporting. There is growing interest in organizations to implement voluntary disclosure of environmental and sustainability data in financial management and accounting (Aras & Crowther, 2008; Bowden et al., 2010; Epstein, 2008, 2010; Hopwood et al., 2010; Gray, 2006, 2010; Unerman et al., 2007; Vann and White, 2004). The sustainable economic, social and environmental practices and activities of organizations have integrated corporate reports and established public accounting standards into verified and certified information.

Table 1. Indicators of sustainability in environmental accounting

Type of indicator	Disclosure score
Environmental index	• Carbon footprint, • Effects of climate change, • Waste disposal, Resource depletion, • Energy consumption • Renewable energy, • Water consumption • Biodiversity Air quality
Social index	• Human rights, • Supply chain, • Discrimination, • Defense of diversity, Poverty
Corporate Government index	• Relations with interest groups, • Executive compensation, Shareholder rights
Financial index	• Economic profitability (ROA) • Financial profitability (ROE) • Result of the exercise • Liquidity, • Debt, Size,
Environmental budget execution indicators	• Fixed assets • Environmental costs • Environmental financing

Global sustainability reports are being self-imposed by companies constrained by voluntarily agreed measures (Press & Arnould, 2014). Some of the guidelines for sustainability auditing and reporting include the fundamentals of good governance and sustainability for integrated corporate performance of social and environmental performance through financial reporting. Corporate governance and sustainability accountability address economic, financial, social, and environmental performance reporting on a voluntary basis, diluting data and information disclosure.

An important element that transposes the European directive on non-financial information and diversity is the accounting audit law, Law 11/2018, which indicates that the regulations are mandatory for companies that exceed 500 employees, that exceed 40 billion euros, that their asset items are above 20 million euros and that they have a headquarters in a community country (Iturrioz del Campo *et al.*, 2019). They must prepare a report containing the following information:

- Current and foreseeable effects of the impact of the environment such as the use of water and renewable energy.
- Measures to guarantee gender equality in working conditions, respect for union rights, health and safety at work.
- Information on internal control procedures in the fight against corruption and bribery.
- Measures applied in terms of professional training, age, so that there is gender equity and there is a balanced number of men and women in decision-making. Below is the table 2 showing the interest group and the relevant measurement indicator:

Integration of sustainability in accounting and management control to improve the performance of economic, social, and environmental practices and activities. Sustainability reporting of performance indicators is related to economically, socially, and environmentally sustainable objectives. The triple bottom line includes external reports containing sustainable economic, social, and environmental information and data, including sustainability standardized to include social and environmental information.

Table 2. Economic efficiency and growth

Stakeholder	Relevant indicator
Stockholders	Economic performance Ethical behavior Zero tolerance fraud policy Corporate governance
Collaborators	Work well-being Health and safety of collaborators Development of human talent
Organizations linked to the care of the environment, society.	Recycling Paper handling Water Energy Climate change management
Clients	Customer satisfaction Responsible marketing Innovation Non-conformity attention

Sustainable reporting on economic, social, and environmental performance, human and environmental management with respect to organizations, including profitability and shareholder value creation. Economic, social, and environmental indicators from sustainable accounting reporting that lead to commitment to sustainability (Adams, 2010; Aras and Crowther, 2008; Hubbard, 2008; Lamberton, 2005; Pava, 2007). Sustainability accounting reporting guidelines define methodologies, processes and procedures, performance indicators and corrective actions for wealth creation and welfare management aimed at reducing imbalance in the community (Savitz and Weber, 2006; Schaltegger and Burritt, 2006). The analysis of sustainability accounting reports defines and identifies the value chain of stakeholders.

Corporate citizenship philanthropy, human capital development and social reporting labor practices, and industry criteria are factors that lead to documenting social performance indicators on sustainability, environment, social reporting, health and safety, financial support, and other related factors.

METHODOLOGY

This study draws on the meta-analysis and descriptive methods to elaborate a reflective proposition based on the conceptual, theoretical, and empirical literature review.

DISCUSSION

As mentioned in this article, since 2022 scientists and experts have already exposed the limits exceeded in the plant in a harmful way. This puts us in a perspective of the urgency of not only changing our vision of the environment, but also consumption habits, business, and public policies. The present provides different concepts, models, and theories of which the ideas of sustainability are strengthened, which is produced by the interaction of the 3 factors: environment, economic and social. Concepts are well exposed, biocultural, bodybuilding, ethnoecology, environmental ethics, liberation ecology, all the above have in common the relationship of people with the environment.

However, and considering the sustainability factor, it already directly involves organizations and companies. This is due to its relationship with the generation of employment which impacts the economy of the region, province, municipality, state, or country. For the above, actions were implemented that increase residence to reduce harmful effects, mainly aimed at the most vulnerable or unattended areas, which are commonly the original villages of each region of the world. On the part of organizations or companies their action is the application of green marketing which exposes the consumption of products or services which do not negatively impact the environment.

While all concepts, theories and models have their foundation, the reality that is current marketing is the worst enemy of green marketing, coupled with the policies and beliefs of each country. This is because many countries are based on consumer economies which have little respect for the environment. If it is possible to connect or harmonize green marketing with a responsible consumer, it will be possible to help sustainability in a prominent way.

The challenge becomes more complex when you want to serve the native peoples of each country since they have a very particular point of view of the environment, although some do live in harmony with it, others are not so much. One might think that progress will not be significant if there is no global policy which impacts consumer habits, companies' production processes, public policies, and the correct application of green marketing.

CONCLUSION

This study analyzes, from the anthropological perspective, the socio-ecological implications between the environment and sustainability, if sociocultural anthropology contributes to more sustainable societies and their environment capable of reproducing changes indefinitely and incrementally. However, from this anthropocentric assumption the human separability of the environment in modernity is justified. This analysis concludes that socio-ecosystem dynamics focus on human interactions with the environment from the ecological perspective aimed at fostering sustainability.

Socio-ecosystem dynamics focus on human interactions from the perspective of historical ecology with the aim of fostering sustainable changes in relationships. Economic, social, and environmental performance integrated into sustainability dimensions are determining factors in resource allocation decisions. Sustainability has implications for population growth and greenhouse gas emissions and issues related to sensitive climate change. A sustainable community can reproduce itself by offering jobs, services, safety, and security.

Non-modern societies transformed local socioecologies, but they were not always sustainable without undermining human living conditions at today's scale and speed. Sustainability is social reproduction that recalls the equilibrium models of social anthropology such as structural functionalism. Changes in sustainability research have led to a rethinking of social and anthropological theory. Societies develop mechanisms that allow them to reproduce indefinitely. Sustainability is the quality that allows sustaining and reproducing without undermining the conditions of existence.

Ecological sustainability is the subject of policy and research. Research should focus on approaches aimed at developing models of sustainability between sustainable relationships, with behaviors and attitudes with natural environments that reflect the ethics of caring for natural processes.

Research on the circulation theory of resource renewal offers a theory of the sustainable neo-animist system as an alternative subjectivity for sustainable eco economics inscribed in the social imaginary. Neo animism is an alternative to the dominant paradigm in social exchange to address challenges to environmental sustainability. A critique of social exchange theory leads to the necessary changes in production, marketing, and consumption to move toward a more sustainable circulation of resources aimed at preserving marketing exchange.

REFERENCES

Adams, R. (2010). *Accounting Report*. Association of Chartered Accountants., www.accaglobal.com/

Aras, G., & Crowther, D. (2008). Developing sustainable reporting standards. *Journal of Applied Accounting Research*, *9*(1), 4–16. doi:10.1108/09675420810886097

Argyrou, V. (2005). *La lógica del ecologismo: Antropología, ecología y poscolonialidad*. Libros Berghahn.

Asad, T. (1986). El concepto de traducción cultural en la antropología social británica. In J. Clifford & G. E. Marcus (Eds.), *Writing culture: The poetics and politics of ethnography*. Prensa de la Universidad de California. doi:10.1525/9780520946286-009

Asamblea General de las Naciones Unidas. (2017). Resolución 71/312 Nuestro océano, nuestro futuro: Llamamiento a la acción. In *Resolución adoptada por la Asamblea General el 6 de julio de 2017*. Naciones Unidas.

Barkin, D. (1998). Sostenibilidad: La economía política del desarrollo autónomo. *Organización y Medio Ambiente*, *11*(1), 5–32.

Bartiaux, F. (2008). ¿La información ambiental supera la compartimentación de las prácticas y cambia los comportamientos de los consumidores? *Revista de producción más limpia*, 16(11), 1170–1180. doi:10.1016/j.jclepro.2007.08.013

Bateson, G. (1972). Metalogue: ¿Por qué las cosas se meten en un lío? En Pasos hacia una ecología de la mente; Bateson, G., Ed.; Chandler: Nueva York, NJ, EE.UU., 1972; págs. 13–18

Belz, F.-M., & Peattie, K. (2010). Marketing de sostenibilidad: Una concepción innovadora del marketing. *Marketing Review St. Gallen*, *27*(5), 8–15. doi:10.1007/s11621-010-0085-7

Berry, J. W. (2004). Una perspectiva ecocultural sobre el desarrollo de la competencia. en Cultura y Competencia. Contextos de Éxito en la Vida; Sternberg, R., Grigorenko, E., Eds.; Asociación Americana de Psicología: Washington, DC, EE.UU.

Bhuian, S. N., Amyx, D. A., & Shamma, H. M. (2014). Una extensión de la investigación del comportamiento ambiental del consumidor entre los expatriados. *Revista Internacional de Comercio y Gestión*, *24*(1), 63–84. doi:10.1108/IJCoMA-01-2012-0005

Biehler, D. D. (2013). *Plagas en la ciudad: moscas, chinches, cucarachas y ratas*. University of Washington Press.

Bonita, J. (2011). Progreso interdisciplinario en enfoques para abordar los sistemas socio-ecológicos y ecoculturales. *Entorno. Conservar*, *38*, 127–139.

Borel-Saladin, J. M., & Turok, I. N. (2013). La economía verde: ¿cambio incremental o transformación? *Environmental Policy and Governance*, *23*(julio-agosto), 209–220. doi:10.1002/eet.1614

Bowden, A. R., Lane, M. R., & Martin, J. H. (2010). *Triple Bottom Line Risk Management: Enhancing Profit. Environmental Performance, and Community Benefit*. John Wiley.

Bradshaw, A., & Zwick, D. (2016). El campo de la sostenibilidad empresarial y el impulso de muerte: Una intervención radical. *Journal of Business Ethics*, *136*(junio), 267–279. doi:10.1007/s10551-014-2443-x

Briassoulis, H. (2002). El turismo sostenible y la cuestión de los bienes comunes. *Anales de Investigación Turística*, *29*(4), 1065–1085. doi:10.1016/S0160-7383(02)00021-X

Brightman, M., & Lewis, J. (Eds.). (2017). *La antropología de la sostenibilidad. Más allá del desarrollo y el progreso; Palgrave*. Reino Unido.

Buffel, T., & Phillipson, Ch. (2016). ¿Pueden las ciudades globales ser ciudades amigables con las personas mayores? *Desarrollo urbano y envejecimiento de la población. Ciudades*, *55*, 94–100.

Cahen-Fourot, L. (2020). Los capitalismos contemporáneos y su relación social con el medio ambiente. *Economía ecológica*, *172* (junio), 106634. doi:10.1016/j.ecolecon.2020.106634

Callison, C. (2014). *Cómo importa el cambio climático: la vida comunitaria de los hechos*. Prensa de la Universidad de Duke.

Campbell, S. K., & Butler, V. L. (2010). Evidencia arqueológica de la resiliencia de las poblaciones de salmón del noroeste del Pacífico y el sistema socioecológico en los últimos ~ 7,500 años. *Ecology and Society*, *15*, 1–17.

Chen, Y.-W., Milstein, T., Anguiano, C., Sandoval, J., & Knudsen, L. (2012). Desafíos y beneficios de la investigación participativa basada en la comunidad para la justicia ambiental: Un caso de examen colaborativo de las luchas ecoculturales. *Entorno. Comuna*, *6*, 403–421.

Cherrier, H., Szuba, M., & Özçağlar-Toulouse, N. (2012). Barreras a la reducción de las emisiones de carbono: Explorando el consumo sostenible frente al piso de vidrio. *Revista de Gestión de Marketing*, *28*(3–4), 397–419.

Creager, A. N. H. (2013). Paisajes de tiempo de trazadores radiactivos en bioquímica y ecología. *Historia y filosofía de las ciencias de la vida*, *35*(1). https://www.jstor.org/stable/43862145

Crutzen, P. J., & Eugene, F. S. (2000). El Antropoceno. *IGBP Newsl*, *41*, 17–18.

Dahake, S. (2020). La creación de identidades ecoculturales fluidas en la India urbana. En Routledge Handbook of Ecocultural Identity; Milstein, T., Castro-Sotomayor, J., Eds.; Routledge: Abingdon, Reino Unido; Nueva York, NY, EE.UU.

Daryl Slack, J. (2008). Resistiendo los estudios ecoculturales. *Cultural Studies*, *2008*(22), 477–497. doi:10.1080/09502380802012575

Del Campo, J. I., Lorenzo, C. M., & Castaño, A. M. (2019). La transparencia sobre Responsabilidad Social Corporativa y su implicación con el comportamiento económico-financiero: efectos en las empresas de Economía Social. *REVESCO: revista de estudios cooperativos,* (132), 122-143.

Di Giminiani, P. y Haines, S. (2020). Introducción: Traducción de entornos. *Ethnos*, *85*(1), 1-16. doi: 10.1080/00141844.2018.1542409

Edelman, M. (2013). Messy hectares: Questions about the epistemology of land grabbing data. *The Journal of Peasant Studies*, *40*(3), 485–501. doi:10.1080/03066150.2013.801340

Epstein, M. J. (2008). *Making Sustainability Work: Best Practices in Managing and Measuring Corporate Social, Environmental and Economic Impacts*. Greenleaf Publishing y Berrett-Koehler Publishers.

Epstein, M. J. (2010). El desafío de mejorar simultáneamente el rendimiento social y financiero: nuevos resultados de investigación, en Epstein, M.J., Manzoni, J.-F. y Davila, A. (Eds), Política pluriversal: lo real y lo possible. Duke University Press: Durham.

Comisión Europea. (2017). Informe sobre la estrategia de crecimiento azul hacia un crecimiento más sostenible y empleo en la economía azul. *Bruselas: Comisión Europea.*

Fiske, S. J., Crate, S. A., Crumley, C., Galvin, K., Lazrus, H., Lucero, L., Oliver-Smith, A., Orlove, B., Strauss, S., & Wilk, R. R. (2014). *Cambiando la atmósfera. Antropología y cambio climático. Informe final del grupo de trabajo sobre el cambio climático global de la AAA.* Asociación Americana de Antropología.

Fox, N. J. y Alldred, P. (2020). La sostenibilidad, el humanismo feminista y las capacidades inusuales de los humanos (post). *Sociología ambiental*, *6*(2), 121–131. doi:10.1080/23251042.2019.1704480

Goldfarb, K. E. (2016). Coming to Look Alike: Materializing Affinity in Japanese Foster and Adoptive Care. *Social Analysis*, *60*(2), 47–64. doi:10.3167/sa.2016.600204

Gómez Hinojosa, J. F. (1990). De la ecología a la ecofilía. *Apuntes para una ecología liberadora. Pasos (El Sauzal)*, *30*(July–August).

Google Académico Gray, R. (2006). ¿Informes sociales, ambientales y de sostenibilidad y creación de valor organizacional? ¿Valor de quién? ¿De quién es la creación? *Accounting, Auditing & Accountability Journal*, *19*(6), 793–819.

Gray, R. (2010). ¿Es la contabilidad de la sostenibilidad realmente responsable de la sostenibilidad? ¿Y cómo lo sabremos? Una exploración de las narrativas de las organizaciones y el planeta. *Contabilidad, organizaciones y sociedad*, *35*(1), 47–62

Guxens, M., Lubczyńska, M. J., Muetzel, R. L., & Dalmau-Bueno, A., Jaddoe, Vincent W., Hoek, G., et al. (2018). Exposición a la contaminación del aire durante la vida fetal, morfología cerebral y función cognitiva en niños en edad escolar. *Psiquiatría Biológica : Publicación Oficial de la Sociedad Española de Psiquiatría Biológica*, *84*(4).

Hamraie, A. (2017). *Acceso a edificios: diseño universal y la política de la discapacidad.* University of Minnesota Press.

Haraway, D., Ishikawa, N., Gilbert, S. F., Olwig, K., & Tsing, A. L. (2016). Los antropólogos están hablando – sobre el Antropoceno. *Ethnos*, *81*(3), 535–564. doi:10.1080/00141844.2015.1105838

Haraway, D. J. (2016). *Staying with the trouble: making kin en el chthulucene.* Duke University Press.

Hoffmann, J. A. (2020). Memoria, vías fluviales e identidad ecocultural. En The Routledge Handbook of Ecocultural Identity; Milstein, T., Castro-Sotomayor, J., Eds.; Routledge: Abingdon, Reino Unido.

Hoke, M. K., & Schell, L. M. (2020). Haciendo antropología biocultural: *Continuidad y cambio. American Journal of Human Biology*, *32*, E23471. doi:10.1002/ajhb.23471 PMID:32681558

Holt, D. B. (2012). Construyendo un consumo sostenible: De los valores éticos a la transformación cultural de mercados insostenibles. *Los ANNALS de la Academia Americana de Ciencias Políticas y Sociales*, *644*(1), 236–255. doi:10.1177/0002716212453260

Hopwood, A. G. (2009). Contabilidad y medio ambiente Contabilidad. Organizaciones y Sociedad, 2009, vol. 34, números 3-4, 433-439

Hopwood, A. G. (2010). In memoriam, Anthony G. Hopwood, 1944-2010. *Accounting, Organizations and Society, 35*(4), 496–497. doi:10.1016/j.aos.2010.05.002

Hornborg, A. (2019). *Naturaleza, sociedad y justicia en el Antropoceno: desentrañando el complejo dinero-energía-tecnología; Cambridge University Press.* Reino Unido., doi:10.1017/9781108554985

Hubbard, G. (2008). Beyond Accounting: Assessing the Impact of Sustainability Reporting on Tomorrow's Business. *University of Adelaide Business School, mimeo.*

Hult, G., & Tomás, M. (2011). Sostenibilidad centrada en el mercado: ¡Orientación al mercado plus! *Revista de la Academia de Ciencias de Marketing, 29*(febrero), 1–6. doi:10.1007/s11747-010-0223-4

Índice de sostenibilidad (2010), índices de sostenibilidad Dow Jones en colaboración con Robeco SAM, disponible en: *J. Peasant Stud., 44*(6). doi:10.1080/03066150.2017.1308353

Kempera, J. A., & Ballantine, P. W. (2019). ¿Qué entendemos por marketing de sostenibilidad? *Journal of Marketing Management, 35*(3–4), 277–309. doi:10.1080/0267257X.2019.1573845

Koehler, T., Weber, O., Fenchel, M., & Scholtz, R. (2005). Principles for sustainability rating of investment of funds. *Business Strategy and the Environment, 14*(1), 54–70. doi:10.1002/bse.423

Kramer, J. M. (2021). De depredador a parásito: En la propiedad privada y nuestro desastre ecológico. *Journal of Economic Issues, 55*(junio), 416–422. doi:10.1080/00213624.2021.1908804

Kuhn, T. S. (1962). *La estructura de las revoluciones científicas.* Prensa de la Universidad de Chicago.

Kulick, D. (2019). Una muerte en la selva: cómo un idioma y una forma de vida llegaron a su fin en Papúa Nueva Guinea. Algonquin Books: Chapel Hill, NC, USA.

Lambert, A. (2019). ¿Psicótico, acrítico y precario? Una exploración lacaniana del sujeto del consumidor neoliberal. *Teoría del marketing, 19*(3), 329–346. doi:10.1177/1470593118796704

Lamberton, G. (2005). Contabilidad de sostenibilidad: Una breve historia y marco conceptual. *Accounting Forum, 29*(1), 7–26. doi:10.1016/j.accfor.2004.11.001

Latour, B. (2018). *Con los pies en la tierra: la política en el nuevo régimen climático. Política.* Reino Unido.

Lewthwaite, G.R. (1966). Ecologismo y determinismo: Una búsqueda de aclaración. *Ann. Assoc. Am. Geogr. 56*, 1–23.

Li, T.M. (2017). Desplazamiento intergeneracional en la zona de plantaciones de palma aceitera de Indonesia.

Lira, A. (2018). Introducción al concepto de sostenibilidad. *Unidades de Apoyo para el Aprendizaje. CUAED/Facultad de Arquitectura-UNAM.* https://uapa.cuaieed.unam.mx/sites/default/files/minisite/static/693ee8e8-f02c-43c2-8222-498e1e8b8814/ConceptoSostenibilidad/index.html

Livi Bacci, M. (2017). *Nuestro planeta menguante (Il Pianeta Stretto); Política.* Reino Unido.

Macdonald, E. K., & Uncles, M. D. (2007). Conocimiento del consumidor: Conceptualización y medición. *Revista de Gestión de Marketing, 23*(5–6), 497–517. doi:10.1362/026725707X212793

Malik, C., Singhal, N., & Tiwari, S. (2017). Antecedentes de la actitud ambiental del consumidor y la intención de comprar productos verdes: Papel moderador de la necesidad percibida del producto. *Revista Internacional de Tecnología y Gestión Ambiental*, *20*(5/6), 259–279. doi:10.1504/IJETM.2017.091290

Mann, C. C. (2011). 1493: Descubriendo el Nuevo Mundo que Colón creó. Vintage: Nueva York.

Marcuse, H. (1972). *Contrarrevolución y revuelta*. Boston Beacon Press.

Masco, J. (2006). *The Nuclear Borderlands: The Manhattan Project in Post-Cold War New Mexico*. Princeton University Press. doi:10.1515/9781400849680

Mathews, D., & Turner, N. J. (2017). *Culturas oceánicas: ecosistemas de la costa noroeste y sistemas de gestión indígena. Levin PS, Poe MR editores. Conservación para el Océano Antropoceno: Ciencia Interdisciplinaria en Apoyo de la Naturaleza y las Personas*. Academic Press.

Mathur, N. (2017). La tarea del traductor climático. *Economic and Political Weekly*, *LII*(31), 77–84.

McDonagh, P., & Prothero, A. (2014). Investigación de mercados de sostenibilidad: Pasado, presente y futuro. *Revista de Gestión de Marketing*, *30*(11–12), 1186–1219. doi:10.1080/0267257X.2014.943263

McElroy, A. (1990). Modelos bioculturales en estudios de salud humana y adaptación. *Anthropological Quarterly*, *4*, 243–265. doi:10.1525/maq.1990.4.3.02a00010

McNeill, J., & Peter, E. (2016). La Gran Aceleración: Una Historia Ambiental del Antropoceno desde 1945. Harvard University Press.

Milstein, T., & Castro-Sotomayor, J. (Eds.). (2020). *Routledge Handbook of Ecocultural Identity. Routledge*. Reino Unido. doi:10.4324/9781351068840

Mittelstaedt, J. D., Shultz, C. J., Kilbourne, W. E. y Peterson, M. (2014). La sostenibilidad como megatendencia: Dos escuelas de pensamiento de macromarketing. *Revista de macromarketing*, *34*(3), 253–264. doi:10.1177/0276146713520551

Mog, J. M. (2004). Luchando con la sostenibilidad. Un marco comparativo para evaluar los programas de desarrollo sostenible. *World Development*, *32*, 2139–2160. doi:10.1016/j.worlddev.2004.07.002

Moore, J. (Ed.). (2016). *Antropoceno o Capitaloceno? Naturaleza, historia y crisis del capitalism*. Kairos Press.

Morimoto, R. (2022) Una persecución de jabalíes: Ecología del daño y la política de HalfLife en la costa de Fukushima. *Antropología cultural*, *37*(1), pp. 69-98.

Moss, M. L. (2011). *Costa Noroeste: Arqueología como Historia Profunda*. Society for American Archaeology Press.

Nozawa, S. (2015). Phatic Traces: Sociality in Contemporary Japan. *Anthropological Quarterly*, *88*(2), 373–400. https://muse.jhu.edu/article/582266. doi:10.1353/anq.2015.0014

Nsamenang, A. B. (2015). Teorías ecoculturales del desarrollo. En la Enciclopedia Internacional de las Ciencias Sociales y del Comportamiento. Elsevier.

Nunn, P. (2017). Dejando de lado a Dios: Por qué los proyectos climáticos seculares en las islas del Pacífico están fallando. La conversación. https://theconversation.com/sidelining-god-why-secular-climate-projects-in-the-pacific-islands-are-failing-77623

OCDE. (2001). *The DAC Journal: Development Co-operation. 2000 Report*. OCDE.

Organización de las Naciones Unidas para la Agricultura y la Alimentación. (2017). *Iniciativa de crecimiento azul de la Organización de las Naciones Unidas para la Alimentación y la Agricultura. Asociación con los países para alcanzar los objetivos de desarrollo sostenible*. Naciones Unidas.

Orlove, B., Lazrus, H., Hovelsrud, G. K. y Giannini, A. (2014). Sobre los orígenes y las consecuencias de la atención desigual al cambio climático en todo el mundo: Sobre los orígenes y las consecuencias de la atención desigual al cambio climático en todo el mundo. *Antropología actual*, *55*(3), 249-275. doi:10.1086/676298

Pava, M. (2007). A response to getting to the bottom of triple bottom line. *Business Ethics Quarterly*, *17*(1), 105–110. doi:10.5840/beq200717116

Peattie, K. (2001). Hacia la sostenibilidad: La tercera era del marketing verde. *The Marketing Review*, *2*(2), 129–146. doi:10.1362/1469347012569869

Peattie, K., & Crane, A. (2005), Green marketing: Legend, myth, farce, or prophesy? *Investigación cualitativa de mercado: una revista internacional*, *8*(4), 357–370. doi:10.1108/13522750510619733

Peet, R., & Watts, M. (1996) Ecología de la liberación: desarrollo, sostenibilidad y medio ambiente en una era de triunfalismo de mercado. Richard Peet & Michael Watts (Eds) Liberation Ecologies: Environment, Development, Social Movements. Routledge.

Petterson, M., Kensek, K., & Noble, D. (2017). Pieles flexibles: Considerando la relevancia, escalabilidad y estrategias de diseño para la resiliencia del sistema de fachadas. *Revista de Educación Arquitectónica*, *71*(1), 34–45.

Pfaffenbichler, P., Emberger, G., Shepherd, S. (2010) Un enfoque de dinámica de sistemas para el modelado de interacción de transporte de uso de la tierra: el modelo estratégico MARS y su aplicación. *Revisión de dinámica de sistemas*, 26, 262 -282. DO - doi:10.1002/sdr.451

Pimenta-Bueno, J. A. (2002). Parque de Innovación Tecnológica y Cultural de Gávea. *Documento interno*.

Polanyi, K. (2001). La gran transformación: Los orígenes políticos y económicos de nuestro tiempo. *Beacon Books*. (Trabajo original publicado en 1947)

Por Melo, M.A.C. (2002). *Innovación y modernización tecnológica y organizacional en las MIPYMES: el dominio interorganizacional*.

Press, M., & Arnould, E. J. (2014). Transparencia narrativa. *Journal of Marketing Management*, *30*(13–14), 1353–1376. doi:10.1080/0267257X.2014.925958

Programa de las Naciones Unidas para el Medio Ambiente. (2012). *Green economy in a blue world: Synthesis report*. Naciones Unidas.

Radcliffe-Brown, A. R. (1940). Sobre la estructura social. *J. R. Antropol. Inst. Gran H. Irel., 1940*(70), 1–12.

Rawes, E. (2019). Los mejores secadores para 2019. *Tendencias digitales.* https://www.digitaltrends.com/home/best-dryers/

Raworth, K. (2017). *Economía del donut. Siete maneras de pensar como un economista del siglo 21; Random House*. Estados Unidos.

Rivers, W. H. R. (Ed.). (1922). *Ensayos sobre la despoblación de Melanesia; Cambridge University Press*. Reino Unido.

Rockström, J., Steffen, W., Noone, K., Persson, Å., Chapin, F. S. III, Lambin, E., ... Foley, J. (2009). Límites planetarios: Explorando el espacio operativo seguro para la humanidad. *Ecology and Society, 14*(2).

Rudiak-Gould, P. (2011). Cambio climático y antropología: La importancia de los estudios de recepción. *Antropología Hoy, 27*(2), 9–12. doi:10.1111/j.1467-8322.2011.00795.x

Sahlins, M. (1976). *Razón práctica*. Aldine.

Sahlins, M. (1996). La tristeza de la dulzura: La antropología nativa de la cosmología occidental. *Current Anthropology, 37*(junio), 395–428. doi:10.1086/204503

Savitz, W., & Weber, K. (2006). *The Triple Bottom Line*. Jossey-Bass.

Schaltegger, M. B., & Burritt, R. (2006), Sustainability Accounting and Reporting. Development, Linkages and Reflection: An Introduction. Springer. doi:10.1080/13511610304521

Shove, E. (2003a). Convenciones convergentes de comodidad, limpieza y conveniencia. *Journal of Consumer Policy, 26*(diciembre), 395–418. doi:10.1023/A:1026362829781

Shove, E. (2003b). *Usuarios, tecnologías y expectativas de confort, limpieza y conveniencia. Innovación: The European Journal of Social Slater, D., & Tonkiss, F. (2001). Sociedad de mercado: mercados y teoría social moderna*. Prensa Política.

Sisaye, S. (2012), An ecological analysis of four competing approaches to sustainability development. *Integration of industrial ecology and ecological anthropology literature.*

Slater, D. y Tonkiss, F. (2001). Sociedad de mercado: mercados y teoría social moderna. Prensa Política.

Sörlin, S. (2007). *El Antropoceno. Un ensayo sobre las edades del hombre; Weylers*. Estocolmo, Suecia.

Stoner, A. M. (2021). Las cosas están empeorando en nuestro camino hacia la catástrofe: el ecologismo neoliberal, la desublimación represiva y el ecoconsumidor autónomo. *Sociología crítica, 47* (mayo), 491-506. doi:10.1177/0896920520958099

Stoner, A. M. (2021). Las cosas están empeorando en nuestro camino hacia la catástrofe: El ecologismo neoliberal, la desublimación represiva y el ecoconsumidor autónomo. *Sociología crítica, 47* (mayo), 491-506.

Tarfaoui, D., & Salah Zkim, S. (2017). Brecha actitud-comportamiento ecológico: Un análisis teórico. [ESMB]. *International Journal of Economics & Strategic Management of Business Process*, *8*(1), 33–38.

Taylor, C. (2002). Imaginarios sociales modernos. *Cultura pública, 14*(1), 91–124. doi:10.1215/08992363-14-1-91

Thomas, N. J. R. (2018). Marketing de sostenibilidad. La necesidad de un enfoque realista de sistemas completos. *Journal of Marketing Management*, *34*(17–18), 1530–1556. doi:10.1080/0267257X.2018.1547782

Toledo, V. M. (1994). *Biodiversity islands in a sea of pasturelands: indigenous resource management in the wet tropics of Mexico Etnoecológica*. Etnoecológica. http://www.etnoecologica.org.mx/Etnoecologica_vol2_n3/art_toledo.htm

Toledo, V. M. (1996). Principios etnoecológicos para el desarrollo sustentable de comunidades campesinas e indígenas. *Documentos Red Latino Americana y Caribeña de Ecología Social.*

Toledo, V. M. (1997) Sustainable development at the village community level: a Third World perspective in: Fraser Smith (Ed.) Environmental Sustainability Boca Raton St. Lucie Press (pp. 233–250). doi:10.1201/9781420048742.ch11

Trosper, R. L. (2009). *Resiliencia, reciprocidad y economía ecológica: sostenibilidad de la costa noroeste*. Routledge. doi:10.4324/9780203881996

Trujillo, M. L. N. (2021). Apuntes para una visión crítica y ecológica del capitalismo patriarcal en la red de la vida. *Capital and Class*, *45*(marzo), 21–32. doi:10.1177/0309816820929115

Unerman, J., O'Dwyer, B., & Bebbington, J. (Eds.). (2007). *Contabilidad de sostenibilidad y rendición de cuentas*. Routledge.

Vann, J.W. & White, G.B. (2004). *Sustainability reporting in the accountingcurriculum*. Research Gate.

Vetlesen, A. J. (2020). Cosmologías del Antropoceno. Panpsiquismo, animismo y los límites del posthumanismo. Routledge.

Weston, A. (1995). Habilitación de la práctica Ambiental. Christine Pierce & Donald VanDeVeer (Eds) People, Penguins and Plastic Trees Belmond Wadsworth.

Wolford, W. (2021). El Plantationocene: Una contribución lusotrópica a la teoría. *Annals of the American Association of Geographers*, *111*, 1622–1639.

Wymer, W., & Polonsky, M. J. (2015). Las limitaciones y potencialidades del marketing verde. *Journal of Nonprofit & Public Sector Marketing*, *27*(3), 239–262. doi:10.1080/10495142.2015.1053341

Yngfalk, C. (2019). Subvertir la sostenibilidad: Trabajos de mantenimiento del mercado y reproducción de la irresponsabilidad corporativa. *Journal of Marketing Management*, *35*(17–18), 1563–1583. doi:10.1080/0267257X.2019.1682031

ZanL.LusianiM. (2011). Managing Machu Picchu: Institutional Settings, Business Model and Master Plans - A Virtual Appendix (22 de enero de 2011). Disponible en SSRN: https://ssrn.com/abstract=1745322 o doi:10.2139/ssrn.1745322

Zedler, J. B., & Stevens, M. L. (2018). Conocimiento ecológico occidental y tradicional en la restauración ecocultural. *San Francisco Estuary and Watershed Science*, *16*, 1–18.

Zeng, X., Radford, S., & Mourali, M. (2017). Antecedentes y consecuencias de las creencias ambientales: Un meta-análisis. En *AMA Marketing & Public Policy Academic Conference Proceedings* (*Vol. 27*, pp. 55–56). Asociación Americana de Marketing.

Zhang, Z. H., Jhaveri, D. J., Marshall, V. M., Bauer, D. C., Edson, J., & Narayanan, R. K. (2014). Un estudio comparativo de técnicas para el análisis de expresión diferencial en datos de RNA-Seq. *PLoS One*, *9*(8), e103207. doi:10.1371/journal.pone.0103207 PMID:25119138

Zoomers, A., Gekker, A., & Schäfer, M. T. (2016). Entre dos exageraciones: ¿Ayudará el big data a desentrañar los puntos ciegos en la comprensión de la fiebre global de la tierra? *Geoforum*, *69*, 147–159. doi:10.1016/j.geoforum.2015.11.017

Chapter 3
Detecting Ecological Changes for Sustainable Futures

Harish Chandra
https://orcid.org/0000-0002-0448-0174
Department of Botany and Microbiology, Gurukula Kangri University (Deemed), Haridwar, India

Ashok K. Rathoure
https://orcid.org/0000-0001-9131-1346
Saraca Research Incorporation, India

ABSTRACT

The chapter delves into the imperative task of detecting ecological changes to ensure the sustainability of our planet's ecosystems. It emphasizes the multidisciplinary approach required, combining field observations, remote sensing technologies, data analytics, and community engagement. Through tracking key indicators such as biodiversity, species abundance, and environmental parameters, scientists unveil subtle shifts and overarching trends in ecosystems worldwide. From forests to oceans, diverse habitats undergo scrutiny, aided by sophisticated analytical techniques that sift through vast datasets to identify patterns and predict future trajectories. Understanding the causes and consequences of ecological changes is essential for effective management and conservation, with human activities and natural factors often playing significant roles. The chapter concludes with a call to action for collective efforts towards sustainability, highlighting the role of individuals, communities, governments, and organizations in safeguarding the health and integrity of Earth's ecosystems.

INTRODUCTION

The intricate web of life on our planet is constantly in flux, shaped by both natural forces and human activities. Yet, amidst this dynamic tapestry, subtle shifts or abrupt alterations in ecosystems can signal significant changes with far-reaching consequences. Detecting these ecological changes is paramount for understanding the health of our planet and steering toward sustainable futures. In this introduction, we explore the critical importance of detecting ecological changes, the methods employed in this endeavor,

DOI: 10.4018/979-8-3693-3330-3.ch003

and the implications for shaping a more resilient and harmonious relationship between humanity and the natural world. Our planet is facing unprecedented challenges, from climate change and habitat destruction to pollution and species loss. These pressures are altering the delicate balance of ecosystems, threatening biodiversity, ecosystem services, and human well-being. Recognizing the urgency of the situation, scientists, conservationists, policymakers, and communities worldwide are turning their attention to detecting and understanding ecological changes (Sala et al., 2000).

Detecting ecological changes involves a multidisciplinary approach that integrates various methods and technologies. From traditional field surveys to cutting-edge remote sensing techniques, researchers employ a range of tools to monitor key indicators such as biodiversity, species abundance, habitat quality, and environmental parameters. By collecting and analyzing data over time, scientists can identify trends, patterns, and anomalies that reveal the state of ecosystems and the drivers of change. Understanding the causes and consequences of ecological changes is essential for effective conservation and management. Human activities, such as deforestation, urbanization, overexploitation of natural resources, and pollution, are often major drivers of ecological changes. Climate change exacerbates these pressures, leading to shifts in temperature, precipitation patterns, and habitat suitability. In addition to human impacts, natural factors such as disease outbreaks, natural disasters, and ecological succession can also contribute to ecological changes (IPBES, 2019).

The implications of ecological changes extend far beyond the realm of nature. They affect the provision of ecosystem services, such as clean air, fresh water, and fertile soil, which are vital for human well-being and economic development. Moreover, they impact vulnerable communities that depend directly on ecosystems for their livelihoods, food security, and cultural identity. In response to these challenges, there is a growing recognition of the need for concerted action to detect, understand, and address ecological changes. By harnessing the power of science, technology, and community engagement, we can pave the way toward more sustainable futures. Through informed decision-making, conservation initiatives, sustainable land-use practices, and policy interventions, we can mitigate the negative impacts of ecological changes and promote ecosystem resilience (Steffen et al., 2015).

NEED FOR VIGILANCE

Our planet's ecosystems are facing unprecedented challenges, from climate change and habitat destruction to pollution and invasive species. These pressures are altering the delicate balance of ecosystems, threatening the stability and resilience of natural systems. To effectively address these challenges, we must first detect and understand the ecological changes occurring in our environments. In an era marked by accelerating environmental transformations, the imperative to detect ecological changes has emerged as a cornerstone for steering humanity towards sustainable futures. Ecosystems, the intricate tapestries of life interwoven with habitats, species, and natural processes, serve as the life support systems for our planet. However, these ecosystems are increasingly under siege from a barrage of anthropogenic pressures and natural forces, resulting in disruptions that reverberate across ecological, social, and economic realms (Millennium Ecosystem Assessment, 2005).

The urgency to detect ecological changes stems from a recognition of the critical role that ecosystems play in sustaining life on Earth. Ecosystems provide a myriad of services essential for human well-being, including clean air and water, fertile soils, pollination, climate regulation, and the provision of food and medicine. Yet, these services are under threat as ecosystems undergo unprecedented alterations at local, regional, and global scales. At the heart of the need for vigilance lies the understanding that ecological changes are not isolated events but interconnected manifestations of broader environmental trends. Climate change, driven by human-induced greenhouse gas emissions, is altering temperature patterns, precipitation regimes, and the frequency and intensity of extreme weather events. Habitat destruction and fragmentation, primarily driven by land conversion for agriculture, urbanization, and infrastructure development, are eroding biodiversity and disrupting ecological processes. Pollution, from industrial effluents to plastic waste, is contaminating air, water, and soil, posing threats to human health and ecosystem integrity (Devictor et al., 2012).

The consequences of failing to detect ecological changes are profound and far-reaching. Biodiversity loss, driven by habitat destruction, overexploitation, and invasive species, undermines the resilience of ecosystems and reduces their capacity to provide essential services. Declines in pollinator populations jeopardize agricultural productivity and food security, while disruptions in marine ecosystems threaten the livelihoods of coastal communities dependent on fisheries and tourism. Climate change-induced shifts in temperature and precipitation patterns exacerbate water scarcity, crop failures, and natural disasters, exacerbating social inequities and exacerbating conflicts over dwindling resources. To address these challenges and pave the way for sustainable futures, vigilance in detecting ecological changes is imperative. It requires a concerted effort to monitor ecosystems, collect data on key indicators, and analyze trends over time. Remote sensing technologies, such as satellite imagery and aerial drones, offer valuable tools for monitoring changes in land cover, deforestation rates, and habitat fragmentation. Citizen science initiatives empower communities to contribute observations of wildlife sightings, phenological events, and environmental changes, enhancing the spatial and temporal resolution of ecological monitoring efforts.

Moreover, detecting ecological changes necessitates interdisciplinary collaboration and the integration of traditional ecological knowledge with modern scientific approaches. By engaging stakeholders from diverse backgrounds, including scientists, policymakers, indigenous communities, and civil society organizations, we can harness collective wisdom and expertise to identify emerging threats, assess their impacts, and develop effective response strategies. The need for vigilance in detecting ecological changes is paramount for safeguarding the health and resilience of ecosystems and ensuring the well-being of present and future generations. By recognizing the interconnectedness of ecological, social, and economic systems, and embracing a holistic approach to environmental stewardship, we can navigate the complexities of global change and chart a course towards a more sustainable and equitable future for all.

In an era marked by unprecedented environmental challenges, the need for vigilance in detecting ecological changes has never been more pressing. Ecosystems, the complex networks of life that sustain biodiversity and provide essential services to humanity, are under increasing threat from a myriad of stressors including climate change, habitat destruction, pollution, and invasive species. In this context, the ability to swiftly and accurately detect ecological changes is essential for effective conservation, sustainable resource management, and the preservation of Earth's natural heritage for future generations.

UNVEILING NATURE'S SECRETS

Detecting ecological changes requires a multidisciplinary approach, integrating field observations, remote sensing technologies, data analytics, and community engagement. By tracking key indicators such as biodiversity, species abundance, habitat quality, and environmental parameters, scientists can unveil the subtle nuances and overarching trends shaping our ecosystems. The intricate tapestry of nature holds invaluable secrets, woven into the fabric of ecosystems worldwide. Amidst the bustling symphony of life, subtle shifts and profound transformations unfold, shaping the destiny of our planet. In the pursuit of sustainable futures, uncovering these secrets and detecting ecological changes are paramount endeavors. Through the lens of scientific inquiry, technological innovation, and interdisciplinary collaboration, humanity endeavors to decipher nature's language, unlocking insights essential for informed decision-making and responsible stewardship of Earth's precious resources. At the heart of this quest lies the recognition of ecosystems as dynamic entities, subject to a myriad of influences both natural and anthropogenic. Climate change, habitat loss, pollution, and the encroachment of human activities are among the forces driving ecological change at an unprecedented pace. Yet, amidst the challenges lie opportunities for discovery and enlightenment, as scientists, conservationists, and communities unite in a shared mission to monitor, understand, and respond to these changes (Millennium Ecosystem Assessment, 2005).

Remote sensing technologies offer a bird's-eye view of the landscape, allowing us to track changes in land cover, vegetation health, and ecosystem dynamics on a global scale. Satellites orbiting high above capture snapshots of Earth's surface, revealing the intricate patterns of life and the subtle nuances of change. From deforestation in the Amazon to the bleaching of coral reefs in the Pacific, these images bear witness to the unfolding drama of ecological transformation, serving as a powerful tool for monitoring and assessment. But the true essence of nature's secrets lies not just in pixels on a screen, but in the rich tapestry of life that thrives on the ground. Field biologists venture into the wilderness, armed with notebooks and binoculars, to observe and document the intricate interactions of plants, animals, and their environment. From the delicate dance of pollinators in a flower meadow to the majestic migration of wildebeest across the Serengeti plains, these observations offer a window into the soul of nature, revealing its resilience, its fragility, and its boundless beauty (Hansen & Loveland, 2012; Pettorelli et al., 2014).

Yet, amidst the wonder and awe of the natural world, lies a sobering truth: that humanity's actions hold the power to shape the destiny of ecosystems far and wide. As we continue to unravel nature's secrets, we must do so with humility, recognizing our role as stewards of the Earth and custodians of its future. Through sustainable practices, conservation efforts, and a commitment to environmental stewardship, we can forge a path towards a brighter tomorrow, where nature's secrets are preserved, cherished, and celebrated for generations to come (Steffen, 2015).

In the grand tapestry of life, every thread is precious, every stitch a testament to the intricate web of existence that binds us all. As we unveil nature's secrets and detect ecological changes, let us do so with reverence, with wonder, and with a steadfast commitment to ensuring a future where humanity and nature thrive in harmony (IPBES, 2019).

MONITORING DIVERSE HABITATS

No corner of our planet is untouched by ecological changes. From the towering forests of the Amazon to the vast expanses of coral reefs in the ocean depths, ecosystems of all shapes and sizes are under scrutiny. Satellite imagery provides a bird's-eye view of changes in land cover and deforestation rates, while underwater sensors reveal shifts in ocean temperature and acidity levels. On the ground, citizen science initiatives empower communities to contribute valuable data on local wildlife sightings and habitat changes. The Earth's ecosystems span a vast array of habitats, from lush forests teeming with life to expansive oceans brimming with biodiversity. These diverse environments are not only interconnected but also face increasing pressures from human activities and environmental changes. Detecting ecological changes across these varied habitats is critical for understanding the health of ecosystems and charting a course towards sustainability. Through innovative monitoring techniques and interdisciplinary approaches, scientists are uncovering the secrets of nature and guiding efforts to protect these vital ecosystems for future generations (Asner et al., 2009).

Forests: Forests cover approximately 31% of the Earth's land area and play a crucial role in regulating the climate, conserving biodiversity, and providing essential ecosystem services. Monitoring changes in forest ecosystems involves a combination of field surveys, satellite imagery, and remote sensing technologies. For example, LiDAR (Light Detection and Ranging) technology allows researchers to obtain detailed 3D measurements of forest structure and biomass, helping to assess forest health and detect changes such as deforestation, degradation, and disturbances due to natural events or human activities. Additionally, satellite-based monitoring tools provide valuable information on forest extent, fragmentation, and land-use changes over large spatial scales, aiding in conservation planning and management efforts (Costanza et al., 1997).

Oceans: Oceans cover over 70% of the Earth's surface and harbor a vast array of marine life, from microscopic plankton to majestic whales. However, marine ecosystems are increasingly threatened by overfishing, habitat destruction, pollution, and climate change. Monitoring changes in marine habitats requires a combination of field observations, underwater surveys, and satellite remote sensing. Techniques such as marine acoustic surveys and underwater drones allow scientists to study marine biodiversity, habitat structure, and ecosystem dynamics in diverse ocean environments. Satellite imagery, including multispectral and synthetic aperture radar (SAR) data, provides valuable insights into ocean temperature, sea surface height, chlorophyll concentration, and other key parameters, helping to detect changes such as coral bleaching, algal blooms, and sea level rise (Halpern et al., 2008).

Interdisciplinary Approaches: Effective monitoring of diverse habitats requires interdisciplinary collaboration among scientists, policymakers, conservationists, and local communities. By integrating data from multiple sources, including field observations, remote sensing, ecological models, and socio-economic surveys, researchers can gain a comprehensive understanding of ecological changes and their drivers. This holistic approach enables informed decision-making and the development of evidence-based conservation strategies tailored to specific habitats and ecosystems. Moreover, engaging stakeholders and fostering community participation are essential for ensuring the success and sustainability of monitoring efforts, as well as promoting local stewardship of natural resources. From forests to oceans, monitoring diverse habitats is essential for detecting ecological changes and safeguarding the health of ecosystems for present and future generations. By harnessing the power of technology, collaboration, and interdisciplinary approaches, we can unravel the complexities of nature, protect biodiversity, and promote sustainable management of Earth's precious natural resources (Turner et al., 2003; Pettorelli et al., 2014).

POWER OF DATA ANALYTICS

In the age of big data, sophisticated analytical techniques offer unprecedented insights into ecological patterns and processes. Time-series analysis, machine learning algorithms, and spatial modeling tools can sift through vast datasets to identify trends, anomalies, and potential drivers of change. By harnessing the power of data analytics, scientists can unlock hidden patterns and predict future trajectories of ecological systems (Chape et al., 2005; Devictor et al., 2012; Hilty et al., 2018).

UNDERSTANDING CAUSES AND CONSEQUENCES

Detecting ecological changes is only the first step; understanding their underlying causes and consequences is essential for effective management and conservation. Human activities such as deforestation, urbanization, overfishing, and pollution are often major drivers of ecological changes, but natural factors such as climate variability and natural disturbances also play significant roles. By untangling these complex interactions, researchers can inform policies and interventions aimed at mitigating negative impacts and promoting ecosystem resilience (McKinney, 2008; Pimm et al., 2014)).

EMPOWERING ACTION FOR SUSTAINABILITY

Ultimately, detecting ecological changes is not just an academic pursuit but a call to action for individuals, communities, governments, and organizations worldwide. By raising awareness, fostering collaboration, and implementing evidence-based solutions, we can chart a course towards a more sustainable future for all life on Earth. Whether through protected area management, sustainable land-use practices, or conservation education initiatives, each of us has a role to play in safeguarding the health and integrity of our planet's ecosystems (Secretariat of the Convention on Biological Diversity, 2014; World Wildlife Fund, 2022).

CASE STUDY OF COAL MINING PROJECT

Ecological change detection involves assessing alterations in the attributes of a particular area across different time periods. This process typically entails the comparison of aerial photographs or satellite imagery captured at various points in time. Ecological change detection serves several purposes, including:

- Monitoring annual or decadal fluctuations in soil moisture or vegetation health.
- Tracking changes in the depth of the winter snowpack.
- Assessing the impact of human activities on a large scale within the landscape.
- Identifying statistically significant temporal shifts in various physical and ecological parameters.
- Observing alterations in mangrove cover.
- Studying the influence of climate change on ocean ecosystems.

Remote sensing techniques play a crucial role in detecting changes caused by human activities without further disrupting ecologically sensitive areas. The insights gained from remote sensing analyses can inform efficient conservation management strategies for the future. Additionally, remote sensing excels in capturing instantaneous changes, such as those occurring between consecutive image acquisitions.

For decadal changes at study area of 10 km radius from an underground coal mining, we have taken 23 years data considering base year 2000. The satellite imagery of y.2000; 2010; 2020 and 2023 have been procured and false color composite (FCC) and land use maps were prepared using GIS software ArcGIS v.10.8. The FCC and land use maps were prepared on the basis of 23 years data (10 years of interval) as given in table 1 below.

Table 1. Land-use pattern in study area showing changes over a period of 23 years

Class/es	Area (ha)			
Type	y.2000	y.2010	y.2020	y.2023
Agriculture	6014.07	10791.18	19687.86	21613.5
Barren	2731.14	5888.34	2487.69	2411.46
Settlement	748.89	966.69	1587.69	1897.2
Vegetation	32731.2	24401.88	18155.7	16153.65
Water	171.09	348.3	477.45	320.58
Total Area (Ha)	**42396.39**	**42396.39**	**42396.39**	**42396.39**

Table 1 presents the land-use pattern in the study area, showcasing changes over a period of 23 years from 2000 to 2023. The table displays the area (in hectares) for different land-use classes for each year as follows:

Agriculture: The area dedicated to agriculture has steadily increased over the years, starting from 6014.07 hectares in 2000 and reaching 21613.5 hectares by 2023. This indicates significant expansion in agricultural activities within the study area.

Barren: The area classified as barren initially increased from 2731.14 hectares in 2000 to 5888.34 hectares in 2010 but then decreased notably to 2411.46 hectares by 2023. This suggests fluctuations in barren land, possibly due to natural processes or human intervention.

Settlement: The area designated as settlement has shown a consistent increase over the years, rising from 748.89 hectares in 2000 to 1897.2 hectares in 2023. This indicates urbanization or the establishment of residential areas within the study area.

Vegetation: The area covered by vegetation experienced a decline over the years, starting from 32731.2 hectares in 2000 and decreasing to 16153.65 hectares by 2023. This indicates a significant loss of vegetative cover within the study area.

Water: The area occupied by water bodies fluctuated over the years, with slight increases and decreases. It started at 171.09 hectares in 2000, peaked at 477.45 hectares in 2020, and then decreased to 320.58 hectares by 2023.

The total area of the study remained constant at 42396.39 hectares throughout the 23-year period. The interpretation of these changes suggests dynamic shifts in land use, including expansion in agriculture and settlement areas, fluctuations in barren land, significant loss of vegetation cover, and varying patterns in water bodies within the study area over the specified timeframe. The analysis of the land-use pattern in the study area over a 23-year period reveals significant changes and trends in various land-use classes. The key findings include:

Expansion of Agriculture: There has been a notable increase in the area dedicated to agriculture over the years, indicating agricultural expansion and possibly changes in land management practices.

Fluctuations in Barren Land: The area classified as barren has shown fluctuations, with an initial increase followed by a decrease. This suggests dynamic processes affecting barren land within the study area.

Urbanization and Settlement Growth: The steady increase in settlement areas indicates ongoing urbanization or the establishment of residential areas, reflecting demographic and societal changes.

Loss of Vegetation Cover: A significant decline in vegetative cover is observed over the 23-year period, highlighting potential environmental degradation or land-use changes affecting natural ecosystems.

Variability in Water Bodies: The area occupied by water bodies has shown fluctuations, indicating dynamic changes in hydrological processes or human interventions affecting water resources.

These findings underscore the dynamic nature of land use and the complex interactions between natural processes and human activities within the study area. Understanding these changes is crucial for effective land management and conservation efforts to ensure sustainable use of resources and mitigate adverse environmental impacts.

CONCLUSION

As stewards of the Earth, we have a responsibility to monitor, protect, and preserve the rich tapestry of life that sustains us all. By tracking nature's pulse and detecting ecological changes, we can illuminate the path towards a more harmonious relationship with the natural world. Through collective action and a commitment to sustainability, we can ensure that future generations inherit a planet teeming with life, diversity, and beauty. In the end, it is not just about detecting ecological changes; it is about shaping a future where humanity and nature thrive together in harmony. Further research and monitoring are essential to track ongoing trends and inform adaptive management strategies for the future.

REFERENCES

Asner, G. P., Knapp, D. E., Kennedy-Bowdoin, T., Jones, M. O., Martin, R. E., Boardman, J., & Hughes, R. F. (2009). Carnegie Airborne Observatory: In-flight fusion of hyperspectral imaging and waveform light detection and ranging for three-dimensional studies of ecosystems. *Journal of Applied Remote Sensing*, *3*(1), 033543. doi:10.1117/1.3223675

Chape, S., Harrison, J., Spalding, M., & Lysenko, I. (2005). Measuring the extent and effectiveness of protected areas as an indicator for meeting global biodiversity targets. *Philosophical Transactions of the Royal Society of London. Series B, Biological Sciences*, *360*(1454), 443–455. doi:10.1098/rstb.2004.1592 PMID:15814356

Costanza, R., d'Arge, R., De Groot, R., Farber, S., Grasso, M., Hannon, B., & Paruelo, J. (1997). The value of the world's ecosystem services and natural capital. *Nature*, *387*(6630), 253–260. doi:10.1038/387253a0

Devictor, V., van Swaay, C., Brereton, T., Brotons, L., Chamberlain, D., Heliölä, J., Herrando, S., Julliard, R., Kuussaari, M., Lindström, Å., Reif, J., Roy, D., Schweiger, O., Settele, J., Stefanescu, C., van Strien, A., van Turnhout, C., Vermouzek, Z., WallisDeVries, M., & Jiguet, F. (2012). Differences in the climatic debts of birds and butterflies at a continental scale. *Nature Climate Change*, *2*(2), 121–124. doi:10.1038/nclimate1347

Halpern, B. S., Walbridge, S., Selkoe, K. A., Kappel, C. V., Micheli, F., D'Agrosa, C., & Watson, R. (2008). A global map of human impact on marine ecosystems. *Science*, *319*(5865), 948–952. doi:10.1126/science.1149345 PMID:18276889

Hansen, M. C., & Loveland, T. R. (2012). A review of large area monitoring of land cover change using Landsat data. *Remote Sensing of Environment, 122*, 66–74. doi:10.1016/j.rse.2011.08.024

Hilty, J., Merenlender, A. M., & Daily, G. C. (Eds.). (2018). *The Ecology of Climate Change: Responses and Consequences of Natural and Managed Ecosystems*. Academic Press.

IPBES. (2019). *Global assessment report on biodiversity and ecosystem services of the Intergovernmental Science-Policy Platform on Biodiversity and Ecosystem Services* (E. S. Brondizio, J. Settele, S. Díaz, & H. T. Ngo, Eds.). IPBES Secretariat.

McKinney, M. L. (2008). Effects of urbanization on species richness: A review of plants and animals. *Urban Ecosystems, 11*(2), 161–176. doi:10.1007/s11252-007-0045-4

Millennium Ecosystem Assessment. (2005). *Ecosystems and Human Well-being: Synthesis*. Island Press.

Pettorelli, N., Safi, K., & Turner, W. (Eds.). (2014). *Satellite Remote Sensing for Conservation Action*. Oxford University Press.

Pimm, S. L., Jenkins, C. N., Abell, R., Brooks, T. M., Gittleman, J. L., Joppa, L. N., Raven, P. H., Roberts, C. M., & Sexton, J. O. (2014). The biodiversity of species and their rates of extinction, distribution, and protection. *Science, 344*(6187), 1246752. doi:10.1126/science.1246752 PMID:24876501

Sala, O. E., Chapin, F. S., Armesto, J. J., Berlow, E., Bloomfield, J., Dirzo, R., & Wall, D. H. (2000). Global biodiversity scenarios for the year 2100. *Science, 287*(5459), 1770–1774. doi:10.1126/science.287.5459.1770 PMID:10710299

Secretariat of the Convention on Biological Diversity. (2014). *Global Biodiversity Outlook 4*. Secretariat of the Convention on Biological Diversity.

Steffen, W., Richardson, K., Rockström, J., Cornell, S. E., Fetzer, I., Bennett, E. M., Biggs, R., Carpenter, S. R., de Vries, W., de Wit, C. A., Folke, C., Gerten, D., Heinke, J., Mace, G. M., Persson, L. M., Ramanathan, V., Reyers, B., & Sörlin, S. (2015). Planetary boundaries: Guiding human development on a changing planet. *Science, 347*(6223), 1259855. doi:10.1126/science.1259855 PMID:25592418

Turner, W., Spector, S., Gardiner, N., Fladeland, M., Sterling, E., & Steininger, M. (2003). Remote sensing for biodiversity science and conservation. *Trends in Ecology & Evolution, 18*(6), 306–314. doi:10.1016/S0169-5347(03)00070-3

World Wildlife Fund. (2022). Living [Bending the Curve of Biodiversity Loss. World Wildlife Fund International.]. *The Planetary Report*.

Chapter 4
Assessing Ecological Risk: Methodologies and Approaches for Biodiversity Loss Evaluation

Bindurani L. G. P. Ram
SGMSPM's Dnyanvilas College of Pharmacy, India

Sachin A. Fegade
SJVPM's Rasiklal M. Dhariwal College of Pharmacy, Chinchwad, India

ABSTRACT

Assessing ecological risk and evaluating biodiversity loss are essential components of conservation planning and management. This chapter examines methodologies and approaches used in these assessments to understand the complex interactions between human activities and ecosystems. The chapter explores various techniques, including measuring species richness and abundance, assessing habitat fragmentation and connectivity, evaluating threats to biodiversity, and modeling ecosystem dynamics under different scenarios. Additionally, the importance of stakeholder engagement and participatory approaches in conservation decision-making is highlighted. By integrating these methodologies and approaches, researchers and policymakers can gain valuable insights into ecological risk and develop effective strategies to mitigate biodiversity loss and safeguard ecosystems for future generations.

INTRODUCTION

The approach for biodiversity loss evaluation is a comprehensive study of various methodologies and approaches used in evaluating ecological risks. The field of ecological risk assessment has evolved over the years, shifting its focus from human health assessments to ecological risk assessments. The risk quotient method is one of the most widely used approaches in ecological risk assessment. It is a practical approach used in the context of environmental protection legislation. The method involves assessing the adverse effects of human interactions with the environment. Ecological risk assessment is a diagnostic tool used to address negative effects of pollutants on the environment and living organisms. It involves using chemical and biological techniques to measure and assess the impacts of external disturbances

DOI: 10.4018/979-8-3693-3330-3.ch004

such as anthropogenic stress, environmental pollution, or natural disasters on regional ecosystems (Magurran, 2004).

The Landscape Pattern Index is a widely used method to assess ecological risks. It is used to quantitatively identify and directly assess ecological risks from the perspective of spatial landscape patterns caused by land-use changes. Land use is considered a comprehensive reflection of the direct impact of human economic and social activities on surface resources and the natural environment. The spatiotemporal heterogeneity of land use is affected by regional terrains and geomorphic features closely related to landscape patterns and landscape ecological risks (IUCN, 2019). The ecological risk of land-use change has obvious stages. Analyzing and studying the evolution of the ecological risks of land-use changes on medium and long timescales can provide a scientific and reasonable basis for planning regional ecological protection schemes. The ecological risk assessment is a vital tool for understanding and managing the impacts of human activities on the environment (Millennium Ecosystem Assessment, 2005). It helps us to identify and mitigate potential risks to ecosystems and biodiversity, ensuring the long-term health and sustainability of our planet (Fahrig, 2003).

Biodiversity loss represents one of the most significant environmental challenges of our time, with far-reaching implications for ecosystems, economies, and human well-being. Understanding the ecological risk associated with biodiversity loss is crucial for effective conservation planning and management. This article explores various methodologies and approaches employed in assessing ecological risk and evaluating biodiversity loss.

Ecological risk assessment (ERA) is a crucial tool for understanding and managing the potential negative impacts of human activities on ecosystems. It is a systematic process that involves identifying, evaluating, and characterizing the likelihood and severity of ecological harm. Here's a breakdown of the methodologies and approaches used in ERA:

PHASES OF ECOLOGICAL RISK ASSESSMENT

A comprehensive ERA typically follows a four-phase approach:

- **Problem Formulation:** This initial phase defines the scope of the assessment. It involves identifying the potential stressors (e.g., pollutants, habitat destruction), the ecological resources of concern (e.g., endangered species, sensitive habitats), and the assessment endpoints (measurable indicators of ecological health).
- **Exposure Assessment:** This phase focuses on determining the likelihood and extent of exposure of ecological receptors to the identified stressors. Factors like environmental fate and transport of pollutants, habitat use patterns of species, and potential co-occurrence with stressors are considered.
- **Effects Assessment:** This phase evaluates the potential effects of the stressors on the identified ecological receptors. It involves toxicity data, field studies, and ecological modelling to predict the magnitude and severity of harm.
- **Risk Characterization:** The final phase integrates the findings from the previous phases. It involves characterizing the overall ecological risk by combining the likelihood and severity of potential effects. Risk communication is crucial at this stage, clearly presenting the findings to stakeholders and decision-makers.

Table 1. Phased approach in the environmental risk assessment

Stage in regulatory evaluation	Stage in risk assessment	Objective	Method	Data Requirement
Phase I	Pre-screening	Estimation of exposure	Action limit	Consumption data, logKow
Phase II Tier A	Screening	Initial prediction of risk	Risk Assessment	Base set aquatic toxicology and fate
Phase II Tier B	Extended	Substance and compartment-specific refinement and risk assessment	Risk Assessment	Extended data set on emission, fate and effects

Table 1 provide the phase wise approach in the environment risk assessment.
Here's a breakdown of the possible pathways of medicines:

Entry Points

- **Prescription** - This is the most common pathway, where a doctor prescribes a medication to a patient.
- **Over-the-counter (OTC) Sales** - Certain medications are available for purchase without a prescription.

Pathways

- **Excretion:** Medications are processed by the body and eliminated through waste products like urine or feces.
- **Storage:** Medications may be stored at home or in a healthcare facility before use.
- **Disposal:** Unused or expired medications should be disposed of properly to avoid environmental contamination. This could involve take-back programs or specific disposal methods depending on the medication.

Possible Subsequent Steps

- **Sewage treatment plant:** If medications are flushed down the drain, they may enter the sewage treatment system.
- **Landfill site:** Improperly disposed of medications may end up in landfills.
- **Incineration:** In some cases, incineration might be used for specific medication disposal. This can introduce air pollutants if not properly managed.
- **Soil and Surface Water Contamination:** Medications from wastewater treatment plants, landfills, or improper disposal can contaminate soil and surface water.
- **Groundwater Contamination:** Contaminated soil and surface water can eventually reach groundwater sources.
- **Drinking Water:** In the worst-case scenario, contaminated groundwater may be used as a source of drinking water, potentially exposing people to residual medications.

METHODOLOGIES FOR ECOLOGICAL RISK ASSESSMENT

Within these phases, various methodologies are employed depending on the specific stressors and ecosystems under study:

- **Site-Specific Assessments:** These assessments focus on a specific location and potential contaminant sources. They may involve detailed field studies, monitoring programs, and modeling to assess exposure and effects.
- **Landscape-Level Assessments:** These assessments consider the broader ecological context and potential cumulative effects of multiple stressors across a larger landscape. They often utilize Geographic Information Systems (GIS) and landscape ecology principles.
- **Species-Specific Assessments:** These assessments focus on the potential impacts of stressors on specific species of concern, particularly threatened or endangered populations. They may involve detailed ecological studies of species life history and habitat requirements.

Considerations and Challenges

- **Data Availability:** Reliable data on ecological receptors, stressors, and their interactions may be limited, requiring estimations and use of best available science.
- **Uncertainty:** Ecological systems are complex, and predicting precise outcomes can be challenging. It is important to acknowledge uncertainties and incorporate them into the risk characterization.
- **Standardization:** Standardized methodologies ensure consistency and comparability of assessments across different ecosystems and projects.
- **Stakeholder Involvement:** Effective communication and collaboration between scientists, risk assessors, policymakers, and stakeholder communities are crucial for successful ERA implementation.

Ecological risk assessment provides a valuable framework for evaluating potential environmental harm and informing risk management decisions (Carpenter et al., 2009). By employing various methodologies and approaches, and addressing the inherent challenges, ERA can be a powerful tool for protecting our ecosystems and ensuring sustainable practices.

Species Richness and Abundance

One fundamental approach to assessing ecological risk is through the measurement of species richness and abundance. Species richness refers to the number of different species present in a given area, while abundance measures the number of individuals within each species. Long-term monitoring of changes in species richness and abundance provides insights into ecosystem health and potential risks to biodiversity (Cash et al., 2003; Peterson et al., 2003; Reed et al., 2009).

Species richness refers to the number of different species present in a given area or ecosystem. High species richness indicates a diverse and healthy ecosystem, while low species richness may indicate disturbances or environmental stressors. Ecological risk assessments often consider changes in species richness as indicators of ecosystem health or degradation. A decrease in species richness could signal ecosystem disruption due to factors such as pollution, habitat loss, invasive species, or climate change (Peterson et al., 2003; Reed et al., 2009).

Abundance refers to the number of individuals of a particular species within a given area. High abundance of species can indicate a healthy population, while low abundance may indicate population decline or other stressors. Assessing changes in species abundance over time can provide insights into the impacts of environmental stressors and human activities on populations within an ecosystem. For example, a decline in the abundance of a keystone species could have cascading effects on other species and ecosystem functions (Daniels & Walker, 2001).

In ecological risk assessments, monitoring and analyzing changes in both species richness and abundance help scientists and policymakers understand the overall health and resilience of ecosystems. These assessments may involve field surveys, data analysis, modeling, and other techniques to quantify changes in species composition and population dynamics over time. Additionally, assessing species richness and abundance allows for the identification of vulnerable or sensitive species that may be at risk from environmental stressors, helping to prioritize conservation efforts and inform management strategies (Fischer, 2000).

Habitat Fragmentation and Connectivity

Habitat fragmentation, caused by human activities such as urbanization and infrastructure development, poses a significant risk to biodiversity. Assessing habitat fragmentation involves quantifying the degree to which natural habitats are broken up into smaller, isolated patches. Connectivity analysis examines the ability of species to move between habitat patches, which is essential for maintaining gene flow and population viability.

Habitat fragmentation and connectivity are critical considerations in ecological risk assessment, particularly when evaluating the impacts of human activities on ecosystems. Here's how they factor into the assessment process:

i. *Habitat Fragmentation*: This refers to the process by which large, continuous habitats are broken up into smaller, isolated patches. Human activities such as urbanization, agriculture, and infrastructure development often lead to habitat fragmentation. Fragmentation can have several ecological consequences, including:
    - Reduction of habitat area: Smaller habitat patches may not be able to support viable populations of certain species, leading to local extinctions.
    - Increased edge effects: Fragmented habitats have a greater proportion of edge habitat, which can experience different environmental conditions than the interior of the habitat. Edge effects can include increased exposure to invasive species, predation, and changes in microclimate.
    - Disruption of species interactions: Fragmentation can disrupt ecological processes such as dispersal, migration, and gene flow, leading to changes in species composition and ecosystem function.

ii. *Connectivity*: Connectivity refers to the degree to which habitats are connected or linked through corridors, dispersal routes, or other means. Maintaining connectivity is essential for facilitating the movement of species between habitat patches, allowing for genetic exchange, population dispersal, and species recolonization. Connectivity helps mitigate the negative effects of habitat fragmentation by:

- Allowing for species movement: Connected habitats provide corridors for species to disperse and migrate between patches, helping to maintain genetic diversity and population viability.
- Enhancing resilience: Connected habitats are more resilient to disturbances such as climate change, as species can move to more suitable habitats as conditions change.
- Supporting ecosystem functions: Connectivity facilitates the flow of energy, nutrients, and species interactions across landscapes, contributing to ecosystem stability and health.

In ecological risk assessments, evaluating habitat fragmentation and connectivity involves mapping habitat patches, assessing landscape features that affect connectivity (such as roads, rivers, and barriers), and modeling species movement and population dynamics. By understanding the spatial configuration of habitats and the degree of connectivity between them, scientists can assess the potential impacts of habitat fragmentation on species populations, biodiversity, and ecosystem functioning. This information is crucial for informing land-use planning, conservation strategies, and habitat restoration efforts aimed at mitigating the negative effects of habitat fragmentation on ecosystems.

Threat Assessment and Species Vulnerability

Identifying and assessing threats to biodiversity is crucial for understanding ecological risk. Threats may include habitat destruction, pollution, climate change, invasive species, and overexploitation of resources. Evaluating species vulnerability involves considering factors such as population size, reproductive rate, habitat specialization, and sensitivity to environmental changes (Leach, 2001).

Threat assessment and species vulnerability are central components of ecological risk assessment, providing insights into the potential impacts of various stressors on individual species, populations, and ecosystems. Here's how they are typically addressed:

i. Threat Assessment:
   - Identifying Stressors: The first step in threat assessment involves identifying and characterizing the stressors or threats that may affect species and ecosystems. These stressors can include natural events (e.g., climate change, natural disasters) as well as human activities (e.g., habitat destruction, pollution, overexploitation).
   - Quantifying Exposure: Assessing the degree of exposure of species and ecosystems to identified stressors is crucial. This involves evaluating the spatial and temporal distribution of stressors and the extent to which species and habitats are affected.
   - Evaluating Effects: Understanding the potential effects of stressors on species and ecosystems is essential for assessing risk. This may involve empirical studies, modeling, and expert judgment to predict how stressors may impact species populations, biodiversity, and ecosystem functions.

ii. Species Vulnerability:
   - Assessing Sensitivity: Species vulnerability refers to the susceptibility of species to adverse effects from stressors. Sensitivity assessments consider intrinsic biological traits (e.g., reproductive rate, habitat specialization, dispersal ability) that influence a species' ability to cope with environmental changes.
   - Evaluating Adaptive Capacity: In addition to sensitivity, species vulnerability assessments consider the adaptive capacity of species to respond to changing environmental conditions.

Species with higher adaptive capacity may be better able to withstand or recover from stressors compared to those with lower adaptive capacity.
- Integrating Multiple Factors: Vulnerability assessments often integrate multiple factors, including species' ecological traits, population status, and exposure to stressors, to provide a comprehensive understanding of species' vulnerability to environmental change.

In ecological risk assessments, threat assessment and species vulnerability are used to prioritize management actions, inform conservation strategies, and guide decision-making processes aimed at minimizing the impacts of stressors on species and ecosystems (Wondolleck et al., 2000). By identifying key threats and vulnerable species, scientists and policymakers can develop targeted interventions to mitigate risks and enhance the resilience of ecosystems in the face of environmental change. Additionally, ongoing monitoring and adaptive management strategies are essential for assessing the effectiveness of conservation measures and adjusting management actions as new information becomes available.

Ecosystem Services and Resilience

Biodiversity loss can compromise the provision of ecosystem services, such as pollination, water purification, and carbon sequestration, which are vital for human well-being. Assessing ecosystem services helps quantify the benefits that ecosystems provide to society and evaluate the risks associated with their degradation. Resilience assessment examines the ability of ecosystems to withstand and recover from disturbances, providing insights into their long-term viability.

Ecosystem services and resilience are essential considerations in ecological risk assessment, as they provide a framework for understanding the benefits that ecosystems provide to human well-being and the capacity of ecosystems to withstand and recover from disturbances. Here's how they are typically addressed:

i. Ecosystem Services:
   - Identification and Classification: Ecosystem services are the benefits that people obtain from ecosystems, such as food provision, water purification, climate regulation, and cultural and recreational opportunities. In ecological risk assessment, identifying and classifying ecosystem services helps to understand the linkages between ecosystem health, human activities, and well-being.
   - Valuation and Quantification: Quantifying the value of ecosystem services helps to make their importance tangible for decision-makers. Economic valuation methods can be used to estimate the monetary value of ecosystem services, while other approaches, such as ecological or social valuation, focus on non-monetary aspects.
   - Trade-off Analysis: Assessing trade-offs between different ecosystem services and between ecosystem services and other societal goals (e.g., economic development) is crucial for sustainable decision-making. Understanding trade-offs helps to identify potential conflicts and synergies among ecosystem services and inform management strategies that optimize benefits while minimizing negative impacts.

ii. Resilience:
   - Definition and Assessment: Resilience refers to the capacity of ecosystems to absorb and recover from disturbances while maintaining essential structure, function, and feedbacks.

Assessing ecosystem resilience involves evaluating the ability of ecosystems to resist, absorb, adapt to, and recover from changes and disturbances.

- Identifying Drivers of Resilience: Understanding the factors that contribute to ecosystem resilience is essential for assessing and managing risks. These factors may include biological diversity, connectivity, habitat heterogeneity, and adaptive management practices.
- Enhancement and Restoration: Ecological risk assessments can inform strategies to enhance and restore ecosystem resilience by identifying key vulnerabilities and implementing measures to strengthen ecosystem functions and services. This may involve habitat restoration, conservation planning, and adaptive management approaches that build on natural ecosystem processes.

In ecological risk assessment, integrating ecosystem services and resilience provides a holistic understanding of the relationships between ecosystems, human well-being, and environmental change. By considering the values that ecosystems provide to society and their capacity to adapt to disturbances, decision-makers can develop more effective strategies for sustainable management and conservation that safeguard both ecological integrity and human welfare.

Ecological Modeling and Scenario Analysis

Ecological modeling allows researchers to simulate the impacts of different scenarios on biodiversity and ecosystem dynamics. Scenario analysis involves developing alternative future scenarios based on different socio-economic and environmental drivers, such as land use change and climate change. These models and scenarios help assess ecological risk under different conditions and inform decision-making processes.

Ecological modeling and scenario analysis are integral components of ecological risk assessment, providing tools for understanding complex ecological systems and predicting the potential impacts of stressors. Here's how they are typically used:

i. Ecological Modeling:
   - Understanding Ecosystem Dynamics: Ecological models simulate the dynamics of ecosystems, including the interactions between species, populations, and their environment. These models can range from simple conceptual models to complex mathematical representations of ecological processes.
   - Predicting Population Dynamics: Ecological models can be used to predict changes in population dynamics in response to environmental stressors. For example, population models may stimulate the growth, reproduction, and mortality rates of species under different scenarios.
   - Assessing Habitat Suitability: Habitat suitability models predict the distribution and abundance of species based on environmental variables such as temperature, precipitation, and habitat characteristics. These models help assess the potential impacts of habitat loss, fragmentation, and degradation on species populations.
   - Quantifying Ecological Interactions: Models of trophic interactions, competition, and mutualism help quantify the relationships between species and understand how changes in one species or environmental variable can cascade through ecological communities.

ii. Scenario Analysis:
   - Exploring Alternative Futures: Scenario analysis involves developing and evaluating different plausible future scenarios to assess their potential impacts on ecosystems. Scenarios may vary in terms of land-use changes, climate conditions, pollution levels, or other factors relevant to ecological risk assessment.
   - Identifying Drivers of Change: Scenario analysis helps identify the key drivers of change that may affect ecosystems and species populations. By examining alternative scenarios, scientists can assess the relative importance of different stressors and prioritize management actions accordingly.
   - Evaluating Management Options: Scenario analysis can be used to evaluate the effectiveness of different management options in mitigating risks to ecosystems. By comparing outcomes across scenarios, decision-makers can identify strategies that enhance ecosystem resilience and promote sustainable management practices.
   - Informing Adaptive Management: Scenario analysis provides a framework for adaptive management by exploring how ecosystems may respond to different management interventions over time. By iteratively updating models and scenarios based on new data and feedback, managers can improve their understanding of ecosystem dynamics and make informed decisions in the face of uncertainty.

In ecological risk assessment, ecological modeling and scenario analysis help integrate scientific knowledge, uncertainty, and stakeholder perspectives to assess the potential impacts of stressors on ecosystems and inform decision-making processes aimed at minimizing risks and promoting ecosystem resilience. These tools are particularly valuable for addressing complex, dynamic systems where traditional empirical approaches may be limited.

Cumulative Effects Assessment (CEA)

Ecological risk assessment should consider the cumulative effects of multiple stressors on biodiversity and ecosystems. Cumulative effects assessment evaluates the combined impacts of various human activities, including land development, pollution, and habitat loss, on ecological systems. Understanding these cumulative effects is essential for prioritizing conservation actions and mitigating biodiversity loss.

Cumulative effects assessment (CEA) in ecological risk assessment involves evaluating the combined impacts of multiple stressors on ecosystems, species, and ecological processes over time. Unlike traditional risk assessments, which typically focus on single stressors or activities, CEA considers the additive, synergistic, or antagonistic effects of multiple stressors operating concurrently or sequentially. Here's how CEA is typically conducted:

i. *Identifying Stressors:* The first step in CEA is to identify and characterize the various stressors affecting the ecosystem. These stressors can include natural processes (e.g., climate change, natural disturbances) as well as human activities (e.g., land use change, pollution, resource extraction).

ii. *Assessing Interactions:* CEA evaluates how different stressors interact with one another and with the ecosystem. Stressors may have additive effects (where the combined impact is the sum of individual effects), synergistic effects (where the combined impact is greater than the sum of individual effects), or antagonistic effects (where the combined impact is less than the sum of individual effects).

iii. *Evaluating Cumulative Effects:* CEA quantifies the cumulative effects of multiple stressors on ecosystem components such as biodiversity, habitat quality, water and air quality, and ecosystem services. This may involve conducting spatial and temporal analyses to assess how stressors accumulate and interact across landscapes and over time.
iv. *Considering Thresholds and Tipping Points*: CEA examines whether cumulative impacts exceed ecological thresholds or trigger tipping points, beyond which ecosystems may undergo rapid and potentially irreversible changes. Identifying these thresholds is crucial for understanding the limits of ecosystem resilience and for informing management decisions aimed at preventing or mitigating undesirable outcomes.
v. *Integrating Stakeholder Input:* CEA often involves engaging stakeholders, including resource managers, policymakers, Indigenous communities, and the public, to identify relevant stressors, set priorities, and incorporate local knowledge and values into the assessment process.
vi. *Informing Decision-making:* The results of CEA are used to inform decision-making processes related to land use planning, natural resource management, environmental regulation, and conservation. By considering the cumulative effects of multiple stressors, decision-makers can develop more holistic and effective strategies for protecting and managing ecosystems and their services (Webler, 2001; NRC, 2008).

CEA is particularly important in landscapes and ecosystems where multiple stressors are present, and where individual risk assessments may underestimate the true extent of environmental impacts. By considering the cumulative effects of stressors, CEA helps improve the sustainability of human activities and promotes the long-term health and resilience of ecosystems.

Stakeholder Engagement and Participatory Approaches

Engaging stakeholders, including local communities, indigenous groups, and policymakers, is essential for effective ecological risk assessment and biodiversity conservation. Participatory approaches involve incorporating local knowledge, values, and preferences into decision-making processes, fostering community ownership and support for conservation initiatives (Beierle, 1999; Rowe, 2000; Webler, 2001; NRC, 2008). Stakeholder engagement and participatory approaches play a crucial role in ecological risk assessment by involving relevant stakeholders in the assessment process, incorporating their knowledge, values, and perspectives, and fostering collaboration and consensus-building. Here's how stakeholder engagement and participatory approaches are typically applied:

i. *Identifying Stakeholders:* The first step in stakeholder engagement is to identify and map relevant stakeholders who have an interest in or may be affected by the ecological risk assessment. This can include government agencies, industry representatives, environmental organizations, Indigenous communities, local residents, scientists, and other stakeholders with expertise or interests related to the assessment.
ii. *Building Relationships:* Establishing and maintaining relationships with stakeholders is essential for effective engagement. This may involve outreach efforts, communication strategies, and build-

ing trust through transparent and inclusive processes. Building relationships helps ensure that stakeholders feel valued, respected, and empowered to contribute to the assessment.

iii. *Involving Stakeholders in Design and Planning:* Stakeholders are often involved in the design and planning of the ecological risk assessment, including defining assessment goals and objectives, selecting assessment methods, and identifying key issues and priorities. Involving stakeholders from the outset helps ensure that the assessment is relevant, credible, and responsive to stakeholder needs and concerns.

iv. *Sharing Information and Knowledge:* Stakeholder engagement involves sharing relevant information, data, and knowledge about the ecological system under assessment, as well as the potential risks and uncertainties associated with different stressors. This helps stakeholders understand the assessment process, make informed decisions, and contribute their own expertise and insights.

v. *Facilitating Dialogue and Collaboration:* Participatory approaches facilitate dialogue and collaboration among stakeholders, enabling them to exchange ideas, perspectives, and concerns, and work together to address complex ecological issues. This may involve workshops, focus groups, stakeholder meetings, and other interactive forums for discussion and deliberation.

vi. *Incorporating Local and Traditional Knowledge:* Indigenous communities and local residents often possess valuable traditional and local ecological knowledge that can complement scientific data and enhance the ecological risk assessment. Participatory approaches involve incorporating local and traditional knowledge into the assessment process, respecting Indigenous rights and sovereignty, and recognizing the importance of diverse perspectives in understanding and managing ecological risks.

vii. *Evaluating and Responding to Feedback:* Stakeholder engagement is an iterative process that involves soliciting feedback on assessment findings, methods, and conclusions, and incorporating stakeholder input into the assessment process. This helps ensure that the assessment is robust, credible, and responsive to stakeholder concerns, and that assessment findings are effectively communicated and used to inform decision-making.

By engaging stakeholders and incorporating participatory approaches into ecological risk assessment, decision-makers can enhance the relevance, credibility, and legitimacy of the assessment process, build consensus around assessment findings, and promote more effective and sustainable management of ecological risks.

CONCLUSION

In conclusion, assessing ecological risk and evaluating biodiversity loss require a multidisciplinary approach that integrates various methodologies and approaches. By combining data-driven assessments, ecological modeling, stakeholder engagement, and participatory approaches, researchers and policymakers can gain a comprehensive understanding of the threats facing biodiversity and develop targeted strategies to mitigate ecological risk and safeguard our natural heritage for future generations. The future of ecological risk assessment will likely involve interdisciplinary collaboration, innovation, and a greater emphasis on sustainability and resilience in managing environmental risks. By embracing these trends and developments, ERA can continue to evolve as an essential tool for protecting ecosystems and biodiversity in a rapidly changing world.

REFERENCES

Beierle, T. C. (1999). Using social goals to evaluate public participation in environmental decisions. *The Review of Policy Research, 16*(3-4), 75–103. doi:10.1111/j.1541-1338.1999.tb00879.x

Carpenter, S. R., Mooney, H. A., Agard, J., Capistrano, D., DeFries, R. S., Díaz, S., & Lubchenco, J. (2009). Science for managing ecosystem services: Beyond the Millennium Ecosystem Assessment. *Proceedings of the National Academy of Sciences of the United States of America, 106*(5), 1305–1312. doi:10.1073/pnas.0808772106 PMID:19179280

Cash, D. W., Clark, W. C., Alcock, F., Dickson, N. M., Eckley, N., Guston, D. H., Jäger, J., & Mitchell, R. B. (2003). Knowledge systems for sustainable development. *Proceedings of the National Academy of Sciences of the United States of America, 100*(14), 8086–8091. doi:10.1073/pnas.1231332100 PMID:12777623

Daniels, S. E., & Walker, G. B. (2001). *Working through environmental conflict: The collaborative learning approach.* Praeger Publishers.

Fahrig, L. (2003). Effects of Habitat Fragmentation on Biodiversity. *Annual Review of Ecology, Evolution, and Systematics, 34*(1), 487–515. doi:10.1146/annurev.ecolsys.34.011802.132419

Fischer, F. (2000). Citizen participation and the democratization of policy expertise: From theoretical inquiry to practical cases. [DOI: 10.1023/A:1004819900491]. *Policy Sciences, 33*(3-4), 283–320.

IUCN. (2019). *Guidelines for Using the IUCN Red List Categories and Criteria. Version 14.* Standards and Petitions Subcommittee. https://www.iucnredlist.org/resources/redlistguidelines

Leach, W. D., & Pelkey, N. W. (2001). Making watershed partnerships work: A review of the empirical literature. [DOI: 10.1061/(ASCE)0733-9496(2001)127:6(378)]. *Journal of Water Resources Planning and Management, 127*(6), 378–385. doi:10.1061/(ASCE)0733-9496(2001)127:6(378)

Magurran, A. E. (2004). *Measuring Biological Diversity.* Blackwell Publishing.

Millennium Ecosystem Assessment. (2005). *Ecosystems and Human Well-being: Synthesis.* Island Press.

NRC. (National Research Council). (2008). Public participation in environmental assessment and decision making. Washington, DC: National Academies Press.

Peterson, G. D., Cumming, G. S., & Carpenter, S. R. (2003). Scenario planning: A tool for conservation in an uncertain world. *Conservation Biology, 17*(2), 358–366. doi:10.1046/j.1523-1739.2003.01491.x

Reed, M. S., Kenter, J., Bonn, A., & Borad, K. (2009). Participatory tools for assessing the impact of land use scenarios on ecosystem services. [DOI: 10.1016/j.jenvman.2009.01.001]. *Journal of Environmental Management, 90*(5), 1933–1940. doi:10.1016/j.jenvman.2009.01.001 PMID:19231064

Rowe, G., & Frewer, L. J. (2000). Public participation methods: A framework for evaluation. [DOI: 10.1177/016224390002500101]. *Science, Technology & Human Values, 25*(1), 3–29. doi:10.1177/016224390002500101

Schultz, P. W. (2001). Assessing the structure of environmental concern: Concern for self, other people, and the biosphere. [DOI: 10.1006/jevp.2001.0227]. *Journal of Environmental Psychology*, *21*(4), 327–339. doi:10.1006/jevp.2001.0227

Webler, T. (2001). Achieving convergence on local environmental disputes: The hard work of soft alternatives. [DOI: 10.2307/2693982]. *The Western Political Quarterly*, *54*(1), 21–47.

Wondolleck, J. M., & Yaffee, S. L. (2000). *Making collaboration work: Lessons from innovation in natural resource management*. Island Press.

Chapter 5
Stress on Biodiversity and the Crucial Role of Environmental Impact Assessment:
Biodiversity Loss Assessment

Ashok Kumar Rathoure
https://orcid.org/0000-0001-9131-1346
Saraca Research Inc., India

ABSTRACT

The chapter explores the various stressors impacting biodiversity and emphasizes the pivotal role of environmental impact assessment (EIA) in mitigating these threats. Biodiversity faces challenges such as habitat loss, climate change, pollution, invasive species, overexploitation, and disease outbreaks. These stressors, driven by human activities, endanger ecosystems and species worldwide. The chapter underscores how EIA promotes sustainable practices by evaluating the environmental implications of development projects. It encourages the adoption of eco-friendly technologies, conservation-oriented designs, and compliance with environmental regulations. Public participation in the EIA process ensures that local communities and stakeholders contribute to decision-making, fostering awareness about biodiversity.

BACKGROUND

The article explores the various stressors impacting biodiversity and emphasizes the pivotal role of Environmental Impact Assessment (EIA) in mitigating these threats. Biodiversity faces challenges such as habitat loss, climate change, pollution, invasive species, overexploitation, and disease outbreaks. These stressors, driven by human activities, endanger ecosystems and species worldwide. Environmental Impact Assessment emerges as a vital tool to address these challenges. Biodiversity experts play a key role in the EIA process, contributing their specialized knowledge to predict and assess potential impacts on ecosystems, species, and habitats. EIA facilitates the identification of effective mitigation measures, such as habitat restoration and conservation programs.

DOI: 10.4018/979-8-3693-3330-3.ch005

The article underscores how EIA promotes sustainable practices by evaluating the environmental implications of development projects. It encourages the adoption of eco-friendly technologies, conservation-oriented designs, and compliance with environmental regulations. Public participation in the EIA process ensures that local communities and stakeholders contribute to decision-making, fostering awareness about biodiversity. Legal compliance is another crucial aspect of EIA, ensuring adherence to environmental laws and regulations. The continuous monitoring and adaptive management approach of EIA allow for real-time adjustments, mitigating unforeseen impacts on biodiversity during the operational phase of projects.

The article advocates for prioritizing and implementing effective Environmental Impact Assessments. By integrating biodiversity considerations into development planning, EIA becomes a powerful tool for achieving sustainable development while safeguarding the intricate web of life on Earth. The cooperation of governments, industries, and communities is essential to ensuring a harmonious coexistence between development and biodiversity conservation.

INTRODUCTION

Biodiversity, the variety of life on Earth, is facing unprecedented stress due to human activities, climate change, and habitat degradation. Environmental Impact Assessment (EIA) emerges as a critical tool in understanding, mitigating, and managing the stress on biodiversity. Recognizing the intricate relationships between human activities and the environment, the integration of biodiversity considerations into Environmental Impact Assessment (EIA) studies has become paramount. EIA serves as a systematic process to evaluate the potential environmental consequences of proposed projects, policies, or programs before they are implemented. Understanding the role of biodiversity within this framework is essential for fostering sustainable development, conserving ecosystems, and mitigating adverse impacts on the planet's rich biological diversity. Biodiversity is a key driver of ecosystem functioning. Diverse ecosystems are more resilient to environmental changes, providing essential services such as pollination, water purification, and nutrient cycling. The health and vitality of ecosystems can be gauged by the diversity and abundance of species present. Biodiversity serves as an indicator of the overall well-being of an environment (Slootweg et al., 2009). EIA studies play a crucial role in identifying and assessing potential impacts on habitats. Understanding the importance of various habitats in supporting biodiversity is essential for effective conservation. Biodiversity assessments within EIAs consider the intricate interactions between different species. This includes predator-prey relationships, symbiotic associations, and the overall ecological balance. Diverse ecosystems tend to be more resilient to disturbances such as climate change, pollution, and habitat destruction. EIA helps identify potential stressors that can affect biodiversity and provides strategies for resilience. Biodiversity contributes to economic activities and cultural practices. EIA recognizes the importance of biodiversity in supporting livelihoods, tourism, and traditional knowledge systems. Certain species and ecosystems play a vital role in mitigating climate change. Forests, for example, sequester carbon dioxide (Akçakaya, 2001). EIAs consider the impacts of projects on these climate-regulating functions. Many countries are signatories to international agreements like the Convention on Biological Diversity (CBD). EIA ensures that projects align with these global commitments to preserve and sustainably use biodiversity. Urbanization and infrastructure projects often lead to habitat fragmentation. EIA studies aim to address this challenge by proposing mitigation measures to maintain or restore connectivity between fragmented habitats. The unintentional introduction of invasive species can have detrimental effects on native biodiversity. EIA assesses the risks associated with invasive species and proposes measures to prevent their spread (Elith et al., 2002). EIA seeks to strike a balance between development objectives and the preservation of biodiversity. This involves identifying areas for conservation, implementing mitigation measures, and exploring opportunities for biodiversity enhancement. The role of biodiversity in EIA studies is multi-faceted, encompassing ecological, social, and economic dimensions (Fernandes, 2000). By systematically assessing and integrating biodiversity considerations into the decision-making process, EIAs contribute to the responsible and sustainable development of projects, ensuring the conservation of Earth's invaluable biological diversity for current and future generations.

This article explores the various stressors affecting biodiversity and the indispensable role of EIA in ensuring sustainable development.

STRESSORS ON BIODIVERSITY

Habitat Loss and Fragmentation

Human activities, such as urbanization, agriculture, and infrastructure development, lead to the destruction and fragmentation of natural habitats. This disrupts ecosystems and threatens the survival of numerous species (Sala et al., 2000). Habitat loss and fragmentation are critical ecological issues that have profound implications for biodiversity, ecosystem stability, and the overall health of our planet. These processes result from human activities and environmental changes, altering landscapes and disrupting natural habitats. Understanding the causes, consequences, and potential solutions to habitat loss and fragmentation is essential for effective conservation efforts (CBD, 1992).

Causes of Habitat Loss and Fragmentation

a. Urbanization and Infrastructure Development: - Rapid urbanization, expansion of cities, and the construction of infrastructure such as roads and highways contribute significantly to habitat loss and fragmentation. The conversion of natural landscapes into urban areas disrupts ecosystems and displaces wildlife (CBD, 1992).
b. Agricultural Expansion: The expansion of agricultural activities, including the conversion of forests and grasslands into farmland, is a major driver of habitat loss. Large-scale monoculture and intensive farming practices can lead to the degradation of ecosystems.
c. Logging and Deforestation: Logging and the clearing of forests for timber, paper, and agricultural purposes result in extensive habitat loss. Deforestation disrupts not only the physical habitat but also the complex ecological relationships within forest ecosystems (CBD, 1992).
d. Mining and Extractive Industries: - Extractive industries, such as mining and oil extraction, can lead to habitat destruction. These activities often involve the removal of large areas of vegetation, altering the physical structure of ecosystems and impacting biodiversity (Millennium Ecosystem Assessment, 2005).
e. Climate Change: Changes in climate patterns, including rising temperatures and altered precipitation, can contribute to habitat loss. These changes can affect the distribution of plant and animal species, leading to shifts in ecosystems (Pimm et al, 1995).

Consequences of Habitat Loss and Fragmentation

a. Loss of Biodiversity: Habitat loss is a major driver of species extinction. As natural habitats disappear or become fragmented, many species are unable to survive in the altered landscape. Specialist species with specific habitat requirements are particularly vulnerable.
b. Disruption of Ecosystem Services: Natural ecosystems provide a range of services, including pollination, water purification, and climate regulation. Habitat loss and fragmentation can disrupt these services, impacting both the environment and human communities (Pimm et al, 1995).
c. Genetic Isolation and Inbreeding: Fragmentation can lead to isolated populations, reducing gene flow between individuals. This genetic isolation may result in inbreeding and a loss of genetic diversity, making populations more susceptible to diseases and environmental changes.
d. Altered Ecosystem Dynamics: The removal or fragmentation of key habitats can alter the structure and dynamics of ecosystems. This may lead to changes in species interactions, community composition, and overall ecosystem functioning (CBD, 1992).

Mitigation and Conservation Strategies

a. Protected Areas and Conservation Reserves: Establishing and maintaining protected areas and conservation reserves are crucial for preserving habitats and providing safe havens for biodiversity. These areas can serve as refuges for species facing habitat loss (Pimm et al, 1995).
b. Habitat Restoration: Efforts to restore degraded habitats can involve reforestation, wetland restoration, and other activities aimed at bringing back the natural vegetation and conditions that support diverse ecosystems (CBD, 1992).
c. Sustainable Land Use Practices: Implementing sustainable land use practices, such as agroforestry and sustainable agriculture, can help minimize the impact of human activities on natural habitats while meeting the needs of growing populations.
d. Corridor Creation: Establishing wildlife corridors or ecological corridors that connect fragmented habitats allows for the movement of species between isolated areas. This helps maintain genetic diversity and promotes healthier populations (Pimm et al, 1995).
e. Community Engagement and Education: Engaging local communities in conservation efforts and providing education about the importance of preserving habitats are critical components of successful conservation strategies. Building awareness fosters a sense of responsibility for environmental stewardship (Glasson, 2012).

Ongoing Challenges and Future Directions

a. Global Cooperation: Addressing habitat loss and fragmentation requires international collaboration. Governments, NGOs, and communities must work together to develop and implement effective conservation policies and practices.
b. Climate Change Mitigation: Mitigating climate change is essential for reducing its impact on habitats. Efforts to decrease greenhouse gas emissions and adapt to changing conditions are crucial for preserving ecosystems (Glasson, 2012).

c. Research and Monitoring: Ongoing research and monitoring programs are essential for understanding the effects of habitat loss and fragmentation and for evaluating the success of conservation interventions. This includes studying species responses, ecosystem dynamics, and the effectiveness of different restoration strategies (Canter, 1996).

Habitat loss and fragmentation pose significant challenges to biodiversity and ecosystem health. Implementing comprehensive conservation strategies, addressing the root causes, and fostering a global commitment to sustainable practices are crucial steps toward mitigating the impact of these processes on our planet's rich biological diversity (Geneletti et al., 2003).

Climate Change

The global climate crisis poses a significant threat to biodiversity. Rising temperatures, extreme weather events, and altered precipitation patterns impact the distribution and behavior of species, often leading to habitat loss and ecosystem shifts.

Climate change is one of the most pressing global challenges of our time, and its impacts on biodiversity are profound and far-reaching. The intricate web of life on Earth, from individual species to entire ecosystems, is being significantly affected by changes in temperature, precipitation patterns, sea levels, and other climate-related factors. Understanding the complex relationship between climate change and biodiversity is crucial for developing effective conservation strategies. Here is an overview of the key aspects:

Temperature Changes: As global temperatures rise; the geographical ranges of many species are shifting. Some species are moving towards higher latitudes or altitudes in search of suitable climate conditions. This can lead to changes in community compositions and potential conflicts as species encounter new ecosystems. Climate change influences the timing of biological events, such as flowering, migration, and breeding. Shifts in phenology can disrupt the synchronization between interacting species, affecting pollination, predation, and other ecological processes (Glasson, 2012).

Sea Level Rise and Coastal Habitats: Rising sea levels, driven by the melting of glaciers and ice caps, pose a significant threat to coastal habitats. Mangroves, salt marshes, and other critical ecosystems are at risk of submersion, leading to habitat loss and the displacement of associated species. Ocean acidification, resulting from the absorption of excess carbon dioxide by the oceans, affects marine life, particularly organisms with calcium carbonate skeletons, such as corals and molluscs. This can lead to the degradation of coral reefs and other marine ecosystems (Guisan & Zimmermann, 2000).

Extreme Weather Events: Climate change is associated with an increase in the frequency and intensity of extreme weather events, including hurricanes, droughts, floods, and wildfires. These events can have direct and indirect impacts on biodiversity, causing habitat destruction, population declines, and disruptions to ecological processes (Lehmann et al., 2003).

Changes in Precipitation Patterns: Changes in precipitation patterns can lead to altered water availability in different regions. This affects freshwater ecosystems, leading to changes in river flows, water temperature, and the distribution of aquatic species. Changes in precipitation can also result in shifts in vegetation zones, impacting terrestrial ecosystems. This may lead to the contraction or expansion of certain plant communities, influencing the composition of associated animal species.

Impacts on Ecosystem Services: Climate change can disrupt critical ecosystem services that support human well-being. Changes in pollination patterns, water availability, and disease dynamics can have cascading effects on agriculture, water resources, and public health (Larson et al., 2004).

Threats to Keystone Species: Keystone species, which play a disproportionately large role in maintaining the structure and function of ecosystems, may be particularly vulnerable to climate change. For example, polar bears, as apex predators in the Arctic, are facing habitat loss due to melting sea ice.

Synergistic Effects with Other Stressors: Climate change often interacts with other stressors, such as habitat loss, pollution, and invasive species. The cumulative impacts of these stressors can exacerbate the challenges faced by biodiversity, making ecosystems more susceptible to degradation (Mladenoff, 2004).

Conservation and Adaptation Strategies: Adaptive management of protected areas involves monitoring and adjusting conservation strategies in response to changing climatic conditions. This may include modifying habitat restoration efforts, adjusting species reintroduction plans, and identifying climate-resilient areas. Establishing ecological corridors that connect fragmented habitats allows species to migrate and adapt to changing climates. These corridors facilitate the movement of plants and animals across landscapes, promoting genetic diversity and enhancing resilience. In certain cases, assisted migration or translocation of species to more suitable habitats may be considered. However, this approach is complex and requires careful consideration of ecological, ethical, and legal implications. Agricultural practices that take into account climate change considerations can contribute to the conservation of biodiversity. These practices may include crop diversification, agroforestry, and soil conservation measures. Addressing the impacts of climate change on biodiversity requires international cooperation. Global efforts to reduce greenhouse gas emissions, such as those outlined in international agreements like the Paris Agreement, are essential for mitigating climate change impacts on biodiversity (Mörtberg & Anders, 2005).

Climate change is an overarching threat to biodiversity, affecting ecosystems and species across the globe. Addressing this challenge requires interdisciplinary research, effective conservation strategies, and global collaboration to ensure the resilience and adaptability of Earth's diverse ecosystems.

Pollution

Pollution from industrial, agricultural, and urban sources introduces harmful substances into ecosystems, affecting both terrestrial and aquatic biodiversity. Chemical pollutants, plastics, and contaminants pose direct threats to species and their habitats. Pollution poses a significant threat to biodiversity, affecting ecosystems and species across the globe. Whether in the form of air, water, or soil pollution, anthropogenic contaminants have wide-ranging and often detrimental impacts on the health and functioning of natural environments. Understanding the interactions between pollution and biodiversity is crucial for developing effective conservation strategies. Here are key aspects of how pollution affects biodiversity:

Types of Pollution

a. Air Pollution: Air pollution, primarily from emissions of pollutants like sulfur dioxide, nitrogen oxides, and particulate matter, can have direct and indirect impacts on terrestrial ecosystems. It can damage plant tissues, disrupt nutrient cycles, and contribute to soil acidification.

b. Water Pollution: Industrial discharges, agricultural runoff, and improper disposal of pollutants contribute to water pollution. Chemical contaminants, such as pesticides, heavy metals, and pharmaceuticals, can contaminate aquatic habitats, affecting the health of aquatic species.
c. Soil Pollution: Pollutants from industrial activities, improper waste disposal, and agricultural practices can contaminate soils. Soil pollution affects the health of plants, microorganisms, and soil-dwelling fauna, influencing ecosystem processes.

Impacts on Biodiversity

a. Direct Harm to Species: Many pollutants are toxic to living organisms. Exposure to high concentrations of pollutants can lead to direct harm, causing physiological and behavioral changes, reproductive failure, and, in severe cases, mortality (Slootweg & Kolhoff, 2003; Slootweg et al., 2009).
b. Disruption of Ecological Interactions: Pollution can alter the composition and abundance of species within ecosystems, disrupting trophic interactions and food webs. This can have cascading effects on the entire ecosystem (Glasson, 2012).
c. Habitat Degradation: Pollution, particularly habitat destruction from industrial activities and urbanization, results in the loss of suitable habitats for many species. This habitat degradation can lead to population declines and increased vulnerability to other threats.
d. Genetic and Population Effects: Pollution can lead to reduced genetic diversity within populations, making species more susceptible to diseases and environmental changes.

Conservation and Mitigation Strategies

a. Pollution Prevention: Implementing technologies and practices to reduce emissions of pollutants from industrial processes, transportation, and energy production is crucial for preventing pollution at the source.
b. Waste Management: Adequate waste management practices, including the proper disposal and recycling of industrial and household waste, are essential for preventing the release of pollutants into the environment.
c. Environmental Monitoring: Establishing comprehensive environmental monitoring programs to track pollution levels and their impacts on biodiversity allows for informed decision-making and timely interventions.
d. Legislation and Regulation: Governments play a crucial role in enacting and enforcing environmental laws and regulations to prevent pollution and protect biodiversity. Compliance with standards for air, water, and soil quality is essential.
e. Public Awareness and Education: Raising public awareness about the impacts of pollution on biodiversity fosters a sense of responsibility and encourages environmentally conscious behavior. Education initiatives can contribute to sustainable practices at individual and community levels (Glasson, 2012).

Pollution represents a significant and pervasive threat to biodiversity. Mitigating its impacts requires a holistic approach, involving pollution prevention, effective waste management, conservation strategies, and global cooperation. Recognizing the intricate connections between pollution and biodiversity is crucial for fostering a sustainable and resilient coexistence between human activities and the natural world (Alcamo, 2002; Abaza et al., 2004).

Invasive Species

The introduction of non-native species to new environments can outcompete native species, disrupt ecosystems, and lead to declines in biodiversity. Invasive species can alter habitats and threaten the survival of indigenous flora and fauna. Invasive species pose a significant threat to biodiversity by disrupting ecosystems, outcompeting native species, and altering ecological processes. An invasive species is a non-native species that, when introduced to a new environment, establishes and spreads rapidly, often causing harm to the native ecosystem. Understanding the impacts of invasive species on biodiversity is crucial for developing effective management and conservation strategies (Amoyaw, 1997). Here are key aspects of the relationship between invasive species and biodiversity:

Introduction and Spread

a. Human-Mediated Introductions: Invasive species are often introduced to new environments through human activities, such as trade, transportation, and agriculture. Globalization has increased the frequency and scale of species introductions, leading to a rise in invasive species worldwide (Aschemann, 2004).
b. High Reproductive Rates and Adaptability: Invasive species often possess traits that contribute to their success in new environments. High reproductive rates, rapid growth, adaptability to diverse environmental conditions, and a lack of natural predators or diseases in the new habitat contribute to their invasiveness.

Impacts on Biodiversity

a. Competition with Native Species: Invasive species can outcompete native species for resources such as food, water, and habitat. This competition can lead to declines in native populations and, in some cases, the displacement or extinction of native species (Ashe, 2002).
b. Predation and Herbivory: Invasive species may exhibit predatory or herbivorous behaviours that native species are not adapted to handle. This can result in increased predation pressure or herbivory on native species, affecting their abundance and distribution.
c. Altered Ecosystem Dynamics: Invasive species can alter ecosystem processes and functions. For example, changes in nutrient cycling, fire regimes, or hydrological patterns may occur due to the presence of invasive species, leading to cascading effects on other species and ecosystem health.
d. Hybridization and Genetic Introgression: Hybridization between invasive and native species can occur, leading to genetic introgression. This can result in the loss of unique genetic traits in native populations and impact their ability to adapt to changing environmental conditions.

e. Disease Transmission: Invasive species may introduce new diseases or parasites to native populations. Native species, lacking evolved defences against these pathogens, can experience population declines or increased susceptibility to other stressors.

Case Examples of Invasive Species Impacting Biodiversity

a. Zebra Mussels (*Dreissena polymorpha*): - Zebra mussels, native to freshwater lakes in Eurasia, have become invasive in many parts of North America. They outcompete native molluscs for resources, attach to hard surfaces, and disrupt aquatic food webs.
b. Cane Toads (*Rhinella marina*): Native to South America, cane toads were introduced to Australia in the mid-20th century to control pests. However, they became invasive, negatively impacting native fauna through predation and toxicity.
c. European Starlings (*Sturnus vulgaris*): Introduced to North America in the 19th century, European starlings have become invasive and compete with native bird species for nesting sites and food resources.

Management and Control Strategies

a. Early Detection and Rapid Response: Early detection of invasive species is crucial for preventing their establishment and spread. Rapid response measures, including eradication or containment efforts, can be more effective when implemented early.
b. Biological Control: Introducing natural enemies or predators specific to the invasive species, known as biological control, can help manage populations. Careful consideration and monitoring are essential to avoid unintended consequences.
c. Chemical and Mechanical Control: Chemical and mechanical methods, such as herbicides or physical removal, are used to control invasive species. These methods aim to reduce the population size and limit their impact on native ecosystems.
d. Restoration and Habitat Enhancement: Restoration efforts involve replanting native vegetation, restoring natural hydrological patterns, and implementing other measures to enhance habitat resilience. These actions can support the recovery of native species affected by invasions.

3.4.5. International Collaboration: Invasive species often transcend national borders. International cooperation is crucial for addressing the global movement of invasive species and developing coordinated strategies for prevention, management, and research.

3.4.6. Public Awareness and Education: Raising awareness among the public, policymakers, and industries about the impacts of invasive species is essential. Education programs can promote responsible practices to prevent the unintentional spread of invasive species. Invasive species represent a major challenge to biodiversity conservation. Addressing this issue requires a combination of proactive prevention measures, effective management strategies, and international collaboration. By understanding the ecological impacts of invasive species and implementing targeted interventions, we can work towards preserving the diversity and resilience of natural ecosystems (Bromley, 1995; Bailey & Rention, 1997; Bina, 2003).

Overexploitation

Unsustainable harvesting of wildlife, overfishing, and illegal poaching contribute to the depletion of species populations. Overexploitation disrupts ecological balance and threatens the survival of many economically and ecologically important species. Overexploitation refers to the unsustainable use or extraction of natural resources from ecosystems, leading to the depletion of populations, decline of species, and degradation of ecosystems. The overexploitation of biodiversity poses a significant threat to the health and sustainability of ecosystems, affecting both terrestrial and aquatic environments. Here are key aspects of the overexploitation of biodiversity:

Harvesting of Wild Species

a. Overfishing: Overfishing occurs when the rate of fishing exceeds the natural reproduction and growth rates of fish populations. This can lead to the depletion of fish stocks, affecting marine and freshwater ecosystems and threatening the livelihoods of communities dependent on fisheries.
b. Hunting and Poaching: Unsustainable hunting and poaching can lead to declines in wildlife populations, especially for species targeted for their meat, fur, horns, or other body parts. This includes iconic species like elephants, rhinoceroses, and tigers.

Logging and Deforestation: Logging, especially when conducted unsustainably, contributes to deforestation and habitat loss. The removal of trees can disrupt ecosystems, leading to the loss of biodiversity, changes in microclimates, and alterations in nutrient cycling. Illegal logging exacerbates the impacts of deforestation. It often occurs in protected areas, undermining conservation efforts and leading to the loss of valuable biodiversity and ecosystem services.

Agriculture and Habitat Conversion: The expansion of agriculture, especially through the conversion of natural habitats into croplands, can result in habitat loss and fragmentation. This poses a threat to species adapted to specific ecosystems, leading to population declines and local extinctions. The establishment of palm oil plantations, often involving the clearance of tropical rainforests, contributes to habitat destruction and biodiversity loss. This affects a wide range of species, including primates, birds, and insects (Buselich, 2002).

Commercial Exploitation: The international trade in exotic species, including plants and animals, can pose a significant threat to endangered species. Collectors, pet traders, and the demand for rare species contribute to the overexploitation of populations. The unsustainable harvesting of medicinal plants for traditional medicine or commercial purposes can lead to the depletion of plant populations, affecting both biodiversity and the availability of important resources for local communities (Cherp & Golubeva, 2004).

Impacts on Ecosystem Services: Overexploitation can disrupt the natural balance of ecosystems by removing key species or functional groups. This can lead to altered nutrient cycling, changes in species interactions, and cascading effects on ecosystem services (CSIR, 2002).

Sustainable Management and Conservation: Implementing sustainable fisheries management practices, including setting catch quotas, regulating fishing gear, and establishing marine protected areas, is crucial for maintaining healthy fish populations (Glasson, 2012). Designating and effectively managing protected areas and reserves play a key role in conserving biodiversity and preventing the overexploitation of species. These areas provide refuges for wildlife and contribute to the maintenance of ecosystem functions. Adopting sustainable logging practices, such as those certified by organizations like the Forest Stewardship Council (FSC), helps promote responsible forest management and reduces the impact of logging on biodiversity. Engaging local communities in conservation efforts and sustainable resource management is essential. Empowering communities to participate in decision-making and benefit from conservation initiatives fosters long-term conservation success (DoE, 1994; EEA, 1999; EC 2002).

Global Collaboration: International agreements such as the Convention on International Trade in Endangered Species of Wild Fauna and Flora (CITES) and the Convention on Biological Diversity (CBD) aim to address the overexploitation of biodiversity through regulation and cooperation among nations. Overexploitation of biodiversity is a multifaceted issue that requires a combination of regulatory measures, sustainable management practices, community involvement, and global cooperation to ensure the long-term health and resilience of ecosystems. Recognizing the importance of biodiversity for ecological balance and human well-being is fundamental to addressing the challenges associated with overexploitation (Gow, 1998; Heather-Clark, 2002; Department of Prime Minister and Cabinet, 2003; Harel et al., 2003).

Disease Outbreaks

The spread of diseases, often exacerbated by climate change and habitat loss, poses a threat to wildlife populations. Emerging infectious diseases can devastate susceptible species and impact entire ecosystems. Disease outbreaks can have profound effects on biodiversity, influencing the health and dynamics of ecosystems, as well as the populations of various species. While diseases are a natural component of ecological systems, anthropogenic factors, climate change, and global connectivity can contribute to the emergence, spread, and impact of infectious diseases. Here are key aspects of the relationship between disease outbreaks and biodiversity:

Disease Dynamics in Wildlife: Many infectious diseases affecting wildlife are zoonotic, meaning they can be transmitted between animals and humans. The spillover of diseases from wildlife to domestic animals or humans can have significant implications for both human and wildlife health. Disease outbreaks can lead to population declines or fluctuations in wildlife species. For example, diseases such as chytridiomycosis in amphibians or white-nose syndrome in bats have been associated with declines in affected populations.

Impact on Species Diversity: Different species exhibit varying susceptibility to diseases. Some species may be more resilient or have evolved defenses against specific pathogens, while others may be highly vulnerable, especially if they lack prior exposure to the pathogen. Biodiversity hotspots, areas with high species richness and endemism, may face increased risks from disease outbreaks. The loss of key species in these regions can have cascading effects on ecosystem functioning and the overall diversity of the area.

Effects on Ecosystem Functioning: Diseases can influence the social structure and dynamics of animal groups. For example, in social species like primates, disease outbreaks can disrupt troop cohesion, reproductive patterns, and social behaviors. Disease-induced changes in prey populations can impact predator populations, affecting the balance of predator-prey interactions within ecosystems. Diseases affecting key species, such as pollinators or seed dispersers, can disrupt important ecosystem services. This may have far-reaching consequences for plant reproduction, nutrient cycling, and overall ecosystem health.

Climate Change and Disease Spread: Climate change can influence the distribution and abundance of disease vectors (e.g., ticks and mosquitoes) and alter the geographic range of diseases. This can expose new populations and species to diseases they were not historically exposed to. Changes in temperature, precipitation patterns, and habitat suitability can impact the life cycles of pathogens and their hosts. This, in turn, can alter the dynamics of disease transmission within ecosystems.

Conservation Challenges and Strategies: Endangered species with small and fragmented populations are particularly vulnerable to disease outbreaks. Conservation efforts for endangered species may need to include disease monitoring and management strategies. In areas where human activities, such as ecotourism or research, intersect with wildlife, implementing biosecurity measures is crucial to prevent the introduction and spread of diseases between human and animal populations. Enhancing the resilience of ecosystems and species through habitat restoration, promoting genetic diversity, and addressing other stressors can contribute to their ability to withstand and recover from disease impacts.

Global Health and One Health Approach: Recognizing the interconnectedness of human, wildlife, and domestic animal health is essential. Diseases at the wildlife-human-domestic animal interface highlight the need for a One Health approach that considers the health of all species and their shared environments. Developing and strengthening disease surveillance systems, particularly in regions with high biodiversity, is crucial for early detection of outbreaks. This allows for timely responses to prevent further spread.

Case Examples: Several emerging infectious diseases, such as Ebola and Nipah virus, have been linked to bats. Understanding the dynamics of bat-associated viruses is essential for preventing spillover events and managing potential outbreaks. The chytrid fungus (*Batrachochytrium dendrobatidis*) has been responsible for declines in amphibian populations worldwide. This infectious disease affects amphibian skin, disrupting electrolyte balance and causing population declines.

Research and Surveillance: Ongoing scientific research is essential to understand the ecology of diseases, host-pathogen interactions, and the factors influencing disease dynamics. This knowledge informs conservation strategies and management practices. Collaborative efforts to establish regional and global surveillance networks can help monitor disease trends, detect emerging threats, and facilitate a coordinated response to mitigate the impact of disease outbreaks.

Understanding the complex interactions between disease outbreaks and biodiversity is critical for mitigating their impacts on ecosystems and species. Adopting interdisciplinary approaches, promoting global cooperation, and integrating disease management into conservation strategies contribute to the resilience and sustainability of both ecosystems and human societies.

THE ROLE OF ENVIRONMENTAL IMPACT ASSESSMENT (EIA)

Predicting and Assessing Impacts

EIA is a systematic process that evaluates the potential environmental impacts of proposed projects before they are undertaken. Biodiversity experts play a crucial role in predicting and assessing how a project may impact local ecosystems, species, and habitats. When it comes to predicting and assessing the impacts of developmental activities on biodiversity, several key aspects should be considered:

- **Baseline Assessment:** Before any developmental activities begin, it's important to conduct a baseline assessment of the existing biodiversity in the area. This includes documenting the species present, their abundance, distribution, and the overall ecosystem structure.
- **Ecosystem Services:** Biodiversity contributes to a wide range of ecosystem services, such as pollination, water purification, climate regulation, and disease control. Assessing the impacts of development on these services is crucial to understanding the full extent of the ecological consequences.
- **Habitat Destruction and Fragmentation:** Development often leads to habitat destruction and fragmentation, which can have severe consequences for biodiversity. Fragmented habitats can isolate populations, reduce genetic diversity, and make species more vulnerable to extinction.
- **Invasive Species:** Developmental activities can introduce invasive species, which may outcompete or prey upon native species. Assessing and predicting the impacts of these invaders on local biodiversity is essential for conservation efforts.
- **Climate Change Interactions:** Biodiversity is influenced by climate patterns, and developmental activities can contribute to climate change. Assessing the synergistic impacts of development and climate change on biodiversity is crucial for understanding future ecological dynamics.
- **Ecological Footprint Analysis:** Evaluating the ecological footprint of developmental activities helps estimate the number of resources consumed and waste generated. This analysis can indicate the sustainability of the development and its potential impacts on biodiversity.
- **Cumulative Impacts:** Consideration should be given to the cumulative impacts of multiple developmental activities in an area. The combined effects of various projects can have more severe consequences than individual projects alone.
- **Biodiversity Indicators:** Establishing and monitoring biodiversity indicators can provide valuable insights into the health of ecosystems. These indicators may include species richness, population trends, and ecosystem function assessments.
- **Mitigation and Conservation Measures:** Based on predictions and assessments, proactive measures can be implemented to mitigate the negative impacts on biodiversity. This may include habitat restoration, conservation easements, or the creation of wildlife corridors to connect fragmented habitats.
- **Public Awareness and Participation:** Engaging the public in the assessment process fosters awareness and understanding of the importance of biodiversity. Informed and involved communities are more likely to support sustainable development practices.

Overall, a comprehensive understanding of the role of biodiversity in ecosystems and thorough assessments of potential impacts are crucial for promoting sustainable development that minimizes harm to the environment. This approach ensures that developmental activities are conducted in a way that supports both human needs and the preservation of biodiversity.

Identifying Mitigation Measures

EIA helps identify potential adverse impacts on biodiversity and proposes effective mitigation measures. This may include habitat restoration, conservation programs, and strategies to minimize habitat fragmentation. When it comes to biodiversity, identifying and implementing effective mitigation measures is a critical aspect of the EIA process. Here are key steps and considerations:

- **Impact Identification:** Clearly identify and assess potential impacts of the proposed project on biodiversity. This may include habitat destruction, fragmentation, pollution, introduction of invasive species, or any other factor that could harm local ecosystems.
- **Avoidance and Minimization:** The first step in mitigation is to avoid or minimize impacts wherever possible. This may involve adjusting project design, location, or timing to reduce the footprint on critical habitats or sensitive ecosystems.
- **Habitat Restoration and Enhancement:** If habitat destruction is unavoidable, consider implementing habitat restoration and enhancement measures. This may involve replanting native vegetation, restoring wetlands, or creating artificial habitats to compensate for the loss.
- **Biodiversity Offsetting:** Biodiversity offsetting involves compensating for the loss of biodiversity in one area by enhancing or restoring biodiversity in another. This can be a strategy to achieve a no-net-loss or a net gain of biodiversity.
- **Wildlife Corridors and Connectivity:** Incorporate measures to maintain or establish wildlife corridors and connectivity between fragmented habitats. This helps species move and migrate, reducing the isolation of populations and promoting genetic diversity.
- **Invasive Species Management:** Develop strategies for preventing the introduction and spread of invasive species associated with the project. If invasive species are identified, implement measures to control or eradicate them.
- **Monitoring and Adaptive Management:** Establish a robust monitoring program to track changes in biodiversity over time. This allows for adaptive management, where mitigation measures can be adjusted based on the actual impacts observed during project implementation.
- **Community Involvement:** Involve local communities and stakeholders in the identification and implementation of mitigation measures. Local knowledge can be valuable in understanding the potential impacts and proposing effective solutions.
- **Capacity Building and Training:** Provide training and capacity-building programs for project staff and local communities to ensure effective implementation of mitigation measures. This may include education on sustainable land use practices and biodiversity conservation.

Effective biodiversity mitigation contributes to sustainable projects that minimize environmental harm and promote long-term ecological health.

Enhancing Sustainable Practices

Enhancing sustainable practices in Environmental Impact Assessment (EIA) is crucial for minimizing the negative impacts of development on biodiversity and ecosystems. By evaluating the environmental implications of development projects, EIA encourages the adoption of sustainable practices. This includes the promotion of eco-friendly technologies, conservation-oriented designs, and adherence to environmental regulations. Here are key considerations to integrate biodiversity and enhance sustainability in the EIA process:

- **Holistic Ecosystem Approach:** Adopt a holistic approach that considers the entire ecosystem. Assess the interconnections between different species, habitats, and ecosystem services to understand the broader impacts of the proposed project.

- **Ecosystem Services Assessment:** Evaluate the project's potential impacts on ecosystem services provided by biodiversity, such as pollination, water purification, soil fertility, and climate regulation. Identify measures to maintain or enhance these services.
- **Stakeholder Engagement:** Involve a diverse range of stakeholders, including local communities, indigenous groups, and environmental NGOs, in the EIA process. Their input can provide valuable insights into the local biodiversity, potential impacts, and mitigation measures.
- **Cumulative Impact Assessment:** Consider the cumulative impacts of the proposed project in conjunction with other existing and planned developments in the region. Assess the combined effects on biodiversity and ecosystems over time.
- **Sustainability Criteria and Indicators:** Develop clear sustainability criteria and indicators to evaluate the environmental, social, and economic aspects of the proposed project. This includes criteria related to biodiversity conservation, habitat protection, and ecosystem health.
- **Sustainable Design and Planning:** Integrate principles of sustainable design and planning into the project from its inception. Consider low-impact development techniques, green infrastructure, and land-use planning that minimizes negative effects on biodiversity.
- **Biodiversity Conservation Zones:** Identify and designate biodiversity conservation zones within or near the project area. These zones can act as refuges for key species and habitats and contribute to overall landscape connectivity.
- **Green Infrastructure and Connectivity:** Incorporate green infrastructure elements, such as wildlife corridors, buffer zones, and green spaces, to enhance habitat connectivity and support biodiversity movement across the landscape.
- **Adaptive Management:** Implement an adaptive management approach that allows for flexibility in project implementation based on ongoing monitoring and evaluation. This ensures that mitigation measures can be adjusted as needed to achieve sustainable outcomes.
- **Biodiversity Monitoring Program:** Establish a comprehensive biodiversity monitoring program to track changes in species abundance, diversity, and ecosystem health over time. Regular monitoring informs adaptive management and ensures the effectiveness of conservation efforts.
- **Capacity Building:** Provide training and capacity-building programs for project staff and local communities to promote a better understanding of biodiversity conservation and sustainable practices. This includes promoting sustainable land use and resource management.
- **Best Available Technologies and Practices (BAT/BAP):** Use the best available technologies and practices to minimize environmental impacts. This includes employing eco-friendly construction methods, waste management, and energy-efficient technologies.
- **Restoration and Rehabilitation:** Include plans for habitat restoration and rehabilitation in the project's lifecycle. This can help restore ecosystems that may have been impacted during construction or other project activities.
- **Incentives for Biodiversity Conservation:** Consider implementing incentives for biodiversity conservation, such as tax breaks, conservation easements, or partnerships with local conservation organizations.

By incorporating these elements into the EIA process, project planners and stakeholders can work towards fostering sustainable practices that not only protect biodiversity but also contribute to the long-term health and resilience of ecosystems.

Public Participation and Awareness

EIA involves public consultation, allowing local communities and stakeholders to voice concerns and provide insights. This inclusive approach fosters awareness about biodiversity and ensures that local knowledge is considered in decision-making processes. Public participation and awareness are crucial components of the Environmental Impact Assessment (EIA) process, especially when it comes to biodiversity conservation. Involving the public in decision-making and raising awareness about the importance of biodiversity helps ensure that diverse perspectives are considered and that communities are informed about the potential impacts of development projects (Gow, 1998; Heather-Clark, 2002; Department of Prime Minister and Cabinet, 2003; Harel et al., 2003). Here are key considerations for incorporating public participation and awareness in the context of biodiversity in EIA:

- **Early and Meaningful Engagement:** Initiate public engagement early in the EIA process to allow for meaningful input and collaboration. Provide opportunities for the public to express concerns, share local knowledge about biodiversity, and contribute to the decision-making process.
- **Accessible Information:** Ensure that information about the proposed project, its potential impacts on biodiversity, and mitigation measures is presented in accessible and understandable formats. Use clear language, visuals, and multiple channels to disseminate information to the public.
- **Stakeholder Identification:** Identify and engage a diverse range of stakeholders, including local communities, indigenous groups, non-governmental organizations (NGOs), and experts in biodiversity. Consider the unique perspectives and knowledge each group brings to the discussion.
- **Public Consultation Meetings:** Organize public consultation meetings or workshops to facilitate direct communication between project proponents, experts, and the public. These forums provide opportunities for discussion, clarification, and the exchange of ideas.
- **Participatory Mapping and Surveys:** Use participatory mapping and surveys to involve local communities in documenting biodiversity, identifying important habitats, and expressing their concerns about potential impacts. This can contribute valuable information to the EIA process.
- **Community Workshops and Training:** Conduct workshops and training sessions to enhance the capacity of local communities to understand biodiversity issues, the EIA process, and how they can contribute constructively. This promotes informed and engaged participation.
- **Feedback Mechanisms:** Establish clear feedback mechanisms to receive input from the public throughout the EIA process. Respond to concerns raised by the public and demonstrate how their input has been considered in decision-making.
- **Cultural Sensitivity:** Be culturally sensitive and respectful of local traditions and knowledge related to biodiversity. Recognize and incorporate indigenous and traditional ecological knowledge into the assessment process.
- **Information Dissemination:** Utilize various communication channels, such as public meetings, websites, social media, and local newspapers, to disseminate information about the EIA process and biodiversity impacts. Ensure that information reaches a wide audience.
- **Public Awareness Campaigns:** Launch public awareness campaigns to highlight the importance of biodiversity, ecosystem services, and the potential impacts of the proposed project. Emphasize the role of the community in biodiversity conservation.

- **Collaboration with NGOs and Community Groups**: Collaborate with environmental NGOs, community-based organizations, and other groups actively involved in biodiversity conservation. Their expertise and local connections can enhance public awareness and participation.
- **Transparency and Openness:** Maintain transparency in the EIA process by openly sharing relevant information, methodologies, and findings with the public. Foster trust through open communication and a commitment to addressing concerns.
- **Language and Cultural Considerations:** Consider linguistic and cultural diversity in the affected community. Provide information in local languages and formats that resonate with the cultural context to ensure inclusive participation.
- **Empowerment and Decision-Making:** Empower local communities to actively participate in decision-making processes related to the project's impact on biodiversity. Foster a sense of ownership and responsibility for biodiversity conservation.

By incorporating these considerations, the EIA process can benefit from diverse perspectives, local knowledge, and community engagement, leading to more comprehensive assessments and better-informed decisions regarding the impact of development projects on biodiversity.

Legal Compliance

EIA ensures compliance with environmental laws and regulations. This helps prevent illegal activities that could harm biodiversity and ensures that development projects adhere to established environmental standards. Legal compliance in Environmental Impact Assessment (EIA) plays a crucial role in ensuring that development projects adhere to established laws and regulations related to biodiversity conservation. Here are key aspects of legal compliance and biodiversity in the context of EIA:

- **National Environmental Laws**

Role: National environmental laws set the framework for EIA processes and standards.

Compliance: EIA processes must align with the requirements specified in national environmental legislation. This includes provisions for biodiversity conservation and the assessment of impacts on ecosystems.

- **Biodiversity Conservation Laws**

Role: Specific biodiversity conservation laws may exist to protect flora and fauna.

Compliance: Project proponents must adhere to laws that safeguard biodiversity, which may involve obtaining permits, following conservation plans, and mitigating impacts on protected species.

- **Protected Areas Legislation**

Role: Laws governing protected areas designate certain regions as ecologically significant and may have specific regulations.

Compliance: Projects within or near protected areas must comply with additional regulations to safeguard critical habitats and biodiversity hotspots.

- **International Agreements and Conventions**

Role: Countries may be signatories to international agreements addressing biodiversity conservation, such as the Convention on Biological Diversity (CBD).

Compliance: Project proponents may need to align their activities with international commitments and agreements related to biodiversity.

- **Wildlife Protection Laws**

Role: Laws protecting wildlife set standards for the treatment of and impact on animal species.

Compliance: Projects must comply with regulations that protect wildlife, including those related to habitat disturbance and the conservation of endangered species.

- **Invasive Species Management**

Role: Laws may exist to control the introduction and spread of invasive species.

Compliance: EIA processes may include assessments of invasive species risks, and compliance involves adopting measures to prevent the introduction or spread of invasive species.

- **Public Consultation Requirements**

Role: Many environmental laws mandate public participation in decision-making processes.

Compliance: Public consultation is essential in biodiversity-related EIAs, ensuring that local communities and stakeholders contribute to the assessment and decision-making process.

- **Mitigation and Monitoring Plans**

Role: Legal frameworks often require the development and implementation of mitigation and monitoring plans.

Compliance: Project proponents must adhere to these plans, detailing how biodiversity impacts will be mitigated and monitored throughout the project lifecycle.

- **Adaptive Management and Reporting**

Role: Some laws may require adaptive management strategies and regular reporting.

Compliance: Projects may need to adjust management practices based on monitoring results and provide periodic reports to regulatory authorities to demonstrate ongoing compliance.

- **Penalties for Non-Compliance**

Role: Legal consequences exist for non-compliance with environmental laws.

Compliance: Penalties may include fines, project suspension, or revocation of permits, reinforcing the importance of adhering to established biodiversity conservation measures.

In summary, legal compliance in EIA with regard to biodiversity involves aligning with national and international environmental laws, conservation regulations, and specific measures aimed at protecting ecosystems and wildlife. It encompasses a range of considerations, from public participation to the implementation of mitigation measures and ongoing monitoring and reporting to regulatory authorities. Compliance with these legal frameworks is essential for promoting sustainable development while safeguarding biodiversity (Gow, 1998; Heather-Clark, 2002; Department of Prime Minister and Cabinet, 2003; Harel et al., 2003).

Continuous Monitoring and Adaptation

EIA is an ongoing process that involves continuous monitoring of projects during their operational phase. This adaptive management approach allows for adjustments to be made in real-time, mitigating unforeseen impacts on biodiversity. Continuous monitoring and adaptation are crucial components of an effective strategy for conserving biodiversity in the context of Environmental Impact Assessment (EIA). Here's how these processes contribute to biodiversity conservation within the EIA framework:

Continuous Monitoring for Conserving Biodiversity

- **Baseline Data Collection:**

Role: Establish a baseline understanding of the existing biodiversity in the project area.

Implementation: Conduct thorough surveys to identify species, habitats, and ecosystem services. Document population sizes, distribution, and ecological dynamics.

- **Ecosystem Health Assessment:**

Role: Monitor the health and resilience of ecosystems.

Implementation: Regularly assess indicators of ecosystem health, such as species diversity, habitat structure, and the functioning of key ecological processes.

- **Population Dynamics Tracking:**

Role: Track changes in population sizes of key species.

Implementation: Monitor trends in population dynamics, including births, deaths, and migrations. Identify factors influencing population changes.

- **Invasive Species Detection:**

Role: Early detection and response to invasive species.

Implementation: Regularly survey for the presence of invasive species. Implement measures to control or eradicate invasive species before they harm native biodiversity.

- **Climate Change Impact Assessment:**

Role: Assess the impact of climate change on biodiversity.

Implementation: Monitor climate-related changes and assess how they may affect biodiversity. Adapt management strategies to help species cope with changing conditions.

- **Habitat Quality Assessment:**

Role: Evaluate the quality and integrity of habitats.

Implementation: Regularly assess habitat conditions, including factors like vegetation cover, water quality, and soil health. Identify and address degradation issues.

- **Community Involvement and Education:**

Role: Engage local communities in biodiversity monitoring.

Implementation: Involve local residents in data collection and monitoring efforts. Educate communities about the importance of biodiversity and the role they can play in conservation.

Adaptation for Conserving Biodiversity

- **Identifying Emerging Threats:**

Role: Recognize and respond to emerging threats.

Implementation: Regularly review monitoring data to identify new threats or stressors to biodiversity. Adjust management plans to address these emerging issues.

- **Mitigation Measure Adjustment:**

Role: Modify mitigation measures based on monitoring results.

Implementation: If initial mitigation measures are not achieving the desired outcomes, adapt them to be more effective. This may involve changing the location of mitigation efforts or employing different conservation strategies.

- **Habitat Restoration and Enhancement:**

Role: Prioritize habitat restoration based on ongoing monitoring.

Implementation: Use monitoring data to identify areas with degraded habitats. Prioritize and adapt habitat restoration efforts to maximize the conservation impact.

- **Biodiversity Offsetting Adjustments:**

Role: Adjust biodiversity offsetting efforts as needed.

Implementation: If the impact on biodiversity is greater than anticipated, adapt biodiversity offsetting measures to achieve a net positive outcome.

- **Climate Change Adaptation Strategies:**

Role: Develop strategies to help biodiversity adapt to climate change.

Implementation: Based on ongoing monitoring, refine and implement adaptive strategies that assist species in adapting to changing climatic conditions.

- **Community Engagement:**

Role: Maintain community support for conservation efforts.

Implementation: Regularly communicate monitoring results to local communities. Adapt community engagement strategies based on feedback and changes in the social context.

- **Legislative Compliance**:

Role: Ensure ongoing adherence to environmental laws and regulations.

Implementation: Regularly review and update project plans to ensure continuous compliance with relevant environmental laws and regulations.

- **Adaptive Management Plan Implementation:**

Role: Implement adaptive management plans.

Implementation: Execute strategies outlined in adaptive management plans based on monitoring data. Regularly review and update these plans to reflect the latest information.

- **Integrated Approach:** An integrated and iterative approach that combines continuous monitoring with adaptive management ensures that conservation strategies remain effective and responsive to changing conditions. Regularly updating and refining management plans based on ongoing monitoring results is essential for achieving long-term biodiversity conservation goals within the EIA process.

RECOMMENDATIONS

Biodiversity in EIA

- Biodiversity plays a critical role in maintaining ecosystem health and stability.
- EIA processes must assess and consider the potential impacts of development on biodiversity.
- Baseline assessments, ecosystem services evaluations, and the consideration of habitat destruction and fragmentation are essential components of biodiversity-focused EIAs.

Legal Compliance

- National and international environmental laws set the framework for EIA processes, ensuring that development activities comply with regulations aimed at protecting biodiversity.

- Specific laws, such as those related to biodiversity conservation and protected areas, guide project proponents in their efforts to minimize negative impacts on ecosystems.
- Compliance with legal frameworks is crucial for obtaining permits and approvals while promoting sustainable development practices.

Public Participation and Awareness

- Involving local communities and stakeholders in the EIA process fosters awareness and understanding of the importance of biodiversity.
- Stakeholder engagement ensures that diverse perspectives are considered, and local knowledge contributes to the assessment of potential impacts and proposed mitigation measures.
- Public participation is a key element in creating a sense of ownership and responsibility for biodiversity conservation.

Continuous Monitoring and Adaptation

- Continuous monitoring is essential for assessing baseline conditions, identifying changes in biodiversity, and detecting emerging threats.
- Biodiversity serves as a crucial indicator of ecosystem health, and ongoing monitoring helps track population dynamics, invasive species, and the impact of climate change.
- Adaptation involves adjusting management strategies based on monitoring results, ensuring that conservation efforts remain effective and responsive to changing environmental conditions.

The integration of biodiversity considerations into the EIA process is fundamental for achieving sustainable development goals while safeguarding ecosystems. Legal compliance provides a regulatory framework, ensuring that development activities adhere to established environmental standards. Public participation fosters inclusive decision-making and community support, contributing to the success of biodiversity conservation efforts. Continuous monitoring and adaptation serve as dynamic tools for assessing the health of ecosystems and adapting conservation strategies over time.

CONCLUSION

As biodiversity faces increasing stress from anthropogenic activities, the role of Environmental Impact Assessment becomes paramount. By integrating biodiversity experts and considerations into the planning and decision-making processes, EIA becomes a powerful tool for achieving sustainable development while safeguarding the intricate web of life on our planet. Governments, industries, and communities must prioritize and implement effective EIAs to ensure a harmonious coexistence between development and biodiversity conservation. In conclusion, the intersection of biodiversity and Environmental Impact Assessment (EIA) is pivotal for sustainable development and biodiversity conservation. Throughout the articles, we explored various aspects, including the role of biodiversity in EIA, legal compliance, public participation, awareness, and the continuous monitoring and adaptation strategies for conserving biodiversity. In essence, a holistic and collaborative approach that considers biodiversity, legal frameworks, public engagement, and ongoing monitoring is essential for achieving the delicate balance between development and conservation. This integrated approach ensures that projects not only comply with regulations but actively contribute to the preservation and enhancement of biodiversity, promoting resilience and sustainability for future generations.

REFERENCES

Abaza, H., Bisset, R., & Sadler, B. (2004). *Environmental Impact Assessment and Strategic Environmental Assessment: Towards an Integrated Approach. Economics and Trade Branch*. UNEP.

Akçakaya, H. R. (2001). Linking population-level risk assessment with landscape and habitat models. *The Science of the Total Environment, 274*(1–3), 283–291. doi:10.1016/S0048-9697(01)00750-1 PMID:11453303

Alcamo, J. (2002). *Scenarios as Tools for International Environmental Assessments. Experts Corner Report, Prospects and Scenarios No.5*. European Environment Agency.

Amoyaw-Osei, Y. (1997). *Developing Strategic EA Capacity in a Developing Country: The Ghana Experience*. Paper to 17th Annual Meeting of the International Association for Impact Assessment, New Orleans, USA.

Aschemann, R. (2004). Lessons Learned from Austrian SEAs. *European Environment, 14*(3), 165–174. doi:10.1002/eet.347

Ashe, J. (2002). The Australian Regional Forest Agreement process: A Case Study in Strategic Natural Resource Assessment, PP 156-181. In S. Marsden & S. Dovers (Eds.), *Strategic Environmental Assessment in Australia*. The Federation Press.

Bailey, J., & Rention, S. (1997). Redseigning EIA to Fit the Future: SEA and the Policy Process. *Impact Assessment, 15*(4), 319–334. doi:10.1080/07349165.1997.9726145

Bina, O. (2003). *Re-Conceptualising Strategic Environmental Assessment: Theoretical Overview and Case Study from Chile*. [PhD thesis, University of Cambridge].

Bromley, D. W. (1995). *Natural Resource Issues in Environmental Policy in South Africa*. Land and Agriculture Policy Centre.

Buselich, K. (2002). *An Outline of Current Thinking on Sustainability Assessment. Background paper prepared for the Western Australian State Sustainability Strategy*. Institute for Sustainability and Technology Policy, Murdoch University.

Canter, L. W. (1996). *Environmental Impact Assessment* (2nd ed.). McGraw-Hill.

CBD (Convention on Biological Diversity). (1992). *Convention on Biological Diversity*. CBD. https://www.cbd.int/doc/legal/cbd-en.pdf

Cherp, A., & Golubeva, S. (2004). Environmental Assessment in the Russian Federation. *Impact Assessment and Project Appraisal, 22*(2), 121–130. doi:10.3152/147154604781766030

CSIR. (2002), Strategic Environmental Assessment (SEA) Resource Document Introduction to the Process, Principle and Application of SEA version1. *CSIR Report ENV-S-C- 2002-073, August 2002.*

Department of Prime Minister and Cabinet. (2003). *Sustainable Development for New Zealand: Programme of Action.* Department of Prime Minister and Cabinet.

DoE. (1994). *Environmental Appraisal in Government Departments.* The Stationary Office.

EC. (2002). *Trans-European Transport Network – TEN-T priority projects.* European Commission.

EEA. (1999). Environment in European Union at Turn of the Century. *Environmental Assessment Report No.2.* European Environment Agency, Copenhagen

Elith, J., Mark, A. B., & Helen, M. R. (2002). Mapping epistemic uncertainties and vague concepts in predictions of species distributions. *Ecological Modelling, 157*(2–3), 313–329. doi:10.1016/S0304-3800(02)00202-8

Fernandes, J. P. (2000). Landscape ecology and conservation management – evaluation of alternatives in a highway EIA process. *Environmental Impact Assessment Review, 20*(6), 665–680. doi:10.1016/S0195-9255(00)00060-3

Geneletti, D., Beinat, E., Chung, C. F., Fabbri, A. G., & Scholten, H. J. (2003). Accounting for uncertainty factors in biodiversity impact assessment: Lessons from a case study. *Environmental Impact Assessment Review, 20*(6), 665–680. doi:10.1016/S0195-9255(03)00045-3

Glasson, J., Therivel, R., & Chadwick, A. (2012). *Introduction to Environmental Impact Assessment: Principles and Procedures, Process, Practice and Prospects.* Routledge.

Gow, L. (1998). *New Zealandís approach to strategic environmental assessment, paper to international seminar on strategic environmental appraisal.*

Guisan, A., & Zimmermann, N. E. (2000). Predictive habitat distribution models in ecology. *Ecological Modelling, 135*(2–3), 147–186. doi:10.1016/S0304-3800(00)00354-9

Harel, F., Bourcier, A., & Skriabine, P. (2003). A French Methodological and Exploratory Study of SEA at the Regional Scale [newsletter published by Transport Research Laboratory, UK]. *SEA and Transport Planning, Issue*, (5), 34–35.

Heather-Clark, S. (2002). *Strategic Integrated Port Planning: Moving from EIA to SEA.* Kuwait Conference, Kuwait.

Larson, M. A., Thompson, F. R. III, Millspaugh, J. J., Dijak, W. D., & Shifley, S. R. (2004). Linking population viability, habitat suitability, and landscape simulation models for conservation planning. *Ecological Modelling, 180*(1), 103–118. doi:10.1016/j.ecolmodel.2003.12.054

Lehmann, A., Overton, J. M., & Leathwick, J. R. (2003). GRASP: Generalized regression analysis and spatial prediction. *Ecological Modelling, 160*(1–2), 165–183. doi:10.1016/S0304-3800(02)00354-X

Millennium Ecosystem Assessment. (2005). Ecosystems and Human Well-being: Biodiversity Synthesis. World Resources Institute.

Mladenoff, D. J. (2004). LANDIS and forest landscape models. *Ecological Modelling, 13*(2–3), 147–159.

Mörtberg, U., & Anders, K. (2005). Predicting forest grouse distribution taking account of spatial autocorrelation. *Journal for Nature Conservation, 13*(2–3), 147–159. doi:10.1016/j.jnc.2005.02.008

Pimm, S. I., Russell, G. J., Gittelman, J. L., & Brooks, T. M. (1995). The Future of Biodiversity. *Science, 269*(5222), 347–350. doi:10.1126/science.269.5222.347 PMID:17841251

Roel, S., Rajvanshi, A., Mathur, V. B., & Kolhoff, A. (2009). *Biodiversity in Environmental Assessment Enhancing Ecosystem Services for Human Well-Being*. Cambridge University Press.

Sala, O. E., Chapin, F. S., Armesto, J. J., Berlow, E., Bloomfield, J., & Dirjo, R. (2000). Global biodiversity scenarios for the year 2100. *Science, 287*(5459), 1770–1774. doi:10.1126/science.287.5459.1770 PMID:10710299

Slootweg, R., & Kolhoff, A. (2003). A generic approach to integrate biodiversity considerations in screening and scooping for EIA. *Environmental Impact Assessment Review, 23*(6), 657–681. doi:10.1016/S0195-9255(03)00114-8

Chapter 6
Marine Biodiversity Impact Assessment

Vinayaka M.
https://orcid.org/0009-0001-9988-2490
University BDT College of Engineering, India

Shanmukha N. T.
Government Engineering College, Krishnarajpet, India

Lokeshappa B.
University BDT College of Engineering, India

Shanwaj Nadaf
University BDT College of Engineering, India

ABSTRACT

Human activities, including pollution, invasive species, habitat degradation, overfishing, and climate change, are causing a decline in marine biodiversity and ecosystems. These issues disrupt food webs, habitat loss, and interfere with marine organisms' health and reproduction. Climate change also strains marine life. Conservation efforts and sustainable management techniques are needed to protect marine biodiversity, including pollution management, habitat protection, sustainable fishing methods, and climate change mitigation. Integrated strategies addressing pollution, habitat protection, sustainable fishing, and climate change mitigation are crucial for preserving marine ecosystems.

INTRODUCTION

The ocean covers more than 70% of Earth and affects the planet's temperature. It provides resources for subsistence and the economy and is home to many species, but we still don't know everything about them. Biodiversity is changing in different habitats and ecosystems at unknown rates. Coastal marine habitats, like seagrasses and bivalve beds, offer important ecosystem services that sustain social, economic, and

DOI: 10.4018/979-8-3693-3330-3.ch006

biological activities. They regulate water quality, create habitat, and cycle important nutrients. Seagrasses also sustain coastal food webs, productivity, and biodiversity.

Mangrove forests are important for fish species that live in seagrass meadows and coral reefs. This helps increase fish diversity and abundance in these areas. Mangroves also help coral reefs recover after disturbances. By connecting mangroves, corals, and seagrasses, fish habitat is improved. Mangroves and seagrass meadows also protect coastal areas from erosion and storms. It is important to maintain and increase connections between maritime areas for robust coastlines and profitable fisheries.

Physical Processes in Marine Environment

Marine animals are impacted by things like temperature, tides, ocean currents, and changes in saltiness. These processes help animals move, larvae spread, and nutrients transport. Physical processes also affect temperature, nutrients, ocean circulation, and habitat availability, which are important for forming marine ecosystems. These activities occur on different scales, from local to large ocean currents. Here are some major physical mechanisms in marine environments.

Ocean Circulation: Ocean currents, influenced by wind, temperature, and Earth's rotation, transport saltwater, heat, nutrients, and gases across the seas, impacting regional temperatures and marine production.1 *(Marshall et al., 1997).*

Tides: Tide cycles, influenced by the moon and sun's gravitational attraction, create dynamic ecosystems in coastal habitats, affecting nutrient transfer, sedimentation, and larval dispersal, affecting marine species distribution *(Pugh et al., 2004).*

Temperature and Salinity: Seawater density and stratification, influenced by temperature and salinity gradients, influence vertical mixing and circulation patterns, impacting marine animal dispersal, and driving thermohaline circulation, crucial for global ocean circulation and climate regulation *(Levitus et al., 1998).*

Upwelling and Downwelling: Upwelling and downwelling transport seawater vertically, transferring nutrients from deep ocean to surface. Upwelling maintains primary production and marine biodiversity, while downwelling transports organic matter and dissolved gases to deeper ocean strata.

Wave Action: Waves, created by wind friction, impact coastal processes like sedimentation, habitat formation, and erosion, shaping ecosystems and marine life, including beaches, rocky shorelines, and coral reefs *(Dally et al., 1985).*

Storms and Extreme Events: Extreme weather events such as storms and hurricanes can have a significant effect on marine ecosystems by destroying habitat, eroding coastal areas, and upsetting ecosystem processes. The long-term effects of extreme events on ecosystem resilience and biodiversity can include modifications to species ranges, population dynamics, and community structure *(Hughes et al., 2003).*

MARINE FOOD PYRAMID

Marine ecosystems, comprising animals and their environments, regulate temperature, nutrients, oxygen, food, climate management, and recreation services, ranging from plankton to whales in oceans and coastal regions.

Producers: Primary producers, such as phytoplankton, algae, and seagrasses, use the energy from sunshine to convert carbon dioxide and nutrients into organic matter through a process called photosynthesis. Marine food webs are supported by these organisms, which also provide the energy needed for higher trophic levels.

Consumers: In marine habitats, a vast range of creatures, including fish, seabirds, marine mammals, and zooplankton, are consumers. By eating primary farmers and other consumers, they transfer energy and nutrients up the food chain.

Decomposers: Like fungi and bacteria, decomposers are vital to recycling nutrients, breaking down organic waste, and maintaining the integrity of ecosystems. They restore the ecosystem with nutrients that can be ingested by primary producers by decomposing organic waste and deceased animals.

MARINE ECOSYSTEM SERVICES

Marine ecosystems provide essential services like oxygen production and fish production. However, land-based activities like pesticides, nitrogen, and phosphorus release pollutants and untreated wastewater, making it difficult to understand their impact. These three-dimensional, connected ecosystems are complex to divide into different regions.

MARINE ECOSYSTEMS

India's marine environment is made up of a range of habitats that may be found throughout the coastline that encircles the two main island groups and the Indian peninsula. The length of the Indian coastline is 7,516 km, of which 5,422 km is on the mainland and 2,094 km is around the two main island groups (132 km surrounding Lakshadweep and 1962 km around the Andaman & Nicobar Islands). The West Coast and the East Coast are the two divisions of the continental coast. The geomorphology of these two shores differs dramatically. While the East Coast is typically shelving with beaches, lagoons, deltas, and marshes, the West Coast is typically exposed with strong surf, rocky shorelines, and headlands. India's coastline region is home to a diverse array of coastal ecosystems such as estuaries, lagoons, mangroves, backwaters, salt marshes etc.

Estuarine Ecosystems

Estuaries are protected coastal wetlands that serve as nurseries, shelters, and nutrient traps for marine organisms. India has 14 major estuaries on the East Coast and 16 on the West Coast, covering an estimated 2million hectares. Human activities like urbanization and industrialization, including sewage disposal, waste water input, and industrial effluents, have significantly impacted estuarine ecosystems, causing heavy organic and inorganic pollutants.

Lagoon Ecosystems

A lagoon is a body of shallow water on the low-lying coast that is both connected to an isolated from the ocean by one or more restricted inlets. The East Coast has eight large lagoons, whereas the West Coast has nine. Bende, Chilka, Gulf of Mannar, Muthupet, Muthukadu, Nizampatnam, Pennar, and Pulicat are the eight major lagoons on the East Coast. Asthamudi, 2. Ettikulum, 3. Lagoons of Bombay Coast, 4.

Lagoons of Lakshadweep atolls, 5. Paravur, 6. Murukumpuzha, 7. Talapady, 8. Veli, and 9. Vembanad are the lagoons located on the West Coast. Similar to estuarine environments, lagoon ecosystems are suffering from industrialization, urbanisation, and aquaculture activities.

Seagrass and Seaweed Ecosystems

Seagrasses are shallow water, protected plants that are found submerged in bays, backwaters, gulfs, and lagoons. They are crucial to the preservation of numerous threatened species, including marine turtles and dugongs. There are reports of about 14 species along the Indian coast. There are nine widely distributed species in the Andaman and Nicobar Islands. in the other hand, seaweeds are primarily found in rocky, level coastal wetlands that are exposed during low tides and submerged during high tides. They are less common along the East Coast but more common on the West Coast, the Andaman & Nicobar Islands, and Lakshadweep. In India's coastal regions, there are estimated to be 120 different species of seaweed. The removal or consumption of nutrients such as phosphorous and nitrogen from household sewage and other effluents is a known ability of seaweeds to mitigate eutrophication. In addition to being significant sources of fertilisers and numerous other compounds with commercial significance, they also contain bioactive chemicals.

Mangrove Ecosystems

Mangroves are habitats found in tropical and subtropical intertidal zones that can tolerate salt water. They grow in shallow, muddy sea water or brackish water and cover around 14 million hectares worldwide. The highest concentration of mangroves is in Tropical Australia and the Indo-Pacific tropical zone. Bangladesh and India's Ganga-Brahmaputra-Meghna deltas have the largest concentration of mangroves in the world. India has 69 different species of mangroves, covering approximately 2.6% of the world's total mangrove area. According to India's most recent satellite data-based survey, completed in 2003, India has 4,461 km^2 of mangrove cover. The Nicobar Islands are 671 km^2 in size. Mangroves protect the coast from storms, waves, and cyclones, and prevent erosion. They provide homes for young marine species and many other forms of marine life. People who live near the coast rely on mangroves for wood, fuel, and assistance with fishing, aquaculture, and tourism. Human activity is a major threat to these ecosystems. Coastal development, urbanization, pesticide and insecticide runoff from agriculture, overexploitation, and high tourism are causing deterioration in mangrove habitats.

Coral Reef Ecosystems

Coral reefs are small creatures in the Phylum Cnidaria, specifically the group Anthozoa. They can create colonies by depositing calcium carbonate and secreting a calcareous skeleton. Coral reefs are a complex system with a variety of organisms like plants and corals that grow together in a specific ecosystem. There are three types of coral reef structures: atolls, barrier reefs, and fringing reefs. All three types are found in India. Fringe reefs are in the Gulf of Mannar, the Gulf of Kachchh, and the Andaman and Nicobar Islands. Barrier reefs have been reported on the western side of the Andaman group of islands, and atoll forms are typical in the Lakshadweep islands. In the Indian Ocean, 199 coral species from 71 genera have been identified. Indian scientists found over 200 coral species in the Andaman Islands. Coral reefs are important ecosystems that provide breeding and living grounds for various marine life. They also give jobs and food to local communities. Natural and human activities threaten coral reefs, including storms, diseases, and pollution. Reefs are fragile and easily damaged by environmental changes. Activities such as improper fishing techniques, oil spills, and deforestation also harm reefs.

THREATS TO MARINE ECOSYSTEMS

Many marine species, including certain marine mammals and creatures, are now known to be threatened, endangered, or vulnerable to climate change, and they may also be negatively impacted by increased human activity. A few marine species have already gone extinct (Estes, Burdin, and Doak, 2015), while others, like the polar bear (Ursus maritimus), are in danger of disappearing because of the melting Arctic ocean ice brought on by global warming (Hunter et al. 2010). Regional species diversity is tracked by the World Conservation Union (IUCN). It keeps a list updated. According to Rodriguez et al. (2015), the IUCN Red List classifies plant and animal species according to their relative abundance or scarcity in a given region by using quantifiable criteria to define their occurrence. Biodiversity loss is dangerous for humans (Díaz et al., 2006). The supply of things we get from the ocean is at risk because of this loss. We need to measure how much biodiversity is decreasing. Biodiversity helps ecosystems produce things (Tilman, Reich, and Isbell 2012). Marine ecosystems are more productive when there is more biodiversity (Worm et al., 2006). It's hard to put a value on biodiversity (Nunes and van den Bergh 2001). There are many things ecosystems do that are important, like giving us clean air and storing carbon (Nunes et al., 2000). To measure the loss of products and services, we can value the loss of biodiversity instead. Coral reefs and coastal environments are important for biodiversity. Damage to these areas can harm the ecosystem and the plants and animals that live there. This can reduce the benefits that these areas provide to people. Over the past few decades, invading species, changes in the global climate, and an increase in human activity have all contributed to the remarkable rate of change in the fauna, flora, and habitats of coastal oceans. Humans do, however, already have a significant impact on the oceans, particularly in coastal regions but also more and more in the open ocean. Currently, shallow coastal waters especially those in enclosed areas are the main source of eutrophication. The following are these threats:

- Introduction of alien species.
- Climate change
- Pollution
- Habitat Destruction
- Overexploitation

Introduction of Alien Species

Any species that has been moved and introduced into an environment beyond its historical geographic range, whether on purpose or by accident, is considered alien. If such species have a negative impact on the environment or the economy, they are referred to as "invasive." Because invasive species disrupt biodiversity and remove essential food chain components, they have the potential to significantly alter the composition and dynamics of maritime ecosystems. These species damage native biodiversity in a variety of ways, such as by dispersing disease or acting as rivals, predators, or parasites.

Water is used by ships as ballast to change where they are in the water, which increases their stability and manoeuvrability. Ballast tanks are normally pumped full of water at one port, which is then released when cargo is loaded at another port. Unintentionally, sailors have used ballast water to move native marine creatures to new locations. A significant biological "regime change" in the Black Sea was brought about by the introduction of Mnemiopsis leidyi, an invasive creature that resembles jellyfish, and it had a role in the decline of the area's commercial fisheries. As planktonic larvae in ballast water, numerous different species have been introduced. Numerous alien species exist. several seaweeds, including Caulerpa taxifolia, zebra mussels (Dreisssena polymorpha), lion fish (Aquarium fish), and red alga (Kappaphycus striatum).

Pollution

Pollutants in air, water, and soil harm organisms, causing plant growth changes, reproduction patterns, and even extinction. Coral reefs are harmed by various contaminants from multiple sources. Algae can outcompete coral due to their rapid development. Coral reefs suffer from untreated wastewater, leading to the loss of reefs. Industrial waste is intentionally dumped in designated areas that end up in the sea, causing poison, acid, or alkaline contamination. Mass fish deaths occur due to industrial waste releases. In the past, industrial and sewage waste was thrown directly on the coast, but now it is discharged via pipelines. Heavy metals such as mercury, cadmium, lead, and arsenic should be taken seriously. The ocean has little industrial debris, but coasts have more. Oil pollution comes from accidents, ballast water, washings, and manufacturers. Animals can be hurt or poisoned by ingesting oil, leading to sleepiness and drowning. Hypothermia is common. Mousse is a sticky material from oil spills that sticks to fur or feathers, causing animals to lose insulation and die from lack of nutrition. Filter feeders can accumulate toxins from oil, harming them or moving up the food chain. Shellfish from affected areas can taste greasy and heavy. Coral spawning is severely impacted by oil spills.

Over Exploitation

Excessive use of resources can result in resource depletion and endanger the existence of certain vulnerable and endangered species. Over 400 species are used by humans as food sources from the marine environment, representing a larger diversity of higher trophic level animals that are utilised in the water than on land. The overuse of marine life resources to satisfy the rising demand for food has resulted from the exponential increase in human population over the past few decades. Fishing fleets across the globe are two to three times larger than what is required to take current fish and other marine species catches. The overuse of resources has increased due to the employment of contemporary methods for gathering, transporting, and storing resources. Thus, over 80% of the world's fisheries are either completely overfished, severely depleted, or in the process of collapsing. According to a recent study, 29% of fish and seafood species have collapsed, meaning their capture has decreased by 90%. If nothing is done now, it is predicted that these species would vanish by 2048. Ninety percent of the world's major predatory fish stocks have already collapsed. Overexploitation usually has the greatest effect on target species. For the species that are categorised as endangered or at risk of extinction, overfishing is by far the greatest concern. Many other species are under peril from overfishing, especially big fish like swordfish, tuna, and sharks.

Climate Change

Rising sea levels are a result of global warming. Consequently, water's capacity to dissolve oxygen is reduced at higher temperatures. However, because humans use a vast amount of fossil fuels, the amount of CO_2 in the atmosphere has been rising. Because of the high temperatures, 90% of the females who hatch from loggerhead turtle eggs in Florida are already formed; males will not be produced in these nests if temperatures rise by an additional 1° C or more due to global warming. The ideal water temperature range for coral reef formation is between 23 and 29°C. Coral reefs require specific environmental conditions to flourish. Overheating the temperature can lead to the expulsion of zooxanthellae from the coral by the polyps, which can result in coral bleaching. Seawater cooling in coastal power facilities disrupts marine ecosystems, especially with low oxygen levels. If current trends continue, the Arctic will become ice-free in decades, causing polar bears, walruses, and other Arctic species to vanish.

Habitat Destruction

Aquatic biodiversity is decreasing because of habitat loss. Habitat destruction and fragmentation refer to the loss or discontinuity of an organism's surroundings. In the last few decades, 20% of the world's coral reefs were lost, 20% degraded, and 35% of mangrove areas were lost. When a species goes extinct, its genetic material is lost forever and cannot be replaced. Extinction wastes life and potential solutions for future issues, like treating illnesses and adapting to a changing environment. The decrease in biodiversity is a big problem. Sea turtles will be affected by rising sea levels and temperatures. Divers catch fish that live in coral reefs, but they are not trained and catch too many fish. They also use poison to stun fish, which kills coral and 90% of the fish. The reef is destroyed by the poison. Clearing mangrove forests harms biodiversity. Mangrove forests are important for many species in the ocean. High water temperatures can cause sickness and bleaching. Rising ocean levels will flood low-lying islands. This will cause flooding and erosion in coastal areas. Extreme weather patterns will make this worse. Ocean currents may also change, causing disruption to ecosystems.

MARINE BIODIVERSITY ASSESSMENT

Marine biodiversity assessment involves systematically analysing organism diversity, genetic diversity, species richness, abundance, and distribution patterns to establish conservation priorities, understand dynamics, and inform management and policy decisions for marine ecosystem protection.

Conducting assessment of critical habitats and species included the following steps:

Stage 1- Collecting data in field
Stage 2 - Manage and compile the data
Stage 3 -Analyse the data
Stage 4- Interpretation and peer-review of data
Stage 5-Communicating the results and adapting the management planning based on the evaluation
Stage 6- continuously monitor the indicators at a regular frequency

Different Types of Marine Biodiversity Assessment

Mangrove Biodiversity

To ensure applicability across various areas, use representative study sites within the mangrove habitat. Set up multiple transects to track natural variability over time, and delineate a predefined number of locations within the area of interest. To ensure statistical rigour, each site should have at least two transects implemented, with the number of replicates varying based on the size of the estuary and the unique study topics. To document changes and patterns in the mangrove environment, keep a close eye on these transects.

Methods of Sampling

For the purpose of returning to each transect at different times and in subsequent years, it must be permanently marked. This will be accomplished by inserting wooden stakes (about 100 cm ×5 cm×5 cm) into the ground. In an ideal world, these stakes would also have a GPS position and a basic sketch map of the pertinent features, allowing transects to be moved in the event that the markers disappear. Although measuring on transects perpendicular to the shoreline and channels is advised, the precise position will depend on the area of interest or selection.

Placing Transect

To mark the selected site, use a wooden stake that is about 100 cm×5 cm 5 cm. Hammer one at each end of two replicate transects. The features of the site determine the transects length and position relative to the mangrove boundary. It's important to ensure the transect is long enough to include the ecosystems of interest, such as a mangrove forest and mudflat or salt marsh. The transect should be between 20 and 50 meters long. Use permanent markers to indicate where the transect is. Note the GPS location where the stakes are placed in case they are removed or lost. To define the transect, use a marked rope or measuring tape between posts. Always remember the stake that equals zero and keep it consistent on consecutive trips. Make a note of variations in surface features like sediment types or plant groups on the transect line. To avoid trampling, only go along one side of the transect line.

Analysis of Ecosystem

To mark a plot near a mangrove forest, use Figure 11 as an example. The plot size depends on the site's features. A sand flat with sparse shrub and seedling cover needs a 5 m × 10 m plot, while densely packed seedling regions in muddy habitats need a 5 m × 1 m or 2 m × 1 m plot. One edge of the plot must align with the transect, and its location should be recorded accurately. Drawings can help. The plot should include the current mangrove tree boundaries and an unobstructed area for future growth. Measure the total number of trees, saplings, and seedlings inside the plot. If there are too many seedlings, count them in smaller subplots and multiply to get the desired area. It is important to keep track of the methods used. Count the mangroves in the same areas on future visits using the same methods.

Satellite Imagery Analysis

SAC research in 2012 found that all states and territories with mangroves will have their mangroves mapped at the community level. These areas include Andaman and Nicobar Islands, Tamil Nadu, Kerala, Puducherry, Maharashtra, Goa, Odisha, Karnataka, West Bengal, and Andhra Pradesh. Mangroves in Andhra Pradesh, Goa, Maharashtra, Odisha, Karnataka, Kerala, and Puducherry have been identified using data from 2005-2007. Low tide data sets were used because mangrove forests are submerged during high tide. Google Earth services were also used to identify peripheral mangroves and new plantation sites. The main source of data for digital analysis is IRS P6 LISS-III data. Equations and calibration factors are used to transform digital values to spectral radiance values.

$$Lrad = \{[DN / \text{max grey}] \times [Lmax - Lmin]\} + Lmin$$

where DN = digital number of each pixel and max grey = 255 for LISS-III.

The experiment began with taking photos of the land using a special camera. These pictures were sorted using a computer program called ISODATA, which grouped similar parts of the land together by their color. The areas were given different color to make them easier to distinguish. The researchers discovered that the height, salt content, and flooding frequency of the land influenced the plants that grew there. The computer images were combined to create a map that depicted the various land areas. To ensure its accuracy, the map was compared to the actual field by checking different spots. All land areas were included. Finally, the data was classified accurately using the correct information. The mangrove community is analyzed and compared to actual data and professional knowledge. Mangroves and other plants can be identified using the IRS LISS-III shortwave infrared band, which has a resolution of 23.5 m. Contextual editing is used to improve accuracy. More information can be found in SAC (2012).

Methods of Assessing Sea Mammals

Marine mammal monitoring programmes often aim to characterise the species distribution and density in a given area, track population status, track the effects of biological events or human activities, or track the use of habitat in space and time to identify critical areas for marine mammals, such as feeding or breeding grounds. Cetaceans (whales, dolphins, and porpoises) and dugongs are among the marine mammals found in Indian seas.

Basics of Mammal Assessment and Monitoring

Marine mammals are rare sightings and are treated differently. To estimate their population, a survey using distance sampling is conducted. The vessel collects genetic, visual, auditory, and behavioural data about them. The technique involves operating strip transects on one side of the ship. Observations are categorized into intervals of 2, 5, or 10 minutes. Marine animals are always recorded. The active survey focuses on the 180° ahead of the ship. The angle and distance to the mammals are always recorded.

Fish Census

The different fish species that appear together are referred to as a "fish assemblage." Three aspects of reef fish assemblages that can be monitored are:

- Diversity: the variety of species present;
- Structure: the relative number and species composition; and
- Fish of a particular species per unit area is known as population density.

While there have been attempts to create a census technique that can reliably measure each of the three attributes, fish biologists largely concur that there isn't one. Prior to choosing one or more suitable techniques, you must determine which information is most crucial for your management requirements.

Census Methods

The most common methods for visual fish censuses are: stationary counts, belt transects, and random swim techniques. In choosing a method, be sure to consider the behaviour of the relevant fish species (e.g., cryptic, schooling, attracted or repelled by divers).

- All of the species observed at the site are the focus of the stationary census, along with their relative abundance and frequency of occurrence.
- The belt transect approach covers a greater area with each census and produces more accurate estimates of density.
- A more comprehensive breakdown of the overall species richness is given by the random swim approach.

Limitations

All visual census methods have the following limitations:

- Most species are generally underreported in terms of abundance by observers.
- The fauna is only counted in its "observable" section. It is possible that species that are nocturnal, secretive, or pelagic will be undervalued.
- Fish behaviour will be impacted by a diver's presence.

Stationary Fish Census

In the stationary fish census method has been used extensively. Although a version of this method might work better in your specific area, the core technique is provided below. Modifications that are appropriate (e.g., altering the area sampled, the duration of each census, the distribution of time spent underwater) will be contingent upon local circumstances (e.g., visibility, depth), as well as management requirements.

1. Find out how big the sample area is. Write down the different types of material on the bottom (like sand or rocks) and how much of each there is at each randomly picked spot. Stand at the 7.5-meter mark after using a tape measure to mark off a 15-meter area.

2. Make a list of the species you see. Look around inside a 7.5-meter imaginary circle from the bottom to the top of the water. Write down the names of all the different kinds of animals you see in the first five minutes. To make it easier, use abbreviations for the names. Only write down the name of the species, not how many or how big they are. If you see a shark or other fast-swimming fish, write down how many you see.
    - Count the number of fish in each species, starting from the bottom of the list and going up.
    - After reaching 7.5 meters, go back and scan the entire area.
    - Use counting in 10s, 20s, 50s, or 100s to estimate the number of fish in large populations.
    - Measure the fish fork length and group each species into size classes based on the minimum and maximum fork length, with a size range of no more than 5 cm.
    - If a species is no longer present, record data from memory.
    - Any additional fish species observed after the initial 5-minute period are ignored, unless you want to include them on a site species list.

Record under sea data for 15 minutes, tracking depth, site relief, habitat structure, sand/clay, and algae. List species and quantity. Consider including all species and reducing cylinder radius to 5 meters for small, cryptic, stationary fish and more accurate density estimates for territorial species.

Belt Transect Census

Belt transects are thought to be most helpful for counting species that are patchily scattered since they cover a greater area per census than stationary counts. Survey stakes or other randomly chosen markers can be used to mark permanent transects along which they are performed. The species that the census is targeting will determine the length and width of a belt transect; however, all transects that are sampled must have the same dimensions. Small, cryptic species might benefit from a tight transect (2 metres wide), while groupers, snappers, and parrotfish might benefit from a broader transect (4 to 5 metres wide). The fundamental method are as follows:

1. While extending a fibreglass measuring tape 50 or 100 metres down the bottom, swim steadily along the designated path.
2. Count the species, number of individuals, and minimum and maximum lengths of fish within a set space (1 to 5 metres) on either side and above the line, including species that cross in front of you or beneath you, as you swim down the transect and unreel the tape. Don't document fish swimming into the transect area from behind you.

In order to avoid overestimating highly mobile species at slow speeds and missing cryptic species at faster speeds, the swimming speed needs to be standardised for repeated censuses.

Random Swim Technique

Although population density is not covered by the random swim technique, it offers useful information on species richness and relative abundance. Rather than documenting additional fish data, the whole census period is devoted to hunting down unrecorded fish species. Replica sample censuses must be carried out in order to gather trustworthy data.

The following is a basic 50-minute census technique.

1. In the chosen area, start the census at a random spot.
2. There are five intervals of ten minutes during the census time. Note the species name along with the time gap when the species is first observed.
3. The relative abundance of species is determined by assigning a score based on the time span between initial observations, with more common species appearing earlier

Sea Grass Habitat Assessment

Sea grasses near estuaries and reef flats filter nutrients and can undergo changes in species, growth, and productivity. Seasonal changes require understanding through environmental monitoring programs. Mapping sea grass distribution can quantify habitat gains or losses, using aerial imagery and GPS-based surveys. Metrics like biomass, area, and ground cover are commonly used to measure composition and abundance. Scientists monitor physical factors like sediment and depth, temperature, salinity, light, turbidity, and temperature to assess sea grass survival and growth. Monitoring plans are carefully designed, considering natural changes in sea grass habitats, and involving government agencies and coastal administrators. We can tell the difference between regular changes and negative impacts on the environment, like habitat destruction, through monitoring and managing activities in a specific area, such as a watershed.

Mapping the Habitat Using Satellite Imagery and Aerial Photography

Research on sea grass distribution near isolated islands is limited. Satellite images and GIS can help understand changes in coverage. Studies on Rameswaram islands and Fletcher et al. (2008) used aerial photography for monitoring and mapping sea grass ecosystems. Grass beds can be disturbed by nature or humans. Scientists used aerial photography, pixel models, and GIS to monitor changes in 1-meter ground features and grass bed disturbances. They converted digital film into different color spaces to identify areas without vegetation and created models to distinguish between vegetated and bare areas. This method produced maps with 75-100% accuracy. In a study, Nobi et al. used satellite data, GIS, and field observations to evaluate restored sea grass ecosystems in Lakshadweep islands. They created sea grass maps from 2000-2008 and marked restoration sites. Around the world, different site selection models have been used to mitigate sea grass losses.

Table 1. Comparison of different fish census methods

Methods	Advantages	Disadvantages
Stationary Census	Benefits relative abundance and permits large sample sizes in many ecosystems	Trains personnel more slowly and is unlikely to offer an exhaustive list of site species (unless altered).
Belt Transect	Large areas can be sampled for each census, making it possible to target more migratory species and possibly estimating species' densities more accurately, including groupers and snapper.	Lower sample rate per unit of time in comparison to stationary approaches; data on small habitats might not be obtained.
Random Swim	Most likely to offer an exhaustive list of species; characterises a greater percentage of species in each sample	Reduction in the number of samples per unit of time; no density estimations and a decrease in quantitative data.

Transplanted sea grass survival can be impacted by various factors. It is better to restore sea grasses in their original regions to avoid disagreements over where to put them, their habitat needs, and invasions. This has been suggested by several experts.

Monitoring Sea Grass Beds

Sea grass beds are very beneficial to fisheries because they give grazing fish food and cover as well as a place for algae and other invertebrates to thrive, which the fish eat. Juveniles of commercial prawns and fish are raised in sea grass beds. Although sea grass beds are inherently changeable and some experience annual dieback episodes, monitoring the number of sea grasses provides vital information for fisheries. It is important to take this into account and to repeat the monitoring at various points throughout the year. Because of the poor vision, manta towing is rarely feasible in sea grass beds. Three approaches are available:

- The first step is to weigh living material, or biomass, by taking samples of it. Only extensive sea grass beds should be used for this because it is harmful. A trained diver can visually measure the weight of the sea grass, or biomass, using the second method. Training and lab work are required for both procedures.
- Observing the sea grass area year-over-year to determine if it is shrinking or expanding is a less complicated approach. The contours of the sea grass beds can be drawn on a map at very low tide and utilised in conjunction with natural features.
- Aerial photography of shallow water sea grass meadows during low tide on sunny days provides accurate assessments of changes in sea grass beds, revealing conditions and reasons behind these events.

Seaweeds

Seaweeds are creatures found in water and can be small or large. They are not plants or animals, but macro algae in the Protista kingdom. Seaweeds use sunlight for growth and contain chlorophyll to absorb it. Some seaweeds have additional pigments that give them different color. Seaweeds are attached to the seafloor by holdfasts, not roots. Seaweeds have different parts with specific functions. The holdfast serves as an anchor but isn't necessary for water or nutrient intake. The holdfast has finger-like projections called haphtara. The stipe is the seaweed's stem or stalk that supports the plant. There are different types of stipe structures. Blades are the seaweed's leaves that absorb sunlight and support the plant's regenerative structure. Different species have varying numbers of blades. Floats or pneumatocysts are gas-filled structures that keep the seaweed afloat and help it collect solar energy. The entire seaweed is called a thallus. There are many unique marine creatures and plants in seaweed masses. These creatures are important for the food chain. Sargassum is a type of brown algae found in protected areas along the Indian coast. It is both a species of algae and a habitat for marine life. Sargassum can be found washed ashore on beaches in Maharashtra and Goa. It has branching stems and pneumatocysts (floats) and is anchored to hard surfaces with holdfasts. Sargassum is a fragile plant that grows tall in water, forming structures like forests. Its floats hold oxygen, nitrogen, and carbon dioxide, making it buoyant. This provides food and shelter for many young fish and endangered invertebrates, like sea horses.

Assessment and Monitoring

Sargassum can be found in Indian waters from the intertidal zone to 10 meters deep, depending on light. In Goa and Maharashtra, it can be found up to 5 meters deep. To gather data on sargassum beds, parallel transects perpendicular to the beach should be used. The transects should be measured to the edge of the bed where sargassum disappears and recorded with GPS. Sargassum presence should be noted at each station. GPS can be used on a boat or by walking/snorkelling to record poly point data. GIS programs can be used to determine the total area and polygon of each sargassum bed. Sargassum coverage estimates are made every five meters along the transect. A 50×50 cm^2 quadrat divided into 25 squares measuring 10 by 10 cm2 should be used on the substratum. The percentage of sargassum cover in each section is estimated using the average of 25 results. To calculate sargassum production, the equation P = B×A×C is used. On sampling days, samples are collected, weighed, and taken to the lab for biomass estimations. The biomass samples are measured for length and cleaned with water before being dried at 60^0 for a day to achieve a constant dry weight.

Coral Reef

Coral reefs are productive and varied communities found in shallow, warm, and clean tropical oceans around the world. They are made by reef-building corals and calcareous algae, which produce a lot of calcium carbonate. This calcium carbonate makes up most of the reef's structure. Coral reefs are important because they create a lot of three-dimensional shapes and structures in the ocean, and they also form many sea islands.

Manta tow

Programmes that use this method:

- Australian Institute of Marine Science Long-term Monitoring Program (AIMS LTMP);
- GCRMN.

Description of Method

Snorkelers continuously drag boats to collect data on creature distribution, abundance, and disturbances like bleaching, COTS, and cyclones. This method provides a comprehensive description of large reef areas and visible factors over long distances.

- Manta tows are useful for counting gigantic clams or COTS, monitoring changes in coral cover, and determining the presence of impacts like illness and bleaching. In order to guarantee consistency in assessments between observers, it is necessary to educate observers to estimate these abundance categories.
- Moreover, manta tows can offer comprehensive data on benthic populations, particularly on particular impacts like bleaching or harmful fishing methods, as well as important macro-invertebrates including giant clams, COTS, and Diadem.
- 50-60 tows are enough to detect a 20% change in COTS. Dive surveys can be calibrated for better manta ray abundance estimates. Combining manta ray and scuba benthic monitoring is the best option. To get detailed info about benthic communities, assess certain parameters.
- The percentage of sand, rubble, and dead coral; the percentage of hard and soft coral;

General Procedures

- The reef survey involves split two-minute surveys, recording data on a manta board, and recording the boat's position and tow number on an aerial photo. Repeat until the entire reef perimeter is surveyed.
- For a depth of 5 to 10 metres, the tow path runs parallel to the reef crest, making the majority of the slope visible. A steady 3 to 5 km/h (1.5 knots) is the tow speed. It could be necessary to adjust the tow speed according to elements like sea conditions and currents.
- The observer scans a breadth of 10–12 m, depending on the number and dispersion of the creatures being counted, as well as the visibility, reef gradient, and distance from the bottom.
- Elements including wind, currents, and solar angle influence the survey direction. To prevent the need to amend data for follow-up surveys, the survey orientation should be standardised.

Advantages

- Since a vast region is covered quickly, there is less possibility of missing sporadic disturbances or population shifts (such as dynamite fishing, COTS, bleaching, sickness, and storm damage).
- Cheap equipment with easily manufactured manta boards locally;
- Simple to operate with minimal field training.
- Wide distances travelled with little strain to observers;
- Suitable for isolated areas requiring little assistance (may be completed on a snorkel).
- Very good for determining the type of reef and its resources, and reasonably accurate (when calibrated with a scuba search) and economical for detecting the number of non-cryptic COTS and corals across wide areas in clear water.

Limitations

- The tow route is controlled by the boat driver, so parts that aren't suitable like sand or steep reef slopes may be covered. Because cryptic organisms are readily missed, such as young COTS, COTS under plate corals, and small gigantic clams (Tridacna), actual abundances are likely underestimated. Medium-scale surveys should be used to calibrate the results.
- There aren't many variables because the observer has to recall every observation made during each 2-minute tow.
- The inability to visually analyse the main reef creatures' limits precision only shallow reefs can be surveyed, particularly in low visibility.
- Is limited to measuring coral cover in broad categories, such as 0–10%, 11–30%, 31–50%, 51–75%, and 76–100%.
- In murky waters where sharks or other potentially dangerous creatures may be present, caution is required.

Random Swim

Programmes that use this method:

- Community monitoring programmes may use the random swim for monitoring.
- Management or research monitoring programmes may use the random swim for site selection.

Method Description

Site selection involves a snorkel or scuba diver buddy pair selecting suitable sites for monitoring e.g. checking if there is sufficient continuous reef for transects. Random swims can also be used for monitoring various coral reef parameters. Information obtained:

General description of the site with semiquantitative counts of various coral reef variables. Equipment required: No special equipment.

Field personnel:

- 1 boat driver/surface watch;
- 2 trained observers.

Lab personnel:

- Data analysis, interpretation and reporting.

General Procedures

Swim around the general reef area to assess whether the chosen monitoring techniques are appropriate, or create a species list to choose which metrics to count when conducting the monitoring.

Advantages

A useful tool for choosing the right level of detail, such as species or genus level, and the type of monitoring technique to use (transects work best for continuous reef sections, whereas quadrats are more suitable for patch reefs).

Limitations

- Only the area visible to the divers is covered, and this might not be the ideal spot. For site selection, manta tow is advised.
- For further information and to create species lists, random swims are advised.

Benthic Communities

To properly allocate resources, coral reef managers require knowledge of the conditions and patterns in benthic communities. While data on soft corals, algae, sponges, and other invertebrates are also needed by managers, hard corals receive the most of the attention. Thus, the focus of these approaches is on monitoring coral diversity, coral cover, health and illness, growth, and recruitment in order to assess corals.

Coral Species Diversity

When there are less species present, like in Malvan, Goa, or Netrani Island, or when a small number of very dominant species exist, assessing coral diversity is simpler. Assessing coral growth type is frequently required in place of diversity assessments in the Andaman and Nicobar Islands, Lakshadweep, Gulf of Mannar, and Gulf of Kachchh. For fish, there are additional “species diversity” metrics.

Percent Cover

Management most commonly utilise the hard coral cover percentage to evaluate the health of the reef. The percentage cover of different types of benthic plants and animals, as well as rock and debris, is simple to calculate and interpret.

Coral Health

It’s critical to keep an eye on a variety of coral health indicators in order to identify and comprehend the reasons for coral dying. Over the past few years, coral bleaching has grown in importance as a worry for coral reef management, and it appears that the frequency and distribution of coral diseases are rising. Medium-scale surveys can offer more specific information on the type and quantity of coral disease, bleaching, or mortality; permanent fine-scale surveys can offer more information on the effects on individual species and coral colonies; and broad-scale surveys can offer information on the general health of corals and causes of death at a large number of sites. Questions might include:

- Are there coral diseases at our sites, and if so, what is the type and abundance?
- How has coral bleaching affected corals?

Coral Bleaching-Specific Methods

Comprehensive data on the volume and trends of coral mortality due to bleaching are needed in order to advance our comprehension of the ecological ramifications of mass coral bleaching. While medium- and fine-scale monitoring offer more accurate information regarding the percentage and kinds of corals that bleach and either survive or perish, broad-scale surveys will only reveal the extent of bleaching. Bleaching can be observed using any of the previously discussed techniques for % cover or coral health. The techniques that follow, however, were created especially for bleaching.

Disease Specific Methods

Disease seems to be becoming more common in corals. In areas where this is a significant issue, disease-specific questionnaires are helpful.

Structural Complexity (Rugosity)

It’s helpful to keep an eye on a reef’s rugosity to see how the structure of the reef evolves with time. Although more laborious and time-consuming, chain intercept transects methods (CIT) offer a good estimate of rugosity. An alternative to chain methods is the use of “line intercept transect” (LIT) procedures. Reef topography evolves over time, and this can be ascertained from the growth form (coral shape) information gathered utilising LIT approaches. Keep in mind that, in comparison to CIT approaches, growth form data offer less comprehensive topographic information.

Coral Recruitment

To determine a reef's health, more than just coral cover or mortality over time is needed. Young recruits are important for the reef's health, and tracking their recruitment is necessary to find areas that serve as sources or sinks for larvae. This information can help determine a reef's ability to recover after a disturbance. Coral recruitment can be measured with settlement plates or visual searches in quadrats. The goal of field searches is to find successful recruits who have survived their first year, and they can currently be observed by field observers.

OVERVIEW OF DATA ANALYSIS

The analysis of substrate categories such as sand, sponge, zoanthids, algae, dead and live coral cover, and their life forms can be done using established procedures and formulas.

Table 2. Comparison of different coral reef assessment methods

Monitoring Category & Scale	When should you choose this method
Manta tow or video towed diver	This method efficiently covers a wide area through mapping and site selection, estimating coral cover percentages limited to shallow snorkel depths with minimal detail.
General observations	Reef change estimates, coral health monitoring, and coral bleaching surveys provide general information for dive tourism workers, researchers, and recreational divers, with low precision and minimal training required.
Timed swim	Scuba diving measures coral cover percentages, providing a thorough view of invertebrates and coral types, but is less precise than belt transects.
Medium scale	More precise, detailed, and smaller area than large-scale approaches. more costly and time-consuming than procedures used on a large scale.
Timed swim	Species diversity: The highest level of expertise is required.
Line transects LIT & PIT	LIT requires skilled personnel for accuracy, while PIT is quick and easy. Coral data requires extensive knowledge, while specific disease data is detailed but time-consuming.
Belt transect	Coral health requires medium to high knowledge, while coral bleaching is simple but requires specific information on affected corals and bleaching degree. Disease-specific information is also mentioned.
Chain transects	Percent cover: highly skilled personnel, exact data, and a high level of detail; requires more time than LIT. Disease-specific: Hard to execute; experience needed; comprehensive details on the kind of corals impacted and the degree of bleaching. Complexity of structure: Hard to execute; expertise needed.
Video transect	High accuracy, medium detail; permanent record; skilled scientists and divers for data collection and analysis; costly equipment to purchase and maintain; don't use until appropriate resources are available are the percentages covered.
Fine scale	helpful for posing specific, targeted queries. costlier and more time-consuming than surveys of a moderate size.
Visual quadrat	Visual methods are more accurate for coral recruitment, but disease assessment is time-consuming and requires detailed data on bleaching severity and new corals' survival.
Permanent photo quadrat	Percent cover: Extreme precision and detail; a permanent record; skilled scientists and divers to gather and interpret data; costly equipment to purchase and maintain. If appropriate resources are not accessible, do not utilise.
Tagging coral colonies	Specific to a disease: Maximum level of detail and most time commitment; offers accurate details on individual coral colonies and the ways in which diseases impact them.
Recruitment tiles or plates	Coral recruitment: Offers details on recently arrived coral recruits on a reef.

Percent cover = Total length of category/length of transect X 100

The relative abundance of each coral is calculated using the following formula

Relative abundance = cover of the species/total live coral cover X100

Same formula can be adopted for other lifeforms such as macro invertebrates, fish, algae, sea grass etc.

Diversity, evenness and dominance

Species diversity and evenness Shannon Weaver and Simpson diversity indices can be used, as Shannon diversity index emphasizes on rare species and Simpson index on most common species.

The Shannon-Weaver formula for diversity can be calculated as follows, with a projection of 1 cm under the line being considered as an 'individual's

$H = 1\text{-}\Sigma\ pi \ln pi$
$i = 1$

where pi = ni/N, N = total number of centimetres (individuals) of all species under the line, and ni = the number of centimetres of species i under the line.

Evenness can be calculated as J' = H'/H max, which is the diversity of the total number of individuals (centimetres) under consideration divided by the diversity which would exist if all the species present are distributed equally. A value of 1.00 represents maximum evenness. Diversity and evenness for each sampling site can be calculated from the combined data from all transects.

In order to compare Simpson's index to Shannon and Weaver's index, following Simpson's index can be used:

$D = 1\text{-}\ \Sigma\ Pi^2$
$i = 1$
Pi = Proportion of individual species i

where, N = Total number of individuals in the plot

S = Number of species in the plot

Calculation of dominance index

Index of dominance = Abundance of dominant coral species / Total coral cover

The Ward method of linking with the Block distance measure is used in an agglomerative hierarchical analysis to analyse the data on the relative abundance and absolute change in the abundance of corals and other lifeforms for interstation comparison. Arcsine transformation is applied to percent values in ANOVAs. Programmes like Microsoft Excel, SPSS 10.0, Origin, and Statistics are used for data analysis.

MARINE BIODIVERSITY IMPACT MITIGATION MEASURES

Mitigating the impact on marine biodiversity is crucial for the health of ecosystems and the sustainability of marine resources. Here are some measures commonly employed for mitigating such impacts.

Marine Protected Areas (MPAs)

Fisheries, biodiversity, and habitat preservation are the three main goals of managing designated marine areas, or MPAs. They place restrictions on fishing, tourism, and mining in order to save marine environments. Examples include the United States' Papahanaumokuakea Marine National Monument and Australia's Great Barrier Reef Marine Park *(Edgar et al., 2014)*.

Sustainable Fishing Practices

Sustainable fisheries management aims to preserve fish stocks and protect fishing communities by reducing bycatch and discards, protecting habitats, and setting catch limits based on science. This helps stop illegal fishing. The Patagonian toothfish fishery in the Southern Ocean is an example of ecosystem-based fisheries management in action. It recovered after catch restrictions and conservation measures were put in place. This approach has also been used in the Gulf of Maine in the United States. *(Worm et al., 2009; Pikitch et al., 2009)*.

Habitat Restoration

Replanting, building artificial reefs, and reducing pollutant inputs are some of the steps involved in restoring degraded marine habitats, such as coral reefs, mangroves, and seagrass beds. In the Caribbean, the Mesoamerican Reef Restoration Network and the Coral Restoration Foundation both have successful restoration initiatives *(Bayraktarov et al., 2016)*.

Marine Spatial Planning (MSP)

By distributing maritime space for various purposes, MSP aims to protect biodiversity and lessen conflict. Social, economic, and ecological factors are taken into account while creating comprehensive maritime zoning plans. Prominent instances of MSP projects are seen in the EU and the US *(Ehler & Douvere, 2009)*.

Pollution Control and Waste Management

To reduce sources of pollution such industrial discharge, agricultural runoff, and marine trash, cleanup efforts, regulations, and public awareness campaigns are all required. Regional agreements like the International Coastal Cleanup and the Barcelona Convention for the Mediterranean Sea are instances of effective initiatives *(Jambeck et al., 2016)*.

Invasive Species Management

In marine habitats, invasive species are managed through control, eradication, and preventive measures. Examples of tactics include ecological impact assessments, elimination projects, and biosecurity protocols. Two examples include initiatives to halt the spread of lionfish in the Caribbean and invasive green crab in various coastal regions *(Simberloff, 2009)*.

CONCLUSION

It's important to understand the challenges marine ecosystems face to protect marine life and ocean services. Marine biodiversity is crucial for the planet's health. The Marine Biodiversity Impact Assessment helps assess and reduce human impacts. We can preserve ocean health through science, stakeholder involvement, and adaptive management. Changes in one area can affect the whole ecosystem. Climate change is a major challenge for marine ecosystems. Rising temperatures and acidity harm marine life, causing coral reefs to bleach and die. Our oceans face challenges from pollution and habitat degradation. To address these issues, scientists, legislators, business executives, and local communities must work together. By cooperating, we can create solutions that benefit everyone and protect marine ecosystems. Sustainable fishing, reducing waste and pollution, and preserving habitats like estuaries and coral reefs are ways to achieve this.

REFERENCES

Aronson, R. B. (2001). The Limits of Detectability: Short-Term Events and Short-Distance Variation in the Community Structure of Coral Reefs. *Bulletin of Marine Science*, *69*, 331–332.

Bakun, A. (1990). Global climate change and intensification of coastal ocean upwelling. *Science*, *247*(4939), 198–201. doi:10.1126/science.247.4939.198 PMID:17813287

Bass, D. K., & Miller, I. R. (1998). Crown-of-Thorns Starfish and Coral Surveys Using the Manta Tow and Scuba Search Techniques. Long-*term Monitoring of the Great Barrier Reef. Standard Operations Procedure Number 1*. www.aims.gov.au/pages/research/reef-monitoring/methods.html

Bayraktarov, E., Saunders, M. I., Abdullah, S., Mills, M., Beher, J., Possingham, H. P., Mumby, P. J., & Lovelock, C. E. (2016). The cost and feasibility of marine coastal restoration. *Ecological Applications*, *26*(4), 1055–1074. doi:10.1890/15-1077 PMID:27509748

Bohnsack, J. A., & Bannerot, S. P. (1986). A Stationary Visual Census Technique for Quantitatively Assessing Community Structure of Coral Reef Fishes. *NOAA Technical Report NMFS*, *41*, 1–15.

Brock, R. E. (1982). A Critique of the Visual Census Method for Assessing Coral Reef Fish Populations. *Bulletin of Marine Science*, *18*, 276–297.

Bruckner, A. W. (2002). Priorities for Effective Management of Coral Diseases, U.S. Department of Commerce. *NOAA Technical Memorandum NMFS-OPR-22*.

Bunce, L., Townsley, P., Pomeroy, R., & Pollnac, R. (2000). *Socioeconomic Manual for Coral Reef Management*. Global Coral Reef Monitoring Network and Australian Institute of Marine Science Townsville Australia.

Carleton, J. H., & Done, T. J. (1995). Quantitative Video Sampling of Coal Reef Benthos: Large Scale Application. *Coral Reefs, 14*(1), 35–46. doi:10.1007/BF00304070

Colin, P. L., Sadovy, Y. J., & Domeier, M. L. (2003). *Manual for the Study and Conservation of Reef Fish Spawning Aggregations*. Society for the Conservation of Reef Fish Aggregations Special Publication.

Couchman, O., Wulffeld, E., Muurmans, M., Steer, M., & Fanning, E. (Eds.). (2009) Society for Environmental Exploration UK. Sea Turtle Monitoring Methods Manual. LIDER Foundation.

Crosby, M. P., & Reese, E. S. (1996). *A Manual for Monitoring Coral Reefs with Indicator Species: Butterflyfishes as Indicators of Change on Indo-Pacific Reefs*. Office of Ocean and Coastal Resource Management, National Oceanic and Atmospheric Administration.

Edgar, G. J., Stuart-Smith, R. D., Willis, T. J., Kininmonth, S., Baker, S. C., Banks, S., Barrett, N. S., Becerro, M. A., Bernard, A. T. F., Berkhout, J., Buxton, C. D., Campbell, S. J., Cooper, A. T., Davey, M., Edgar, S. C., Försterra, G., Galván, D. E., Irigoyen, A. J., Kushner, D. J., & Thomson, R. J. (2014). Global conservation outcomes depend on marine protected areas with five key features. *Nature, 506*(7487), 216–220. doi:10.1038/nature13022 PMID:24499817

Ehler, C., & Douvere, F. (2009). Marine Spatial Planning: A Step-by-Step Approach toward Ecosystem-Based Management. Intergovernmental Oceanographic Commission and Man and the Biosphere Programme, IOC Manual and Guides No. 53, ICAM Dossier No. 6. UNESCO.

English, S., Wilkinson, C., & Baker, V. (1997). *Survey Manual for Tropical Marine Resources. Townsville, Australia*. Australian Institute of Marine Science.

Green, E. P., Mumby, P. J., Edwards, A. E., & Clark, C. D. (2000). *Remote Sensing: Handbook for Tropical Coastal Management*. UNESCO Publishing.

Hiby, A. R. and Hammond, P. S. (1989). *Survey techniques for estimating abundance of ceta- ceans*. Rep. Int. Whal. Commn.

Hill, J., & Wilkinson, C. (2004). *Methods for Ecological Monitoring of Coral Reefs (Version 1): A Resource for Managers*. Australian Institute of Marine Science.

Hoegh-Guldberg, O., Mumby, P. J., Hooten, A. J., Steneck, R. S., Greenfield, P., Gomez, E., Harvell, C. D., Sale, P. F., Edwards, A. J., Caldeira, K., Knowlton, N., Eakin, C. M., Iglesias-Prieto, R., Muthiga, N., Bradbury, R. H., Dubi, A., & Hatziolos, M. E. (2007). Coral reefs under rapid climate change and ocean acidification. *Science, 318*(5857), 1737–1742. doi:10.1126/science.1152509 PMID:18079392

Hughes, T. P., Baird, A. H., Bellwood, D. R., Card, M., Connolly, S. R., Folke, C., & Nystrom, M. (2003). Climate change, human impacts, and the resilience of coral reefs. *Science, 301*(5635), 929–933. doi:10.1126/science.1085046 PMID:12920289

Jambeck, J. R., Geyer, R., Wilcox, C., Siegler, T. R., Perryman, M., Andrady, A., Narayan, R., & Law, K. L. (2015). Plastic waste inputs from land into the ocean. *Science, 347*(6223), 768–771. doi:10.1126/science.1260352 PMID:25678662

Kaliaperumal, N., & Kalimuthu, S. (1997). *Seaweed potential and its exploitation in India. Sea- weed Res. Utiln. 19(1&2):33-40. Koskimies, P. and Väisänen, R.A. 1991. Monitoring bird populations*. Zoological Museum, Finnish.

Marshall, J., Adcroft, A., Hill, C., Perelman, L., & Heisey, C. (1997). A finite-volume, incompressible Navier Stokes model for studies of the ocean on parallel computers. *Journal of Geophysical Research, 102*(C3), 5753–5766. doi:10.1029/96JC02775

Minton, C. D. T. (1968). Pairing and breeding in mute swans. [Morgan, D.G., Estimating Vertebrate Population Densities by Line Transect Methods, Melbourne College of Advanced Education, Australia.]. *Wildfowl, 19*, 41–60.

Museum of Natural History & Helsinki, P. (1959). Zur Methodik der quantitativen Vogelforschung in den Binnengewässern. *Ornis Fennica, 36*(3–4), 66–78.

O'Connor, R. J. (1985). Long-term monitoring of British bird populations. *Ornis Fennica, 62*(2), 73–79.

OECD. (1996). Guidelines for Aid Agencies for Improved Conservation and Sustainable Use of Tropical and Sub-Tropical Wetlands. Guidelines on Aid and Environment, 9. OECD.

Ohansen, K. L., Boertmann, D., Mosbech, A., & Hansen, T. B. (2012). *Manual for Seabird and Marine mammal survey on Seismic Vessels in Greenland* (3rd Revised edition).

Pikitch, E. K., Santora, C., & Babcock, E. A. (2009). A global overview of the management effectiveness of marine protected areas. *Marine Biology, 156*, 273–283.

Pugh, D. T. (2004). *Tides, surges, and mean sea-level: A handbook for engineers and scientists*. John Wiley & Sons.

Rogers, C. S., Garrison, G., Grober, R., Hillis, Z. M., & Franke, M. A. (1994). *Coral Reef Monitoring Manu- al for the Caribbean and Western Atlantic*. USVI, Virgin Islands National Park.

Schwarz, A. M. (2004). Contribution of photosynthetic gains during tidal emersion to production of Zostera capricorni in a North Island, New Zealand estuary. *New Zealand Journal of Marine and Freshwater Research, 38*(5), 809–818. doi:10.1080/00288330.2004.9517280

Short, F. T., McKenzie, L. J., Coles, R. G., & Vidler, K. P. (2002). Sea grass Net Manual for Scientific Monitoring of Sea grass Habitat. QDPI,QFS, Ciarns.

Simberloff, D. (2009). The Role of Propagule Pressure in Biological Invasions. *Annual Review of Ecology, Evolution, and Systematics, 40*(1), 81–102. doi:10.1146/annurev.ecolsys.110308.120304

Smart, M. (1997). Sehumed. Boletin de la Sedepara el Estudio de los Humedales Mediterraneos.

Worm, B., Hilborn, R., Baum, J. K., Branch, T. A., Collie, J. S., Costello, C., Fogarty, M. J., Fulton, E. A., Hutchings, J. A., Jennings, S., Jensen, O. P., Lotze, H. K., Mace, P. M., McClanahan, T. R., Minto, C., Palumbi, S. R., Parma, A. M., Ricard, D., Rosenberg, A. A., & Zeller, D. (2009). Rebuilding Global Fisheries. *Science*, *325*(5940), 578–585. doi:10.1126/science.1173146 PMID:19644114

Chapter 7
Biodiversity Loss Estimation on Natural Stream Diversion:
Biodiversity Loss Assessment

Ashok Kumar Rathoure
https://orcid.org/0000-0001-9131-1346
Saraca Research Inc., India

ABSTRACT

The variability of life on Earth is called biodiversity. Biodiversity takes into account all the living organisms present on Earth. Utilitarian values include the many basic needs humans obtain from biodiversity such as food, fuel, shelter, and medicine. Further, ecosystems provide crucial services such as pollination, seed dispersal, climate regulation, water purification, nutrient cycling, and control of agricultural pests. Biodiversity has cultural value to humans as well, for spiritual or religious reasons for instance. The intrinsic value of biodiversity refers to its inherent worth, which is independent of its value to anyone or anything else. Biodiversity loss refers to the decline or disappearance of biological diversity, understood as the variety of living things that inhabit the planet, its different levels of biological organization and their respective genetic variability, as well as the natural patterns present in ecosystems. The present case study is taken from mining project for extraction of coal by underground semi-mechanized mining bearing area of 659 ha.

BACKGROUND

The resulting case study pertains to a mining project focused on extracting coal through underground semi-mechanized methods, covering an area of 659 hectares in Madhya Pradesh India. Within this lease area, 12.733 hectares are classified as forest land. The mining plan targets coal deposits with a thickness exceeding 1.20 meters, predominantly comprising high-grade coal. The net proved reserve within this thickness range is estimated at 51.490 million metric tons (MT), with a total mineable reserve of 53.606 million MT. The extractable reserve stands at 22.187 million tons. Water requirements for mining and

DOI: 10.4018/979-8-3693-3330-3.ch007

associated activities are projected at 394 kiloliters per day. The operational workforce, including absentees, is anticipated to consist of 725 individuals.

A natural stream flows through the lease area, spanning 2.6 kilometers, prompting a biodiversity loss assessment due to proposed diversion measures. Decadal changes in the study area over 23 years, from 2000 to 2023, have been analyzed using satellite imagery and GIS software. Vegetation cover has significantly decreased from 32,731.2 hectares in 2000 to 16,153.65 hectares in 2023. Agriculture area has expanded from 14.19% in 2000 to 50.98% in 2023, while vegetation has declined from 77.2% to 38.10% over the same period. Human settlement has increased modestly, alongside fluctuations in barren land and water bodies.

Based on the current scenario in 2023, the average tree density is calculated at 20 trees per hectare, and shrub density stands at 45 per hectare. This translates to approximately 13,180 trees and 29,655 shrubs within the mining lease area. The estimated cost of carbon offset by the enumerated trees is Rs. 1.3 million per annum, while the total cost of carbon offset by all trees within the 659-hectare area is Rs. 16.5 million per annum. Additionally, the cost of carbon sequestration by trees near the natural stream amounts to Rs. 0.5 million per annum in 2023.

BIODIVERSITY STUDY

Plants and animals are particularly vulnerable to environmental pressures, and changes in their distribution, frequency, density, and abundance reflect shifts in the biological communities within ecosystems. These alterations over time can be measured and correlated with environmental factors. Natural flora and fauna are integral components of ecosystems, organized into distinct communities and highly responsive to external influences.

It is imperative to incorporate ecological considerations into planning processes, especially given the ongoing degradation of the natural environment, which is an unfortunate but direct consequence of development activities. The study area encompasses reserved and protected forests, including wildlife sanctuaries. Biodiversity conservation is approached from various perspectives, including scientific, philosophical, economic, ethical, and aesthetic viewpoints.

Scientists emphasize that there is still much to be learned about species and ecosystems worldwide, and the loss of biodiversity would prevent the opportunity for further discoveries. Regardless of whether a species holds economic significance, each possesses intrinsic value, underscoring the importance of conserving biological diversity and ecosystem integrity. An ecosystem comprises a cohesive unit housing both flora and fauna whose existence hinges upon the interplay of biotic and abiotic components. The study area can be categorized into cropland, terrestrial, and aquatic ecosystems, depending on the distribution pattern of organisms and the physical environment. To comprehend the governing factors of the system, both abiotic elements (physical setting) and biotic factors (flora and fauna) have been delineated.

METHODOLOGY & APPROACH

Sampling plots for floral inventory were chosen randomly within suitable habitats, drawing on methodologies outlined by Anderson (1867), Jain and Rao (1983), Dixit (1984), Wilson and Reeder (2005), Kumar (2013), and Kumar et al. (2013). The faunal survey methodology encompassed random surveys, opportunistic observations, diurnal bird observations, active searches for reptiles, assessment of faunal habitats, active searches for scats and footprints, animal calls, and review of past studies. Special attention was given to the presence of endemic and threatened species within the study area.

A comprehensive desktop literature review was conducted to identify a representative spectrum of threatened species, populations, and ecological communities listed by organizations such as IUCN, WCMC, ZSI, BSI, and under the Indian Wildlife Protection Act of 1972 (Bentham and Hooker, 1862-1883; Hunter, 1879; Dixit, 1984; Ghosh et al., 2004; Lushington, 1915; Wilson and Reeder, 1993; BirdLife International, 2000; BirdLife International, 2004a, b; Wilson and Reeder, 2005; BirdLife International, 2010; Kumar and Srivastava, 2012; Kumar, 2013; Kumar et al., 2013; Kumar and Aggarwal, 2013). The status of individual species was evaluated using the revised IUCN/SSC category system, as described by WCMC (1988), IUCN (1994), WCMC (2000), IUCN (2001, 2003, 2008, 2010).

DATA ANALYSIS

The data of primary assessment of floral and faunal species has been analyzed by using standard methods. Following are the standard methods of calculations for floral species:

i. Frequency:

$$\text{Frequency (\%)} = \frac{\text{no. of quadrants in which species occured}}{\text{total no. of sampling units studied}} \text{ x } 100$$

ii. Density:

$$\text{Density (\%)} = \frac{\text{number of individual of the species}}{\text{total area studied}} \text{ x } 100$$

iii. **Abundance:** This is the number of individuals of any species per unit area of occurrence.

$$\text{Abundance} = \frac{\text{total no. of individuals of species in all quadrants}}{\text{no. of quadrants in which species occurred}} \text{ x } 100$$

iv. **Dominance:** Dominance is the area occupied by stems of a species in any given area.
Basal area of a species= Sum of basal areas of all the stems
Basal area of individual stem = $\pi D^2/4$
Where D=Diameter of stem

v. Relative Density:

$$\text{Relative Density} = \frac{\text{density of the species}}{\text{total density of all the species}} \text{ x } 100$$

vi. Relative Frequency:

$$\text{Relative frequency} = \frac{\text{frequency of the species}}{\text{total frequency of all the species}} \text{ x } 100$$

vii. Relative Dominance:

$$\text{Relative dominance} = \frac{\text{dominance (cover) of the species}}{\text{total dominance of all the species}} \text{ x } 100$$

vii. **Importance Value Index (IVI):** The value reflecting the relative importance of individual species in study area.

IVI = relative density+ relative dominance +relative frequency

Species Diversity Index (Shannon-Weiner Index): species diversity in a biological community is based on presence of a wide variety of species. It is in expressed by Shannon-Weiner Index (1948) in the following way:

$$H = -\sum_{i=1}^{S} p_i * \ln p_i$$

Where H = Species diversity index
S = the number of species in a sample
Pi = (ni/N)
N = the total number of individuals of all the species in a sample
ni = the number of individuals of a species.

ESTIMATION OF AVERAGE AGE OF TREE

Basal Area

Basal area is the cross-sectional area of trees at breast height (1.3m or 4.5 ft above ground). It is a common way to describe stand density. In forest management, basal area usually refers to merchantable timber and is given on a per hectare or per acre basis. If cut down all the merchantable trees on an acre at 4 ½ feet off the ground and measured the square inches on the top of each stump (πr*r), added them all together and divided by square feet (144 sq inches per square foot), that would be the basal area on that acre. In forest ecology, basal area is used as a relatively easily-measured surrogate of total forest biomass and structural complexity and change in basal area over time is an important indicator of forest recovery during succession.

Estimation From Diameter at Breast Height

The basal area (BA) of a tree can be estimated from its diameter at breast height (DBH), the diameter of the trunk as measured 1.3m (4.5 ft) above the ground. DBH is converted to BA based on the formula for the area of a circle:

$BA = \pi \, (DBH)^2 \div 4$

Circumference (C)

The circumference of a tree is the measurement of the outer boundary of the stem or trunk of a tree, including the bark. It's measured at the highest possible point of measurement, which is 4.5 feet above ground level or the germination point.

Diameter (d)
Diameter (d) = $C \div \pi$
radius (r) = $d \div 2$
Radius Less Bark
radius-($1 \div 2$)
Average Ring Width
radius ÷ number of rings
Average Age of Tree
radius ÷ average ring width
Floral Diversity of ML Area and Its IVI

The floral diversity should be calculated on the basis of standard formula and present for current scenario.

BIOLOGICAL STRESS

Stress is commonly understood as a physiological or psychological state resulting from environmental or social pressures, potentially affecting the functioning or behavior of individuals. Ecological or biological stress, on the other hand, arises when a physical factor adversely impacts an ecosystem or its living components. Biodiversity serves as a crucial support system within all ecosystems, offering resilience against disturbances and facilitating the provision of essential ecosystem services. These services, vital for our survival, carry significant economic value, highlighting the immediate concern that biodiversity loss may lead to dire consequences for current and future generations.

Human activities have accelerated species extinction rates to unprecedented levels, rivaling those of pre-human eras. Among the numerous ecosystem services bolstered by biodiversity, one crucial aspect is its contribution to human health and well-being through exposure to diverse wildlife. For example, green spaces such as forests, parks, gardens, and other natural environments play a pivotal role in alleviating stress. Research indicates that both short-term and long-term exposure to nature can reduce stress and anxiety, leading to enhanced subjective psychological well-being.

An illustrative example of this phenomenon is the landmark study revealing that patients recovering from surgery showed faster recovery rates and required less pain medication when they had a view of trees outside their hospital window compared to those facing a brick wall.

As environmental stress intensifies significantly (such as by increasing pollution), the following changes are observed:

- Mortality increases, especially of the most vulnerable species.
- Species richness decreases. the stocks of nutrients and biomass become depleted.
- The rate of community respiration exceeds that of production, so the net production becomes negative.
- Sensitive species are replaced by more-tolerant ones
- Top predators and large-bodied species may be lost from the ecosystem
- Previously self-maintaining ecosystems may require active management to sustain their desirable attributes, for example, to maintain declining populations of rare or economically valuable species that have become threatened

A disturbance is an episodic but intense disruption that causes severe biological and ecological damage. An event of disturbance is followed by a sometimes-lengthy period of ecological recovery through the process known as succession. Disturbance is a natural force that affects all ecosystems. There are two broad types of disturbances: *community-replacing disturbances and micro-disturbances.*

- A community-replacing disturbance is extensive in scale and results in a catastrophic destruction of one or more original communities. Natural examples are caused by wildfire, windstorm, avalanche, and glaciation, while anthropogenic ones include clear-cutting and ploughing. These large-scale disturbances may be followed by a successional recovery that eventually regenerates a community similar to what was destroyed. Younger communities in the successional sequence (or sere) are relatively dynamic in their structural and functional properties. They are typically dominated by species that are abundant only during the initial stages of recovery, when competition is not so intense. Community changes in later stages are somewhat less dynamic, until a late-stage community is re-established.
- A micro-disturbance involves a local disruption that only affects a small area within an otherwise intact community. Anthropogenic micro-disturbances include the selective harvesting of individual large trees or particular animals, while leaving the community otherwise intact. Ecological changes are relatively rapid within a habitat patch that has been affected by a recent micro-disturbance, but at the stand level the community is stable. So-called patch- or gap-phase successional dynamics occur in all natural forests but are particularly important during the later stages of succession. This is especially the case in older-growth forest, where individual trees might die from disease, insect attack, or a lightning strike, creating a gap in an otherwise intact canopy.

Ecotoxicology encompasses a broad scope, examining not only the direct toxic effects of chemicals but also their indirect ecological impacts. These indirect effects may include alterations in habitat or changes in food abundance. For example, the application of a herbicide in forestry or agriculture can lead to shifts in vegetation biomass and species composition within treated areas, thereby affecting animal habitats. Even if animals exposed to the herbicide do not suffer direct poisoning, they may still be impacted by changes in their environment.

Various factors contribute to the ecotoxicological risks associated with chemical exposure in the environment. Key factors include:

i. Biological sensitivity: Different organisms exhibit varying degrees of sensitivity to chemical exposure.
ii. Inherent toxicity of the chemical: The inherent toxic properties of the chemical being studied play a significant role in determining its impact.
iii. Intensity of exposure: The level and duration of exposure to the chemical influence the extent of its effects.
iv. Indirect effects: Changes in habitat or food availability caused by chemical exposure can also have indirect consequences on ecosystems and organisms.

These factors collectively shape the ecological risks posed by exposure to environmental chemicals.

The sensitivity to chemical exposures varies significantly among individual organisms and species. Toxicological studies, typically conducted in controlled laboratory settings, often compare the susceptibility of different organisms to toxic substances. Acute toxicity refers to instances where short-term exposure to a chemical at high concentrations leads to biochemical or anatomical damage, or even death—a common acute endpoint. Conversely, chronic toxicity involves prolonged exposure to low to moderate concentrations of a chemical. Over time, such exposures may result in biochemical or anatomical damage, or potentially fatal conditions such as cancer.

The concept of physiological thresholds of tolerance helps distinguish between contamination and pollution, as discussed earlier. Additionally, the notion of thresholds underscores the importance of framing discussions in terms of potential toxic exposures to chemicals, particularly when actual environmental concentrations are unknown, and when the biological risks associated with extremely small doses are not fully understood.

ECOLOGICAL BALANCE

Carbon sequestration refers to the long-term storage of carbon in various natural reservoirs such as plants, soils, geological formations, and oceans. Specifically, it involves capturing waste carbon dioxide (CO2) emissions from significant point sources like fossil fuel power plants, transporting the captured CO2 to designated storage sites, and securely depositing it underground in geological formations to prevent its release into the atmosphere. Terrestrial carbon sequestration occurs as part of this process, whereby CO2 from the atmosphere is absorbed by trees and plants through photosynthesis and subsequently stored as carbon within soil and biomass components such as tree trunks, branches, foliage, and roots.

Method of Calculating Current CO2 Sequestration in Tree

- Broward County (2012) devised a method known as carbon counting method. The equation to estimate a tree's dry weight is based on the physical relationship between tree volume and wood density.
- As tree density figures can vary, even within the same species, using different formula for each tree species does not necessarily provide more accurate CO_2 estimations.
- Therefore, the method we used is based on average wood density figures across species. For example – Hard wood trees, shrubs and ornamental plants.
- The total weight of the tree using the diameter of the trunk and an estimate of the height
- Convert the weight of the tree to the dry weight of organic matter (on average 72.5 per cent of the total weight)
- Estimate the amount of carbon, being a proportion of the dry weight (on average 50 per cent of the dry weight)
- Convert the weight of the tree to the dry weight of organic matter (on average 72.5 per cent of the total weight)
- Estimate the amount of carbon, being a proportion of the dry weight (on average 50 per cent of the dry weight)

Tree Density in ML Area

The density of trees (woody and shrubs) was studied on primary data and calculation done as per standard practice. The average tree (woody) density was calculated as 20/Ha. while average density of shrubs was 45/Ha. as per current scenario (y.2023). The tree density is given in subsequent tables.

As per table presented above, total no. of trees in the ML area would be 20 x 659 = **13,180** no.s

The total no of shrubs in ML area would be 45 x 659 = **29,655** no.s

Carbon Sequestration by Existing Biodiversity in ML Area

The total estimated cost of carbon offset by enumerated trees (3478 Nos.) would be Rs. 1.3 million / annum while total cost of carbon offset by estimated trees (42,835 Nos.) in ML area of 659 Ha would be 16.5 million / annum for y.2023.

As per Terrapass (2023), the maximum cost of 1-ton caron offset upto 2023 is 60 USD i.e., 4980 INR (@83).

The carbon sequestration by all floral species emerged on nala is given below table 5:

It comes to Rs. 0.5 million/annum for y.2023.The cost of carbon sequestration by trees at nala is Rs. 0.5 million/annum.

Concept of Biodiversity Loss

Biodiversity loss refers to the reduction or disappearance of the variety of life forms inhabiting the planet, encompassing different levels of biological organization and their genetic variability, as well as the intricate patterns found within ecosystems. The biodiversity of an area experiences fluctuations through natural cycles. For instance, seasonal changes, like the arrival of spring, create favorable conditions for

Table 1. Tree density in ML area

HS No.	No. of Trees	Area (Ha)	Density
1.	162	9.11	18
2.	19	1.13	17
3.	89	2.81	32
4.	29	1.22	24
5.	53	5.12	10
6.	21	0.5	42
7.	56	4.3	13
8.	111	5.4	21
9.	59	2.73	22
10.	82	4.63	18
11.	28	0.93	30
12.	73	1.30	56
13.	55	3.11	18
14.	76	4.50	17
15.	73	3.92	19
16.	104	2.01	52
Total	1090	52.72	-

Table 2. Density of shrubs in ML area

HS No.	No. of Shrubs/bushes	Area (Ha)	Density
1.	346	9.11	38
2.	64	1.13	57
3.	125	2.81	44
4.	58	1.22	48
5.	89	5.12	17
6.	21	0.5	42
7.	167	4.3	39
8.	213	5.4	39
9.	168	2.73	62
10.	234	4.63	51
11.	56	0.93	60
12.	147	1.30	113
13.	198	3.11	63
14.	134	4.50	30
15.	211	3.92	54
16.	157	2.01	78
Total	**2388**	**52.72**	-

Table 3. Carbon sequestration by enumerated trees

Type of Plantation	Number	Average Age	Estimated diameter	Estimated Height	tCO_2e/year
Tree	1090	21 years	35 inches	32 Feet	220.75
Shrubs	2388	5 years	9 inches	7.5 Feet	52.46
Total	**3478**	-	-	-	**273.21**
Cost of carbon offset		Rate 4980 per ton		**Rs. 1,360,586**	

Table 4. Carbon sequestration by estimated trees in ML area

Type of Plantation	Number	Average Age	Estimated diameter	Estimated Height	tCO_2e/year
Tree	13,180	21 years	35 inches	32 Feet	2669.20
Shrubs	29,655	5 years	9 inches	7.5 Feet	651.51
Total	**42,835**	-	-	-	**3320.71**
Cost of carbon offset		Rate 4980 per ton		**Rs. 16,537,136**	

Table 5. Carbon sequestration by floral species along the Nala

Type of Plantation	Number	Average Age	Estimated diameter	Estimated Height	tCO_2e/year
Tree	409	21 years	37 inches	35 Feet	82.82
Shrubs	903	5 years	9 inches	7.5 Feet	19.83
Total	**1312**	-	-	-	**102.65**
Cost of carbon offset		Rate 4980 per ton		**Rs. 511,197**	

feeding and reproduction, leading to an increase in biodiversity as populations of various species thrive. Conversely, the onset of winter often results in a temporary decline in biodiversity as cold-sensitive insects perish and migratory animals depart.

Moreover, the seasonal fluctuations in plant and invertebrate populations, such as insects and plankton, which serve as crucial food sources for other organisms, also influence an area's biodiversity. Biodiversity loss is commonly associated with more permanent ecological alterations in ecosystems, landscapes, and the global biosphere. Natural ecological disturbances, such as wildfires, floods, and volcanic eruptions, can significantly alter ecosystems by eliminating local populations of certain species and reshaping entire biological communities. Despite their disruptive nature, such disturbances are temporary, as ecosystems have evolved to adapt to the challenges posed by natural events.

Human-Driven Biodiversity Loss

Human-induced disturbances have resulted in more severe and enduring biodiversity losses. The expansion of human settlements, agriculture, and livestock farming has led to a significant appropriation of Earth's land area. Approximately half of the world's habitable land, equivalent to 51 million square kilometers (19.7 million square miles), has been converted to agricultural use. Moreover, about 77 percent of agricultural land, covering approximately 40 million square kilometers (15.4 million square miles), is utilized for grazing livestock such as cattle, sheep, and goats.

This extensive transformation of forests, wetlands, grasslands, and other terrestrial ecosystems has precipitated a substantial decline in vertebrate populations worldwide. On average, there has been a 60 percent reduction in vertebrate numbers since 1970, with the most significant declines observed in freshwater habitats (83 percent) and in regions of South and Central America (89 percent). Over the period from 1970 to 2014, the global human population nearly doubled, growing from approximately 3.7 billion to 7.3 billion individuals.

By 2018, the combined biomass of humans and their livestock greatly outweighed that of wild mammals and birds. Specifically, human and livestock biomass measured 0.16 gigatons, contrasting sharply with the biomass of wild mammals (0.007 gigatons) and wild birds (0.002 gigatons). Researchers estimate that the current rate of species loss is between 100 and 10,000 times higher than the background extinction rate, which typically sees around one to five species lost per year when considering the entire fossil record.

Furthermore, a 2019 report by the Intergovernmental Science-Policy Platform on Biodiversity and Ecosystem Services underscored that up to one million plant and animal species are confronting extinction due to human activities.

Forest clearing, wetland filling, stream channeling and rerouting, and road and building construction are often part of a systematic effort that produces a substantial change in the ecological trajectory of a landscape or a region. As human populations grow, the terrestrial and aquatic ecosystems they use may be transformed by the efforts of human beings to find and produce food, adapt the landscape to human settlement, and create opportunities for trading with other communities for the purposes of building wealth. Biodiversity losses typically accompany these processes.

- Habitat loss and degradation—which is any thinning, fragmentation, or destruction of an existing natural habitat—reduces or eliminates the food resources and living space for most species. Species that cannot migrate are often wiped out.
- Invasive species—which are non-native species that significantly modify or disrupt the ecosystems they colonize—may outcompete native species for food and habitat, which triggers population declines in native species. Invasive species may arrive in new areas through natural migration or through human introduction.
- Overexploitation—which is the harvesting of game animals, fish, or other organisms beyond the capacity for surviving populations to replace their losses—results in some species being depleted to very low numbers and others being driven to extinction.
- Pollution—which is the addition of any substance or any form of energy to the environment at a rate faster than it can be dispersed, diluted, decomposed, recycled, or stored in some harmless form—contributes to biodiversity loss by creating health problems in exposed organisms. In some cases, exposure may occur in doses high enough to kill outright or create reproductive problems that threaten the species' survival.
- Climate change associated with global warming—which is the modification of Earth's climate caused by the burning of fossil fuels—is caused by industry and other human activities. Fossil fuel combustion produces greenhouse gases that enhance the atmospheric absorption of infrared radiation (heat energy) and trap the heat, influencing temperature and precipitation patterns.

Ecological Effects

The impact of biodiversity loss is particularly severe for species experiencing population declines. The decline in gene pools and individual numbers poses a significant threat to the long-term survival of a species, as it reduces the availability of mates and increases the risks associated with inbreeding among closely related survivors. Moreover, the widespread loss of populations heightens the risk of species extinction.

Biodiversity plays a crucial role in maintaining ecosystem health. Diminishing biodiversity results in decreased ecosystem productivity—the amount of food energy converted into biomass—and diminishes the quality of ecosystem services. Additionally, biodiversity loss jeopardizes the structure and functioning of ecosystems. While ecosystems possess some degree of adaptability to stressors associated with reduced biodiversity, such loss diminishes ecosystem complexity. Roles once fulfilled by multiple interacting species or individuals are now assumed by fewer or none, leading to a loss of ecosystem resilience. As components are lost, ecosystems lose their capacity to recover from disturbances, thereby undermining ecological resilience. Beyond a critical threshold of species removal or diminishment, ecosystems can become destabilized and face collapse.

Economic and Societal Effects

Economic shortages within common food crops may attract more attention than biodiversity losses in ecosystems and landscapes distant from global markets. Since 1900, approximately 75 percent of food crop varieties have vanished, primarily due to an overdependence on a small number of high-yielding crop strains. This reduction in crop biodiversity poses a significant risk to food security, as these limited varieties may be susceptible to diseases, pests, invasive species, and the impacts of climate change. Similar patterns are observed in livestock farming, where highly productive breeds of cattle and poultry are preferred over less productive, more natural breeds.

Solutions to Biodiversity Loss

Addressing biodiversity loss is intricately linked to confronting the conservation challenges posed by its underlying drivers. Conservation biologists emphasize the potential for solutions through a combination of public policy and economic measures, supported by ongoing monitoring and educational efforts. Collaboration among governments, non-governmental organizations, and the scientific community is essential to establish incentives for conserving natural habitats and protecting the species within them from overexploitation, while discouraging behaviors that contribute to habitat loss and degradation.

Sustainable development practices, which prioritize economic growth while preserving environmental quality, should be integral to the creation of new farmland and human living spaces. Strengthening and enforcing laws to combat poaching and the illicit wildlife trade, as well as implementing thorough inspections for stowaway organisms at ports, are critical measures.

While solutions tailored to each cause of biodiversity loss can alleviate pressure on species and ecosystems, conservation biologists emphasize the utmost importance of safeguarding remaining species from overhunting, overfishing, and habitat destruction. Preserving habitats and ecosystems from species invasions and land use conversion is crucial. Monitoring the status of individual species, facilitated by tools such as the Red List of Threatened Species by the International Union for Conservation of Nature and Natural Resources (IUCN) and the United States Endangered Species list, is essential for decision-makers to prioritize conservation efforts.

Furthermore, identifying and protecting areas rich in unique species, known as "hot spots," is crucial. These hot spots, characterized by high levels of endemism, particularly occur in tropical environments where species richness and biodiversity are significantly greater than in ecosystems closer to the poles. Loss and gain are calculated as the difference between: (1) the amount of biodiversity that would exist with a change in management (i.e., with development and with the offset) and (2) the amount of biodiversity that would exist under the counterfactual i.e., without development and without the offset.

FLORAL BIODIVERSITY

Across an approximate area of 15.77 hectares (or a length of 2.6 kilometers), a diverse range of species has flourished. The average density of trees and shrubs is reported as 25 per hectare and 57 per hectare, respectively. The cost of carbon sequestration attributed to these floral species has been determined to be Rs. 511,197 per annum.

In a report submitted to the Supreme Court of India panel in February 2021, the value of a single tree is estimated at Rs. 74,500 per year. This valuation comprises various components, with the cost of oxygen alone valued at Rs. 45,000 and the contribution of biofertilizers valued at Rs. 20,000. However, according to the Ministry of Environment, Forest and Climate Change (MoEFCC) Standard Operating Procedure dated July 7, 2021, the cost of a tree is calculated at Rs. 1100 per tree, with compensation set at Rs. 1500 per tree or shrub in cases of tree or shrub cutting, with ten times plantation being a requirement.

Furthermore, the cost associated with habitat loss due to the felling of trees, shrubs, and grasses—critical habitats for avifauna, insects, mammals, and other species including those dependent on water sources such as streams, butterflies, reptiles, and domestic animals—is of significant concern.

Table 6. Existing tree density along the natural stream

Hotspot No.	No. of Trees	Area (Ha)	Density
1.	28	0.93	30
2.	73	1.30	56
3.	55	3.11	18
4.	76	4.50	17
5.	73	3.92	19
6.	104	2.01	52
Total	**409**	**15.77**	-

Average tree density 25/ha.

Table 7. Existing shrubs density along the Nala (both sides)

HS No.	No. of Shrubs	Area (Ha)	Density
1.	56	0.93	60
2.	147	1.30	113
3.	198	3.11	63
4.	134	4.50	30
5.	211	3.92	54
6.	157	2.01	78
Total	**903**	**15.77**	-

FAUNAL DIVERSITY

The faunal biodiversity of nala is assessed and given below:

BIODIVERSITY LOSS ON DIVERSION OF EXISTING NALA

The nallah flows towards north along the western boundary of the ML area. It enters the block from its southwest corner and exits from the northern boundary. The nallah passes through western part of the block taking winding courses following the topography. The loss of biodiversity on diversion of nala is calculated on the standard pro rata basis. The major biodiversity parameters considered for calculation of cost of biodiversity on diversion of nala as follows:

Table 8. Mammals

S.No.	Family	Common Name	*Scientific name*	Density* (individuals/km^2)
1.	Antilopinae	Nilgai (Blue Bull)	*Boselaphus tragocamelus*	16.3
2.	Herpestidae	Common Mongoose	*Herpestes edwardsii*	13.5
3.	Leporidae	Hare	*Lepus nigricollis*	5.5
4.	Muridae	Common House Rat	*Rattus rattus*	25.5
5.	Sciuridae	5 striped Palm Squirrel	*Funambulus pennanti*	17.8

Table 9. Reptiles

S.No.	Family	Common Name	Scientific name	Density* (individuals/km^2)
1.	Agamidae	Common Garden Lizard	*Calotes versicolor*	16.6
2.		Fan-Throated Lizard	*Sitana ponticeriana*	5.5
3.	Bufonidae	Toad	*Bufo bufo*	134.1
4.	Chamaeleonidae	Indian chameleon	*Chameleon calcaratus*	44.5
5.	Colubridae	Common Rat Snake	*Ptyas mucosus*	6.4

Table 10. Avi-fauna

#	Family	Scientific Name	Common Name	Density* (individuals/km^2)
1.	Alcedinidae	*Halcyon smyrnensis*	White-throated Kingfisher	12.34
2.	Apodidae	*Apus affinis*	Little Swift	24.56
3.	Ardeidae	*Bubulcus ibis*	Cattle Egret	23.32
4.	3)	*Egretta garzetta*	Little Egret	12.98
5.	Coraciidae	*Coracias benghalensis*	Indian Roller	26.67
6.	Corvidae	*Corvus splendens*	Crow	23.87
7.	Cuculidae	*Centropus sinensis*	Coucal	22.98
8.	Dicruridae	*Dicrurus macrocercus*	Black drongo	23.78
9.	Gruidae	*Antigone antigone*	Crane	11.54
10.	Megalaimidae	*Megalaima haemacephala*	Coppersmith barbet	23.11
11.	Meropidae	*Merops leschenaultia*	Chestnut-headed Bee-eater	12.14
12.	Motacillidae	*Anthus spinoletta*	Water Pipit	34.34
13.	3)	*Motacilla flava*	Yellow Wagtail	11.24
14.	Muscicapidae	*Ficedula parva*	Red breasted Flycatcher	23.98
15.	Paridae	*Cyanistes caeruleus*	Blue Tit	32.45
16.	Passeridae	*Passer domesticus*	Sparrow	31.24
17.	Picidae	*Picus viridis*	Green Woodpecker	32.39
18.	Psittaculidae	*Psittacula krameri*	Rose-ringed Parakeet	24.51
19.	Pteroclidae	*Pterocles exustus*	Sandgrouse	14.62
20.	Rallidae	Amaurornis phoenicurus	White-breasted Water hen	13.87
21.	3)	*Fulica atra*	Common Coot	16.21
22.	3)	*Philomachus pugnax*	Ruff	17.43

Option One: Biodiversity Cost (Without Spiritual and Religious Value)

The biodiversity cost considering above parameters has been calculated and presented in table 8 below-

The total cost of biodiversity which support various ecosystem services comes to 81.42 Cr for 2.6 km of existing nala covering 15.77 Ha area. The biodiversity cost comes 5.16 Cr/Ha. for existing nala.

Option Two: Biodiversity Cost (With Spiritual and Religious Value)

The total cost of biodiversity which support various ecosystem services comes to 84.46 Cr. for 2.6 km of existing nala covering 15.77 Ha area. The biodiversity cost comes 5.36 Cr/Ha. for existing nala which is maximum out of two options. The average biodiversity cost in terms of ecosystem services is 5.26 Cr/Ha. for existing Nala.

Table 11. Butterflies

Scientific Name and Family	Common Name	Relative Abundance	Density* (individuals/km^2)
Family Asclepiadaceae			
Danaus genutia Cramer	Striped Tiger	Common	26.12
Family Papilionidae			
Papilio polytes	Common Mormon	Common	23.43
Family Pieridae			
Eurema hecabe	Common Grass yellow	Very Common	24.51
Catopsilia crocale	Common Emigrant	Common	23.12
Family: Nymphalidae			
Danaus chrysippus	Plain Tiger	Common	21.37
Phalantha phalantha	Common Leopard	Fairy Common	19.62
Hypolimanas misippus	Danaid egg fly	Common	25.82
Precis lemonias	Lemon pansy	Common	24.35
Junonia orithya Linnaeus	Blue pansy	Fairy Common	

Table 12. Domestic animals who frequently visit the Nala

S.No.	Common Name	Scientific name	Density* (individuals/km^2)
1.	Buffalo	*Bubalus bubalis*	15.43
2.	Goat	*Capra aegagrus hircus*	31.54
3.	Cow	*Bos taurus*	41.23
4.	Oxen	*B. taurus primigenius*	15.21
5.	Dog	*Canis lupus familiaris*	13.42
6.	Chicken	*Gallus gallus*	41.64

BENEFIT COST RATIO

The benefit cost ratio (BCR) is an indicator showing the relationship between the relative costs and benefits of a proposed project, expressed in monetary or qualitative terms. If project has a BCR greater than 1.0, the project is expected to deliver a positive net present value.

The benefit-cost ratio (BCR) is calculated by dividing the expected benefit of a project by the cost of the project. The formula is:

BCR: = PV of Benefit Expected from the Project / PV of the Cost of the Project

The BCR varies across protected areas. It is reported that the estimates of earth's ecological system are worth more than US$ 33 trillion every year, which indicates the great importance and value of the natural resources that our planet offers. Despite the expenditure on nature conservation globally which exceeded US$ 6 billion, little research is done on how successful conservation projects are, and which project provides the best benefit or value for money. Thus,

Table 13. Insects along the Nala

S.No.	Common Name	Scientific name	Density* (individuals/km^2)
1.	Honey Bees	*Apis sp.*	145.45
2.	Grasshopper	*Goniaea australasiae*	300.43
3.	Common Black Ant	*Lasius niger*	342.65
4.	Common Red Ant	*Myrmica rubra*	423.76
5.	Beetle	*Callosobruchus maculatu*	342.34
6.	Lady bug	*Coccinella septempunctata*	634.87
7.	Dragon Fly	*Anisoptera sp.*	265.25
8.	Centipede	*Chilopod sp.*	234.65
9.	Earthworm	*Lumbricus terrestris*	542.98
10.	Spider	*Araneus diadematus*	235.24
11.	Mosquito	*Aedes sp.*	653.94
12.	Firefly	*Photinus sp.*	343.56
13.	Worms	*Lumbricus sp.*	367.73
14.	Crane fly	*Nephrotoma appendiculata*	365.13
15.	Wasp	*Vespula vulgaris*	376.62
16.	Moth/ Caterpillar	*Lepidoptera*	567.49
17.	Pill bug	*Armadillidium vulgare*	834.27
18.	Cricket	*Acheta domesticus*	343.87
19.	Locust	*Locusta migratoria*	546.35
20.	Slug	*Arion hortensis*	212.76
21.	Mantis	*Mantis religiosa*	345.34

*estimated data with SD:±3.57

it is vitally important that the decision makers and project managers make robust decisions and choices about resource allocation for obtaining the preferred outcome which generates benefits that exceed the costs of the project.

Moran et al (2010) mention that decisions in biodiversity conservation are complex and multi-faceted. They believe that evaluating conservation projects face different challenges, mainly the partial ecological understanding of some species, and the difficulty in evaluating the outcomes of a project. Furthermore, Esther et al. (2011) see that it can be debatable and complicated to prioritize conservation management actions, due to having a large number of problems to deal with, coupled with low budgets allocated for them. Such challenge becomes greater when there are uncertainties about the expected benefits that may arise from the actions of conservation projects. Moreover, there are always diverse opinions about where should the resources be allocated, what decisions should be taken and what is obtained from the investment. There are different tools that are used in conservation project management which help conservation managers taking decisions in this regard. Cost effective analysis (CEA) is one of these tools that can help conservation managers prioritize management actions, and fulfill conservation objectives while ensuring an optimum use of available resources. This analysis works on maximizing the benefits that are obtained from different options of

Table 14. Parameters

S.No.	Parameters
1	Cost of Floral Species
2	Cost of Climate Change (Carbon Sequestration)
3	Cost of Habitat Loss
4	Cost of Pollinators Loss
5	Wildlife Conservation Cost
6	Loss of Grazing Land
7	Social Benefits due to Nala (Recreation & Aesthetic)
8	Cost of Noise absorption & Dust collection system
9	Cost of Water Cycling / purification system
10	Cost of Nutrient Cycling
11	Cost of Oxygen purification System
12	Cost of Primary Production (Food, Fuel & Fibre)
13	Cost of Genetic Resources
14	Cost of Spiritual & Religious Value (Optional)
15	Cost of Natural Hazard Protection/ Erosion Regulation
16	Cost of Aquatic Ecosystem
17	Cost of Potential for future use/ Environmentally friendly farming/cropping
18	Loss of Water for Irrigation
19	Cost of Medicinal value of vegetation

actions across various conservation and biological values. However, the cost-benefit analysis (CBA) takes the process a step further by assigning a monetary value to benefits and costs and adjust them for the time value of the money, so the total benefits and costs are represented in common bases (Cellini and Kee. 2010). Therefore, the comparison between benefits and costs becomes clear, and it helps taking the decision that serves the objectives of the institution or the project. CBA facilitates in an efficient allocation of the society's resources; it also helps governments in the allocation of resources and forming public policies.

Criteria of Acceptance of BCR

0-0.99 – Not acceptable
>1.00 – Acceptable
1.01-4.99 – Good
5.00-9.99 – Better
>10.00 – Best

Table 15. Biodiversity cost of Nala (option one)

S.No.	Parameters	Basis of Calculation	Cost (INR)
1.	Cost of Floral Species	10 times of existing tree/shrubs 1312 @Rs.1500/tree as per SOP 7th July 2021 (MoEFCC)	19680000
2.	Cost of Climate Change (Carbon Sequestration)	Carbon Seq. by existing vegetation per year	10735137
3.	Cost of Habitat Loss	Habitat of Avifauna, reptiles, fishes, Insects, krill, etc. @Rs.1000/sq.m	157700000
4.	Cost of Pollinators Loss	$47.2/acre =Rs.3918/acre (@83) ~Re 1/m^2	157700
5.	Wildlife Conservation Cost	20%(max) of Wildlife Conservation cost (2% of project cost)	15600000
6.	Loss of Grazing Land	30% of land is being utilized by animals for grazing, the cost of land as per local market is 1 Cr/ha.	473100000
7.	Social Benefits due to Nala (Recreation & Aesthetic)	Social benefits *viz.* recreational & Aesthetic values (Rs.13/sq.m.)	2050100
8.	Cost of Noise absorption & Dust collection system	Noise barriers @Rs.400/sq.m; dust collection system @Rs.200/sq.m	94620000
9.	Cost of Water Cycling / purification system	Water Area 4.731 ha x 0.75 m water depth during monsoon season (100 days) @Rs.5/m^3	17741250
10.	Cost of Nutrient Cycling	Bacterial & fungal consortium maintain the cycling the nutrient in the soil; cost @Rs. 4/sq.m	630800
11.	Cost of Oxygen purification System	Oxygen from plants, phytoplankton etc. @120kg/tree x 409x365 & 40kg/ shrubs x 903 x 365; the cost of oxygen @Rs.15.22/m^3 (as per National Pharmaceutical Pricing Authority)	473311.56
12.	Cost of Primary Production (Food, Fuel & Fibre)	Food for faunal species, domestic animals, & human, wood for fuel, fibre etc.@Rs.10/sq.m	1577000
13.	Cost of Genetic Resources	Genetic resource to produce new plants, animals, insects etc. @ Rs. 50/sq.m.	7885000
14.	Cost of Natural Hazard Protection/ Erosion Regulation	Control of Heavy rainfall, soil erosion, water level regulation, aquifer recharge, drought etc. @Rs. 75/sq.m	11827500
15.	Cost of Aquatic Ecosystem	Aquatic area 30% @5000 /ha	23650
16.	Cost of Potential for future use/ Environmentally friendly farming/cropping	Nala supporting environment friendly farming/cropping by giving rainwater (renewable) @Rs. 500 /ha.	7885
17.	Loss of Water for Irrigation	Irrigation Water cost Rs. 960 /Ha (max) as per MP as per gazette notification F-35-43/M/31 dtd 29.10.2021. Catchment area is approx. 100 ha.	96000
18.	Medicinal value of vegetation	Anti-inflammatory, antiviral, antitumor, antimalarial, and analgesic properties cost 40,000/ha (50% of 15.77)	315400
19.	Loss of Forest Land	No forest land involved	-
20.	Loss of Agriculture land	No agriculture land involved	-
21.	Cost of Shoreline Stabilisation	No shoreline involved.	-
	Total		814220733.6

Outcome

The cost of Biodiversity min 81.42 Cr (5.16 Cr/ha.) and max 84.46 Cr (5.36 Cr/ha) which comprised the average BCR 1.3 which indicate the positive criteria for acceptance. The diversion of stream can be accepted on adequate mitigation measure like plantation of 10 times trees, shrubs,

Table 16. Biodiversity cost of Nala (option two)

S.No.	Parameters	Basis of Calculation	Cost (INR)
1.	Cost of Floral Species	10 times of existing tree/shrubs 1312 @Rs.1500/tree as per SOP 7th July 2021 (MoEFCC)	19680000
2.	Cost of Climate Change (Carbon Sequestration)	Carbon Seq. by existing vegetation per year	10735137
3.	Cost of Habitat Loss	Habitat of Avifauna, reptiles, fishes, Insects, krill, etc. @Rs.1000/sq.m	157700000
4.	Cost of Pollinators Loss	$47.2/acre =Rs.3918/acre (@83) ~Re 1/m^2	157700
5.	Wildlife Conservation Cost	20%(max) of Wildlife Conservation cost (2% of project cost)	15600000
6.	Loss of Grazing Land	30% of land is being utilized by animals for grazing, the cost of land as per local market is 1 Cr/ha.	473100000
7.	Social Benefits due to Nala (Recreation & Aesthetic)	Social benefits *viz.* recreational & Aesthetic values (Rs.13/sq.m.)	2050100
8.	Cost of Noise absorption & Dust collection system	Noise barriers @Rs.400/sq.m; dust collection system @Rs.200/sq.m	94620000
9.	Cost of Water Cycling / purification system	Water Area 4.731 ha x 0.75 m water depth during monsoon season (100 days) @Rs.5/m^3	17741250
10.	Cost of Nutrient Cycling	Bacterial & fungal consortium maintain the cycling the nutrient in the soil; cost @Rs. 4/sq.m	630800
11.	Cost of Oxygen purification System	Oxygen from plants, phytoplankton etc. @120kg/tree x 409x365 & 40kg/shrubs x 903 x 365; the cost of oxygen @Rs.15.22/m^3 (as per National Pharmaceutical Pricing Authority)	473311.56
12.	Cost of Primary Production (Food, Fuel & Fibre)	Food for faunal species, domestic animals, & human, wood for fuel, fibre etc.@ Rs.10/sq.m	1577000
13.	Cost of Genetic Resources	Genetic resource to produce new plants, animals, insects etc. @ Rs. 50/sq.m.	7885000
14.	Cost of Spiritual & Religious Value	Spiritual & religious value *viz.* strength, individuality and expression, calmness, growth and the interconnectedness of everything @ 74500 per tree (as per SC panel 2021)	30470500
15.	Cost of Natural Hazard Protection/ Erosion Regulation	Control of Heavy rainfall, soil erosion, water level regulation, aquifer recharge, drought etc. @Rs. 75/sq.m	11827500
16.	Cost of Aquatic Ecosystem	Aquatic area 30% @5000 /ha	23650
17.	Cost of Potential for future use/ Environmentally friendly farming/cropping	Nala supporting environment friendly farming/cropping by giving rainwater (renewable) @Rs. 500 /ha.	7885
18.	Loss of Water for Irrigation	Irrigation Water cost Rs. 960 /Ha (max) as per MP as per gazette notification F-35-43/M/31 dtd 29.10.2021. Catchment area is approx. 100 ha.	96000
19.	Medicinal value of vegetation	Anti-inflammatory, antiviral, antitumor, antimalarial, and analgesic properties cost 40,000/ha (50% of 15.77)	315400
20.	Loss of Forest Land	No forest land involved	-
21.	Loss of Agriculture land	No agriculture land involved	-
22.	Cost of Shoreline Stabilisation	No shoreline involved.	-
	Total		844691233.6

84.46 Cr.

herbs etc. no obstruction of natural flow of water, the diversion can be done in non-monsoon season only.

Table 17. BCR calculation for diversion/straightening of Nala (2.6 km)

Particulars	Detail	Nala Length	Cost (unit)	Remarks
Alternative 1	**Straightening of course**			
Meandering Length in ML Area	Total	2600 m	-	
Strengthen Length	Straight	1100 m	-	
Coal Released by Nala Diversion/ Straightening	2,07,760 tons	-	*G-6: Rs. 4085/ton G-7: Rs. 3900/ton	
Revenue Generation	Total cost of Coal G-6 & G-7	-	83 Cr	67 Cr as Govt. Revenue
BCR	a) Biodiversity Loss of 81.42 Cr b) Biodiversity Loss of 84.46 Cr	2.6 km	**1.02** or **0.98**	Cost of Coal/ Biodiversity Loss
Alternative 2	Straightening the nallah at two strategic places			
Meandering length	Total	2600 m		
Strengthen Length	At two places	554 m & 504 m	-	
Coal Released by Nala Diversion/ Straightening	2,35,666 tons	-	*G-6: Rs. 7642/ton G-7: Rs. 4085/ton	
Revenue Generation	Total cost of Coal G-6 & G-7	-	133 Cr	102 Cr as Govt. Revenue
BCR	a) Biodiversity Loss of 81.42 Cr b) Biodiversity Loss of 84.46 Cr	2.6 km	**1.63** or **1.57**	Cost of Coal/ Biodiversity Loss
Average BCR	Taking average of all	2.6 km	**1.3**	

*Representative price of coal (MoC Nov. 2022)

CONCLUSION

For biodiversity ecosystem services, 19 parameters were considered out of 22 as forest, agriculture and shoreline is not applicable for current case study. Total biodiversity cost of existing nala (max) for 2.6 km was estimated as 84.46 Cr and (min) 81.42 Cr. The average biodiversity cost in terms of ecosystem services is 5.26 Cr/Ha. for existing Nala. On the basis of CBA, the BCR (average) comes to 1.3 which shows positive to accept the proposal of diversion. The diversion of stream can be accepted on adequate mitigation measure like plantation of 10 times trees, shrubs, herbs etc. no obstruction of natural flow of water, the diversion can be done in non-monsoon season only. Study recommends the diversion of Existing Nala as BCR is positive (1.3) with proper mitigation measures for diversion without obstructing natural flow of water, hence diversion process can be taken in non-monsoon season especially in the month of Jan-May when meagre water flow.

REFERENCES

Aerts, R., Honnay, O., & Van Nieuwenhuyse, A. (2018). Biodiversity and human health: Mechanisms and evidence of the positive health effects of diversity in nature and green spaces. *British Medical Bulletin*, *127*(5), 5–22. doi:10.1093/bmb/ldy021 PMID:30007287

Allen, E. L., Engeman, R., & Krupa, H. (1996). Evaluation of three relative abundance indices for assessing dingo population. *Wildlife Research*, *23*(2), 197–206. doi:10.1071/WR9960197

Alvarsson, J. J., Wiens, S., & Nilsson, M. E. (2010). Stress recovery during exposure to nature sound and environmental noise. *International Journal of Environmental Research and Public Health*, *7*(3), 1036–1046. doi:10.3390/ijerph7031036 PMID:20617017

Anderson, T. (1867). An enumeration of the Indian species of Acanthaceae. *Journal of the Linnean Society of London, Botany*, *9*(39), 425–454. doi:10.1111/j.1095-8339.1867.tb01308.x

Andersson, R., & Östlund, L. (2004). Spatial patterns, density changes and implications on biodiversity for old trees in the boreal landscape of northern Sweden. *Biological Conservation*, *118*(4), 443–453. doi:10.1016/j.biocon.2003.09.020

APHA. (1971). *Standard methods for the examination of water and waste water*. American Public Health, Association.

Ashok, K. (2013). Butterfly (Lepidoptera: Insecta) Diversity from Different Sites of Jhagadia, Ankleshwar, District-Bharuch, Gujarat, *Oct. Jour. Env. Res.* *1*(1), 09-18

Ashok, K. (2014). Environmental Management Plan for Chemical Industries Especially Resin Manufacturing Unit, *Oct. Jour. Env. Res*, *2*(3), 262–273.

Ashok, K., & Goyal, A. S. (2013a). Ecology and Biodiversity status of Sachin gidc and its surroundings with Special reference to Conservation measures for Indian Peafowl (Pavo cristatus) schedule –I Bird species. *Oct. Jour. Env. Res.*, *2*(1), 82–100.

Ashok, K., & Goyal, A. S. (2013b). Study of Common Property Resources (Cpr) With Special Reference To Water And Biological Resources At Projected Area Near Village Ninat, Bardoli, District-Surat. *Oct. Jour. Env. Res.*, *1*(4), 319–331.

Ashok, K., Meena, S., & Savita, G. (2013). The Biodiversity At Sandi Bird Sanctuary, Hardoi With Special Reference to Migratory Birds. *Oct. Jour. Env. Res*, *1*(3), 173–181.

Batten, S. D., Clarke, R., & Flinkman, J. (2003). CPR sampling: The technical background, material and methods, consistency and comparability. *Progress in Oceanography*, *58*(2-4), 193–215. doi:10.1016/j.pocean.2003.08.004

Begon, M., Harper, J. L., & Townsend, C. R. (1996). *Ecology: Individuals, Populations, and Communities* (3rd ed.). Blackwell Science Ltd. doi:10.1002/9781444313765

Berg, A., Ehnstrom, B., Gustafsson, L., Hallingback, T., Jonsell, M., & Weslien, J. (1994). Threatened Plant, Animal, and Fungus Species in Swedish Forests: Distribution and Habitat Associations. In Conservation Biology, 8(3). doi:10.1046/j.1523-1739.1994.08030718.x

BirdLife International. (2000). *Threatened Birds of the World*. Lynx Edicions and BirdLife International.

BirdLife International. (2004a). *Threatened Birds of the World 2004. CD-ROM*. BirdLife International.

BirdLife International. (2004b). *State of the World's Birds 2004-Indicators for our changing world*. BirdLife International.

BirdLife International. (2010). *The BirdLife checklist of the birds of the world, with conservation status and taxonomic sources*. BirdLife International. http://www.birdlife.info/docs/SpcChecklist/Checklist_v3_June10.zip

Bohil, C. J., Alicea, B., & Biocca, F. A. (2011). Virtual reality in neuroscience research and therapy. *Nature Reviews. Neuroscience*, *12*(December), 752–762. doi:10.1038/nrn3122 PMID:22048061

Brown, P., & Daigneault, A. (2014). Cost-benefit analysis of managing the Papuana uninodis (Coleoptera: Scarbaeidae) taro beetle in Fiji. *Journal of Economic Entomology*, *107*(5), 1866–1877. doi:10.1603/EC14212 PMID:26309277

Busch, J., & Cullen, R. (2009). Effectiveness and cost-effectiveness of yellow-eyed penguin recovery. *Ecological Economics*, *68*(3), 762–776. doi:10.1016/j.ecolecon.2008.06.007

Co-operation on Health and Biodiversity (2010). *Biodiversity and Human Health Policy Brief.* Co-operation on Health and Biodiversity.

Colebrook, J. M. (1960). Continuous Plankton records: Methods of analysis, 1950-59. *Bulletins of Marine Ecology*, *5*, 51–64.

Dedovic, K., Renwick, R., Mahani, N. K., Engert, V., Lupien, S. J., & Pruessner, J. C. (2005). The Montreal Imaging Stress Task: Using functional imaging to investigate the effects of perceiving and processing psychosocial stress in the human brain. *Journal of Psychiatry & Neuroscience*, *30*(5), 319–325. PMID:16151536

Dixit, R. D. (1984). A census of the Indian Pteridophytes. Flora of India Series 4. Botanical Survey of India, Howrah (Calcutta).

Edmondson, W. T. (1974). A simplified method for counting phytoplankton. In R. E. Vollenmeider (Ed.), *A manual on methods for measuring primary production in Aquatic environments* (pp. 14–16). Balckwell Sci. Pub.

FAO. (2010). *The State of World Fisheries and Aquaculture: 2010 Accessed March 2013*. FAO.

Fischer, J., Lindenmayer, D. B., & Manning, A. D. (2006). Biodiversity, ecosystem function, and resilience: Ten guiding principles for commodity production landscapes. *Frontiers in Ecology and the Environment*, *4*(2), 80–86. doi:10.1890/1540-9295(2006)004[0080:BEFART]2.0.CO;2

Galea, S., & Vlahov, D. (2005). URBAN HEALTH: Evidence, Challenges, and Directions. *Annual Review of Public Health*, *26*(1), 341–365. doi:10.1146/annurev.publhealth.26.021304.144708 PMID:15760293

Gamble, J. S. (1924). *The Flora of Presidency of Madras 2*. Botanical Survey of India.

Ghosh, S. R., Ghosh, B., Biswas, A., & Ghosh, R. K. (2004). The Pteridophytic Flora of Eastern India. Flora of India Series 4. Botanical Survey of India, Kolkata.

Groom, M. J., Meffe, G. K., & Carroll, C. R. (2006). *Principles of Conservation Biology* (3rd ed.). Sinauer Associates.

Hedblom, M., Gunnarsson, B., Iravani, B., Knez, I., Schaefer, M., Thorsson, P., & Lundström, J. N. (2019). Reduction of physiological stress by urban green space in a multisensory virtual experiment. *Scientific Reports*, *9*(1), 1–11. doi:10.1038/s41598-019-46099-7 PMID:31300656

Hunter, W. W. (1879). *Statlstlcal Account of Assani* (Vol. II). Trubner and Co.

Hutto, R. L., Pletsechel, S. M., & Hendrick, P. (1986). A fixed radius point count method for non-breeding season use. *The Auk*, *103*(3), 593–602. doi:10.1093/auk/103.3.593

IUCN. (1994). IUCN Red List Categories. IUCN Species Survival Commission.

IUCN. (2001). IUCN Red List Categories and Criteria: Version 3.1. IUCN Species Survival Commission.

IUCN. (2003). Guidelines for Application of IUCN Red List Criteria at Regional Levels: Version 3.0. IUCN Species Survival Commission.

IUCN. (2008). *Red List of Threatened Species*. IUCN. (www.iucnredlist.org)

IUCN. (2010). *Guidelines for Using the IUCN Red List Categories and Criteria, version 8.1 (August 2010), prepared by the Standards and Petitions Subcommittee of the IUCN Species Survival Commission*. IUCN. http://intranet.iucn.org/webfiles/doc/SSC/RedList/ RedListGuidelines.pdf

Jain, S. K. (1968). *Medicinal Plants Nation Book Trust, New Delhi. Jain, S.K. 1983. Rare and Endangered Specles: Observation on rare, imperfectly known endemic plants. In the sacred groves of Western Maharashtra*. Bot. Sur of India.

Jain, S. K. (1991). *Dictionary of Indian folk medicine and ethnobotany*. Deep publications.

Jain, S. K. (1992). The Problem of Endangered Species: Concepts, Problems and Solutions. In Tropical Ecosystems: Ecolosv and Management (Eds. K.P.Singh and J.S.Singh.), Iiley Eastern limited.

Jain, S. K., & Rao, R. K. (1983). *An assessment of threatened plants of India*. Bot. Surv. of India.

Jain, S. K., & Sastry, A. R. K. (1980). *Threatened plants of India - A State of the Alf Report Bot.* Surv. of India.

Jain, S. K., & Sastry, A. R. K. (1984). Safeguarding Plant diversity in threatened Natural Habitats. In A. Hall (ed.) Conservation of Threatened Natural Habitats. African nat. Sci. Prog. Report. 92.

Kholia, B. S., & Bhakuni, K. (2009). Western Himalaya a new range of distribution for a critically endangered fern, Dryopsis manipurensis (Bedd.) Holttum et P. J. Edwards. Nelumbo. *Bulletin of the Botanical Survey of India*, *51*, 245–248.

Krabbendam, L., & Van Os, J. (2005). Schizophrenia and Urbanicity : A Major Environmental Influence — Conditional on Genetic Risk. *Schizophrenia Bulletin*, *31*(4), 795–799. doi:10.1093/schbul/sbi060 PMID:16150958

Kumar Ashok and Srivastava Meena. (2012). Diversity of medicinal Plants in Uttarakhand and their conservation Strategy with special reference to Orchids, *In*: Proceeding of National Conference on Environementla Health: Challaneges and Management. PG College Rajajipuram, Lucknow.

Lackey, J. B. (1938). The manipulation and counting of river plankton and changes in some organisms due to formalin preservation. *Public Health Reports*, *53*(47), 2081–2093. doi:10.2307/4582717

Lindbladh, M., Bradshaw, R., & Holmqvist, B. H. (2000). Pattern and process in south Swedish forests during the last 3000 years, sensed at stand and regional scales. *Journal of Ecology*, *88*(1), 113–128. doi:10.1046/j.1365-2745.2000.00429.x

Lushington, A. W. (1915). *Vernacular list of trees, shrubs and woody cl~rnbers of the Madras Presidency*. Govt. Press.

Magurran, A. E. (1988). *Ecological Diversity and its Measurement*. Princeton University Press. doi:10.1007/978-94-015-7358-0

Mcallister, E., Bhullar, N., & Schutte, N. S. (2017). Into the Woods or a Stroll in the Park : How Virtual Contact with Nature Impacts Positive and Negative Affect. *International Journal of Environmental Research and Public Health*, *14*(786), 1–11. doi:10.3390/ijerph14070786 PMID:28708074

Misra, R. (2013). Ecology Workbook. Scientific Publishers. (Page 31 to 45). Moran, D; Laycock, H; White, P C.L. The role of cost-effectiveness analysis in conservation decision-making. *Biological Conservation*, *143*(4), 826–827.

Nautiyal, D. C., Sharma, S. K., & Pandit, M. K. (2009). Notes on the taxonomic history, rediscovery and conservation status of two endangered species of Ceropegia (Asclepiadaceae) from Sikkim Himalaya. *Journal of the Botanical Research Institute of Texas*, *3*(2), 815–822.

Nautiyal, D. C., Sharmaand, S. K., & Pandit, M. K. (2009). Notes on the taxonomic history of two rare species of Begonia (Begoniaceae) from Sikkim Himalaya and their conservation. *Journal of the Botanical Research Institute of Texas*, *3*(2), 823–830.

Nayar, M. P. (1980). Endemism and patterns of distribution of endemic genera (Angiosperms) in India. *Journal of Economic and Taxonomic Botany*, *I*, 99–110.

Nayar, M. P. (1996). Hotspots of Endemic Plants of India, Nepal and Bhutan. Thiruvanathapuram: Tropical Botanical Garden and Research Institute. 204 pp.

Nayar, M. P., & Sastry, A. R. K. (1987). *Red Data Book of Indian Plants* (Vol. I). Botanical Survey of India.

Nayar, M.P., & Sastry, A. R. K. (1988). *Red Data Book of Indian Plants* (Vol. II). Botanical Survey of India.

Nayar, M. P., & Sastry, A. R. K. (1990). *Red Data Book of Indian Plants* (Vol. III). Botanical Survey of India.

Ohasi, H. (1975). Flora of Eastern Himalaya, Third Report. *University Museum of University of Tokyo Bulletin*, *8*, 1–458.

Oldfield, S., Lusty, C., & MacKinven, A. (1998). *The World List of Threatened Trees*. World Conservation Press.

Rosenzweig, M. L. (1995). *Species Diversity in Space and Time*. Cambridge University Press. doi:10.1017/CBO9780511623387

Rosser, A. M., & Mainka, S. A. (2002). Overexploitation and Species Extinctions. *Conservation Biology, 16*(3), 584–586. doi:10.1046/j.1523-1739.2002.01635.x

Roth, D. S., Perfecto, I., & Rathcke, B. (1994). The effects of management systems on ground-foraging ant diversity in Costa Rica. Ecological Applications 4(3):423-436. Botanical Survey of India.

Shaffer, F., & Ginsberg, J. P. (2017). An Overview of Heart Rate Variability Metrics and Norms. *Frontiers in Public Health, 5*(September), 1–17. doi:10.3389/fpubh.2017.00258 PMID:29034226

Shendage, S. M., & Yadav, S. R. (2010). Revision of the Genus Barleria (Acanthaceae) in India. *Rheedea, 20*(2), 81–230.

Standard, W., Zielasko, D., Dawans, B. Von, & Domes, G. (2020). *Open TSST VR : Trier Social Stress Test in Virtual Reality.* September.

Strengbom, J., Dahlberg, A., Larsson, A., Lindelöw, Å., Sandström, J., Widenfalk, O., & Gustafsson, L. (2011). Introducing intensively managed spruce plantations in Swedish: Forest landscapes will impair biodiversity decline. *Forests, 2*(3), 610–630. doi:10.3390/f2030610

Terrapass. (2023). *Cost of Carbon Offset*. Terrapass. https://terrapass.com/blog/creating-carbon-credits-is-it-profitable/#:~:text=Carbon%20credits%20in%202023%20cost,can%20fluctuate%20greatly%20with%20demand

Thommpson, I. D., Davidson, I. J., O' Donnell, S., & Brazeau, F. (1989). Use of track transects to measure the relative occurrence of some arboreal mammals in uncut forest and regeneration stands. *Canadian Journal of Zoology, 67*(7), 1816–1823. doi:10.1139/z89-258

United Nations Environment Program. (2005). Millennium Ecosystem Assessment. Ecosystems and Human Well-being: Current State and Trends. CH 4. *Biodiversity (Nepean).*

Walker, B. (1995). Conserving Biological Diversity through Ecosystem Resilience. *Conservation Biology, 9*(4), 747–752. doi:10.1046/j.1523-1739.1995.09040747.x

Welch, F. S. (1948). *Limnological Methods*. McGraw Hill Book Co Inc.

Welsh B. (1987). *Monitoring herpetofauna in woodlands of north western California and south west Oregon: a comparative strategy.*

Welsh, H.H. Jr. & Lind, A. (1991). *The structure of the herpetofaunal assemblage in the Douglas-fir/hardwood forests of northwestern California and south western Oregon.*

White, M. P., Yeo, N. L., & Albin, M. (2018). A prescription for "nature" – the potential of using virtual nature in therapeutics. *Neuropsychiatric Disease and Treatment, 14*, 3001–3013. doi:10.2147/NDT.S179038 PMID:30510422

Wilson, D. E., & Reeder, D. M. (Eds.). (1993). *Mammal Species of the World a Taxonomic and Geographic reference* (2nd ed.). Smithsonian Institution Press.

Wilson, D. E., & Reeder, D. M. (Eds.). (2005). *Mammal Species of the World. A Taxonomic and Geographic Reference* (3rd ed.). Johns Hopkins University Press. doi:10.56021/9780801882210

World Conservation Monitoring Centre. (1988). *The Conservation of Biological Diversity. WCMC*. I.U.C.N.

World Conservation Monitoring Centre. (2000). *Global Biodiversity: Earth's living resources in the 21st Century. By: Groombridge B. and Jenkins, M.D.* World Conservation Press.

Chapter 8
Harmonizing Development With Nature:
A Comprehensive Environmental Impact Assessment (EIA) for Sustainable Biodiversity Conservation

Rekha Rani
Indira Gandhi National Tribal University, India

Vineet Kumar
https://orcid.org/0000-0002-4415-4971
Dr. Yashwant Singh Parmar Government P.G. College, India

ABSTRACT

As societies strive for progress and economic growth, the imperative to balance development with the preservation of natural ecosystems becomes increasingly pressing. The concept of harmonizing development with nature emphasizes the need to ensure that human activities are conducted in a manner that sustains biodiversity and ecological integrity. Central to achieving this balance is the implementation of comprehensive environmental impact assessments (EIAs) that incorporate considerations for sustainable biodiversity conservation. By employing innovative tools and methodologies and engaging stakeholders at all levels, EIAs can inform decision-makers about the potential impacts of development projects on biodiversity and ecosystems, thus facilitating informed decision-making and promoting sustainable development practices. Through collaborative efforts and a commitment to environmental stewardship, the harmonization of development with nature can be realized, ensuring the long-term health and resilience of our planet's ecosystems for generations to come.

INTRODUCTION

The increasing human population and the need for development have led to significant changes in the environment. This has resulted in a decline in biodiversity, which is essential for the health and well-being

DOI: 10.4018/979-8-3693-3330-3.ch008

of ecosystems and human societies. In response to this challenge, the concept of sustainable development has emerged, which aims to balance economic growth, social equity, and environmental protection. One of the key tools in achieving this balance is the Environmental Impact Assessment (EIA), a comprehensive approach to understanding and managing the potential impacts of development on the environment and biodiversity (Glasson et al., 2012). This article discusses the importance of harmonizing development with nature through a comprehensive EIA for sustainable biodiversity conservation.

In the pursuit of progress and development, the delicate balance between human activities and the natural environment often comes under strain. As societies strive for economic growth and infrastructural advancement, the repercussions on biodiversity and ecosystems cannot be ignored. However, there exists a paradigm that emphasizes not only development but also the preservation of nature's wealth - the concept of harmonizing development with nature.

Environmental Impact Assessment (EIA) serves as a critical tool for achieving sustainable development by systematically evaluating the potential environmental consequences of proposed projects, policies, or plans. This paper explores the role of EIA in promoting biodiversity conservation within the framework of sustainable development. It examines how EIA processes incorporate biodiversity considerations, assesses their effectiveness in mitigating adverse impacts on biodiversity, and highlights opportunities for enhancing biodiversity conservation through EIA. Drawing on case studies and empirical evidence, the paper underscores the importance of integrating biodiversity concerns into EIA practices to safeguard ecosystems, support ecosystem services, and ensure the long-term well-being of both human societies and the natural world.

At the forefront of this endeavor lies the Environmental Impact Assessment (EIA), a crucial tool employed to evaluate the potential environmental consequences of proposed projects and activities. In recent years, there has been a growing recognition of the need to integrate biodiversity conservation considerations into EIAs, thereby ensuring that development initiatives are carried out in a sustainable manner.

As per Canada Environmental Assessment Act, a comprehensive EIA for sustainable biodiversity conservation encompasses a multifaceted approach that goes beyond mere compliance with regulatory requirements. It involves a thorough assessment of the project's potential impacts on the surrounding ecosystem, including flora, fauna, and their habitats. By identifying sensitive areas and vulnerable species, mitigation measures can be devised to minimize adverse effects and safeguard biodiversity.

One of the key pillars of sustainable biodiversity conservation through EIA is the principle of ecosystem-based management. Rather than focusing solely on individual species or habitats, this approach considers the interconnectedness of various components within an ecosystem. By assessing the project's cumulative impacts on ecosystem structure and function, EIAs can inform decision-makers about the long-term sustainability of development activities (Bojórquez-Tapia & Ezcurra, 2009).

Moreover, stakeholder engagement plays a pivotal role in the EIA process, particularly concerning biodiversity conservation. Local communities, indigenous peoples, and other stakeholders possess valuable traditional knowledge and insights regarding their surrounding environment. By involving them in the decision-making process, EIAs can benefit from their perspectives and enhance the effectiveness of biodiversity conservation measures.

Furthermore, technological advancements have revolutionized the field of EIA, offering innovative tools and methodologies for assessing biodiversity impacts. Geographic Information Systems (GIS), remote sensing, and predictive modeling techniques enable more accurate mapping of biodiversity hotspots and the prediction of potential impacts. Additionally, the incorporation of ecosystem services valuation allows decision-makers to weigh the costs and benefits of development projects from a broader perspective.

THE NEED FOR SUSTAINABLE DEVELOPMENT

Sustainable development is a concept that has gained momentum in recent years as the world becomes more aware of the consequences of unsustainable practices on the environment and human societies (IPCC, 2019). The Brundtland Report (1987) defined sustainable development as "development that meets the needs of the present generation without compromising the ability of future generations to meet their own needs." This definition emphasizes the importance of balancing economic growth, social equity, and environmental protection. Biodiversity conservation is integral to sustainable development for several reasons:

Ecosystem Services: Biodiversity provides essential ecosystem services such as pollination, water purification, and carbon sequestration, which are crucial for human well-being and economic development.

Genetic Resources: Biodiversity serves as a vast genetic resource for agriculture, medicine, and industry. Sustainable use of biodiversity ensures continued access to these resources for future generations.

Ecological Stability: Biodiverse ecosystems are more resilient to environmental changes and disturbances. Conserving biodiversity helps maintain ecological stability, reducing the risk of ecosystem collapse and its associated socio-economic impacts.

Cultural and Spiritual Values: Biodiversity is deeply intertwined with cultural and spiritual practices of many communities worldwide. Preserving biodiversity preserves cultural diversity and traditional knowledge systems.

Tourism and Recreation: Many ecosystems and species attract tourists, generating revenue and employment opportunities. Sustainable biodiversity conservation supports ecotourism and recreational

Climate Change Mitigation: Healthy ecosystems play a vital role in climate change mitigation by sequestering carbon dioxide. Protecting biodiversity helps maintain the integrity of these carbon sinks, aiding in global efforts to combat climate change.

Food Security: Biodiversity supports food security by providing diverse sources of food, including wild edible plants and animals. Maintaining diverse ecosystems safeguards against crop failures and pests, ensuring stable food production.

Health Benefits: Biodiversity contributes to human health through the provision of medicinal plants and ecosystem services that regulate diseases. Preserving biodiversity is crucial for discovering new drugs and preventing disease outbreaks.

Legal and Ethical Obligations: Many nations have legal obligations under international agreements such as the Convention on Biological Diversity to conserve biodiversity. Failure to fulfill these obligations can have diplomatic and legal consequences.

Interconnectedness of Life: Every species, no matter how small, plays a role in the web of life. Loss of biodiversity can lead to cascading effects throughout ecosystems, affecting human societies in unforeseen ways.

In summary, integrating biodiversity conservation into sustainable development efforts is essential for ensuring the long-term well-being of both ecosystems and human societies. By recognizing the intrinsic value of biodiversity and its contributions to economic, social, and environmental sustainability, we can work towards a more harmonious relationship between humans and the natural world.

ENVIRONMENTAL IMPACT ASSESSMENT (EIA) AS A TOOL FOR SUSTAINABLE DEVELOPMENT

An EIA is a systematic process that involves the identification, assessment, and management of potential environmental impacts associated with a proposed project. The purpose of an EIA is to ensure that projects are developed in a manner that minimizes their negative impacts on the environment and promotes sustainable development (Canter, 1996).

As per Intergovernmental Science-Policy Platform on Biodiversity and Ecosystem Services (IPBES), Environmental Impact Assessment (EIA) serves as a critical tool for achieving sustainable development by systematically evaluating the potential environmental consequences of proposed projects, policies, or plans. This paper explores the role of EIA in promoting biodiversity conservation within the framework of sustainable development. It examines how EIA processes incorporate biodiversity considerations, assesses their effectiveness in mitigating adverse impacts on biodiversity, and highlights opportunities for further improvement.

First, let's discuss how EIA processes incorporate biodiversity considerations. EIAs typically involve multiple stages, including site selection, project design, implementation, and monitoring. At each stage, potential impacts on biodiversity are assessed, and mitigation measures are developed to minimize adverse effects. For example, during the site selection stage, EIAs may identify areas with high biodiversity value and avoid or minimize the disturbance of these areas. During the project design stage, mitigation measures such as habitat corridors, creation of protected areas, and careful selection of construction materials can be implemented to reduce the impact on biodiversity. Monitoring and evaluation of the project's progress can also help to identify any further potential threats to biodiversity and inform future decision-making.

However, the effectiveness of EIA processes in mitigating adverse impacts on biodiversity can vary. While some EIAs have been successful in reducing the negative effects of projects on biodiversity, others have been less effective due to a lack of comprehensive information, inadequate participation of stakeholders, or insufficient implementation of mitigation measures. To enhance the effectiveness of EIAs in promoting biodiversity conservation, it is essential to address these challenges and improve EIA processes. The comparative analysis for EIA and biodiversity conservation is listed below in table 1.

As per International Council for Science, one opportunity for improvement is to enhance the availability and quality of information used in EIAs. This can be achieved by strengthening collaboration between researchers, government agencies, and local communities to collect and share accurate data on the distribution and status of species and ecosystems. Additionally, incorporating the latest scientific knowledge and best practices in EIA processes can help to ensure that projects have a lower impact on biodiversity.

Another opportunity is to improve stakeholder involvement in EIA processes. This can be achieved by promoting transparency, providing opportunities for public participation, and encouraging the inclusion of local communities, indigenous peoples, and other affected parties in the decision-making process. This will help to ensure that the needs and concerns of these stakeholders are taken into account, and that the resulting projects are more likely to be sustainable and acceptable to all parties involved (Partidário, 2008).

Finally, it is important to recognize that EIA processes are just one aspect of a broader approach to sustainable development and biodiversity conservation. Policies and regulations that promote sustainable development, as well as public awareness and education campaigns, are also essential components of a comprehensive strategy to protect and preserve biodiversity. By integrating EIA processes with other

Table 1. Comparative analysis of EIA and biodiversity conservation

Aspect	Environmental Impact Assessment (EIA)	Biodiversity Conservation
Definition	EIA is a process for evaluating the potential environmental consequences of proposed projects, policies, or plans.	Biodiversity conservation involves the protection, management, and restoration of ecosystems and species diversity.
Objective	To systematically identify, predict, and evaluate the environmental impacts of development activities and propose mitigation measures to minimize adverse effects.	To safeguard the variety and variability of life on Earth, including ecosystems, species, and genetic diversity.
Focus	Primarily focuses on assessing the overall environmental impacts of proposed projects or activities on various components of the environment.	Specifically targets the conservation and management of biodiversity, including ecosystems, species, and genetic resources.
Process	Involves a systematic process of screening, scoping, impact assessment, mitigation, monitoring, and decision-making, often guided by legal and regulatory frameworks.	Involves various strategies such as protected area management, habitat restoration, species conservation programs, and sustainable use of natural resources.
Scope	Covers a wide range of environmental factors, including air quality, water resources, land use, noise pollution, social impacts, and cultural heritage.	Focuses specifically on biodiversity values, threats to biodiversity, conservation planning, and implementation of conservation measures.
Integration with Sustainable Development Goals	Contributes to achieving sustainable development goals by promoting environmentally sound decision-making and balancing economic development with environmental protection.	Biodiversity conservation is recognized as a key component of sustainable development, contributing to ecosystem resilience, food security, and human well-being.
Tools and Techniques	Utilizes various tools and techniques such as environmental impact matrices, risk assessment, spatial analysis, modeling, and stakeholder engagement.	Employs tools and techniques such as habitat mapping, species inventories, ecological assessments, genetic analysis, and participatory approaches.
Legal and Policy Framework	Often guided by national environmental laws, regulations, and international standards, requiring developers to conduct EIAs for certain types of projects.	Supported by international agreements such as the Convention on Biological Diversity (CBD) and national laws and policies aimed at biodiversity conservation.
Implementation and Effectiveness	Effectiveness may vary depending on factors such as the quality of the assessment, stakeholder participation, enforcement of mitigation measures, and follow-up monitoring.	Effectiveness depends on factors such as the designation and management of protected areas, enforcement of conservation laws, public awareness, and international cooperation.

approaches, we can create a more robust framework for achieving sustainable development and promoting biodiversity conservation (Sadler, 1996).

In conclusion, Environmental Impact Assessment (EIA) plays a crucial role in promoting biodiversity conservation within the framework of sustainable development. By incorporating biodiversity considerations throughout the EIA process and addressing the challenges that hinder their effectiveness, we can ensure that projects and policies have a lower impact on biodiversity and contribute to the long-term health and resilience of ecosystems and the species that depend on them (United Nations Environment Programme, 2002).

A comprehensive EIA for sustainable biodiversity conservation should include the following components:

Scope and Objectives: The first step in an EIA is to define the project's scope and objectives. This includes identifying the project's purpose, location, and key activities, as well as the potential environmental impacts and objectives for sustainable biodiversity conservation.

Site Characterization: A thorough assessment of the site's environmental characteristics, such as geology, hydrology, and ecosystems, is necessary to understand the potential impacts of the project on the environment.

Impact Assessment: This involves the identification and evaluation of potential environmental impacts on biodiversity, including habitat loss, fragmentation, and degradation, as well as the impacts on species and ecosystems.

Mitigation and Enhancement Measures: Based on the impact assessment, appropriate mitigation and enhancement measures should be developed to minimize the negative impacts on biodiversity and promote sustainable development.

Monitoring and Evaluation: Throughout the project, it is essential to monitor and evaluate the effectiveness of the implemented mitigation and enhancement measures to ensure that the project is achieving its intended goals for sustainable biodiversity conservation.

HARMONIZING DEVELOPMENT WITH NATURE THROUGH A COMPREHENSIVE EIA

Harmonizing development with nature through a comprehensive EIA can be achieved by integrating environmental considerations into all stages of the project lifecycle. This includes:

Early Engagement of Stakeholders: Stakeholder involvement from the beginning of the project can help ensure that environmental concerns are identified and addressed throughout the EIA process.

Purpose: Engaging stakeholders early in the project lifecycle allows for the identification of environmental concerns and potential impacts, fostering transparency and inclusivity in decision-making processes.

Approach:

- Conducting stakeholder consultations and public hearings to gather input and feedback.
- Facilitating dialogue between project proponents, regulatory agencies, affected communities, and other stakeholders.
- Incorporating traditional knowledge and indigenous perspectives into the decision-making process.

4.2. Integration of Environmental Considerations: The EIA process should integrate environmental considerations into all aspects of the project, from site selection to project implementation, ensuring that potential impacts on biodiversity are identified and mitigated.

Purpose: Integrating environmental considerations throughout project planning, design, implementation, and monitoring phases helps minimize adverse impacts on ecosystems, species, and natural resources.

Approach:

- Conducting baseline studies to assess the existing environmental conditions and biodiversity.
- Identifying potential environmental impacts and assessing their significance using scientific methods and predictive modeling.
- Developing mitigation measures and alternative solutions to minimize adverse effects on biodiversity and ecosystem services.
- Incorporating principles of sustainable design, such as green infrastructure and low-impact development techniques, to enhance environmental performance.

4.3. Collaboration with Local Communities: Involving local communities in the EIA process can help build support for sustainable development and promote the conservation of biodiversity.

Purpose: Collaborating with local communities fosters mutual understanding, promotes social cohesion, and empowers communities to participate in decision-making processes that affect their environment and livelihoods.

Approach:

- Establishing community advisory groups or committees to facilitate communication and consultation.
- Conducting participatory mapping exercises and community surveys to identify local ecological knowledge and cultural values related to biodiversity.
- Providing capacity-building and training opportunities for local communities to participate effectively in environmental monitoring and management activities.
- Implementing benefit-sharing mechanisms to ensure equitable distribution of project benefits and responsibilities among stakeholders.

4.4. Enforcement of Regulations and Policies: Strong enforcement of environmental regulations and policies can help ensure that projects are developed in a manner that minimizes their impacts on the environment and promotes sustainable development.

Purpose: Enforcement of regulations and policies ensures compliance with environmental standards, promotes accountability, and deters activities that may harm biodiversity or natural ecosystems.

Approach:

- Establishing clear regulatory frameworks and guidelines for environmental impact assessment, monitoring, and enforcement.
- Conducting regular inspections and audits to verify compliance with environmental permits and regulations.
- Imposing penalties or sanctions for non-compliance with environmental requirements, such as fines, suspension of permits, or project shutdowns.
- Strengthening institutional capacity and coordination among government agencies responsible for environmental management and enforcement.

As per United States Environmental Protection Agency, by implementing these strategies, projects can effectively harmonize development with nature, minimize adverse environmental impacts, and promote sustainable development outcomes that benefit both people and the planet.

BIODIVERSITY CONSERVATION & EIA

According to World Bank, Environmental Impact Assessment (EIA) has emerged as a vital tool for integrating environmental considerations into decision-making processes related to development activities. With the growing recognition of the importance of biodiversity conservation for sustainable development, there is a need to explore the role of EIA in safeguarding biodiversity.

Role of EIA in Biodiversity Conservation: EIA serves as a proactive mechanism for identifying, predicting, and evaluating the potential impacts of development projects on biodiversity. By conducting comprehensive assessments of ecosystems and species likely to be affected by proposed activities, EIA helps stakeholders understand the ecological consequences of their decisions. Moreover, EIA enables the identification of alternative approaches and mitigation measures to minimize adverse impacts on biodiversity.

Effectiveness of EIA in Biodiversity Conservation: While EIA has made significant strides in integrating biodiversity considerations, its effectiveness in conserving biodiversity varies across different contexts. Challenges such as inadequate baseline data, limited stakeholder engagement, and insufficient enforcement of mitigation measures can undermine the outcomes of EIA processes. Nonetheless, numerous case studies demonstrate the positive outcomes of EIA in preserving biodiversity and enhancing ecosystem resilience (Morrison-Saunders & Fischer, 2006).

Opportunities for Enhancing Biodiversity Conservation through EIA: To enhance biodiversity conservation through EIA, several opportunities can be explored. These include strengthening legal and institutional frameworks, promoting the use of strategic environmental assessment (SEA) to address cumulative impacts on biodiversity, and fostering interdisciplinary collaboration among stakeholders. Additionally, incorporating indigenous and local knowledge systems into EIA processes can enrich biodiversity assessments and foster community participation in conservation efforts (International Union for Conservation of Nature, European Commission, Convention on Biological Diversity).

CONCLUSION

Environmental Impact Assessment (EIA) plays a crucial role in advancing biodiversity conservation within the framework of sustainable development. By integrating biodiversity considerations into decision-making processes, EIA contributes to the preservation of ecosystems, protection of endangered species, and promotion of ecosystem services essential for human well-being. However, realizing the full potential of EIA in biodiversity conservation requires concerted efforts to address existing challenges and capitalize on emerging opportunities. Harmonizing development with nature through a comprehensive EIA is essential for sustainable biodiversity conservation. By incorporating environmental considerations throughout the project lifecycle and working collaboratively with stakeholders, we can ensure that projects are developed in a manner that minimizes their impacts on the environment and promotes the long-term health and well-being of ecosystems and human societies. Through collaborative efforts and informed decision-making, we can strive towards a future where development and biodiversity conservation go hand in hand, paving the way for a more sustainable and resilient world.

REFERENCES

Bojórquez-Tapia, L. A., & Ezcurra, E. (2009). Assessing environmental impacts of large construction projects: An integrative approach. *Environmental Impact Assessment Review*, *29*(4), 219–227.

Canter, L. (1996). *Environmental impact assessment* (2nd ed.). McGraw-Hill.

Convention on Biological Diversity (CBD). (2004). *Handbook of the convention on biological diversity*. CBD. https://www.cbd.int/doc/handbook/

Glasson, J., Therivel, R., & Chadwick, A. (2012). *Introduction to environmental impact assessment*. Routledge.

Intergovernmental Panel on Climate Change (IPCC). (2019). *Special report on climate change and land*. IPCC. https://www.ipcc.ch/srccl/

Millennium Ecosystem Assessment. (2005). *Ecosystems and human well-being: Synthesis*. Island Press.

Morgan, R. K., & Henrion, M. (1990). *Uncertainty: A guide to dealing with uncertainty in quantitative risk and policy analysis*. Cambridge University Press. doi:10.1017/CBO9780511840609

Morrison-Saunders, A., & Fischer, T. B. (2006). What is wrong with EIA and SEA – a diagnostic approach. *Environmental Impact Assessment Review*, *26*(6), 552–566.

Partidário, M. R. (2008). Strategic environmental assessment and the management of Natura 2000 sites in Portugal: The experience of a research project. *Impact Assessment and Project Appraisal*, *26*(3), 213–222.

Sadler, B. (1996). International study of the effectiveness of environmental assessment. *Impact Assessment and Project Appraisal*, *14*(1), 3–9.

United Nations Environment Programme (UNEP). (2002). *Biodiversity and environmental assessment: Enhancing ecosystem services for human well-being*. UN. https://www.unep.org/

Chapter 9
From Canopy to Concrete:
Assessing Urban Green Space Depletion's Impact on Biodiversity With OSM Data

Froilan Delute Mobo
https://orcid.org/0000-0002-4531-8106
Philippine Merchant Marine Academy, Philippines

Munir Ahmad
https://orcid.org/0000-0003-4836-6151
Survey of Pakistan, Pakistan

ABSTRACT

This study provides a comprehensive analysis of green spaces in Punjab Province, Pakistan, from 2019 to 2023. The OpenStreetMap (OSM) data has been employed to gain insight into green spaces focusing on leisure, natural, and land use tags. These findings indicate a consistent decline in green spaces across all key categories over the five-year period. This decline raises concerns about habitat loss, biodiversity depletion, and the potential impact on ecosystem services. Recommendations for policymakers include strengthening habitat protection, promoting sustainable land use planning, investing in restoration, enhancing urban green infrastructure, and raising public awareness. Collaboration among stakeholders is essential to effectively address these challenges and safeguard biodiversity for future generations.

INTRODUCTION

Urbanization is a process of population concentration (Tisdale, 1942). This process involves different facets including demographic shifts, infrastructural developments, and socioeconomic transitions coupled with big urbanization where rural dwellers are opting to migrate to urban cities for exploitations and amenities. Urbanization forms the basis of significant land use and social pattern changes, affecting culturally entrenched issues such as consumption patterns and identity formation. This concept reflects a holistic reshaping of the spatial organization that is not only concerned with the sustainable utiliza-

DOI: 10.4018/979-8-3693-3330-3.ch009

tion of the existing resources but also about the supply of basic services, thus being among the crucial aspects of global developments in our age. Urbanization is a global process that brings about emerging patterns in land use and land cover distribution, especially within urban settlements where population growth and development pressures are high (Fox, 2012; Zhang, 2008).

Urban green spaces act as a key to sustainable urban environments that safeguard human health and create a robust city as highlighted by (Jabbar et al., 2022). These spaces act as vital links in the process of promoting total health in the urban setting and building urban resilience. These green spaces, which comprise parks, forests, and community gardens, bring about health gains to urban dwellers in terms of psychological and physical well-being as elucidated by (Wolch et al., 2014). Apart from enhancing the popularity and fitness levels of citizens both for exercising and recreation, these spaces contribute extensively to psychological well-being endeavors and stress mitigation in the midst of urban hustle. Besides that, the cities' green spaces are prime ecosystems, sheltering biodiversity and mitigating the pollution impacts, consequently, the cities become more ecological. Such outcomes clearly point to the necessity of preservation and equitable distribution of urban green spaces, being components of the policy tools and vision of urban development, which, in turn, serves as a foundation for healthy and resilient communities.

However, the rapid pace of urbanization is diminishing green spaces within cities. As urban areas expand to accommodate growing populations and infrastructural demands, green spaces are often sacrificed, leading to increased fragmentation and loss of biodiversity. For example, rapid urban expansion in Southeast Asian cities has reduced green space from 45% to 20 as noted by (Nor et al., 2021). Similarly, Ethiopia, experienced a 51.8% decrease in green space at a rate of 3.9% annual loss between 2003 and 2016, driven by factors such as population growth, high land values, and lax enforcement of planning regulations as endorsed by (Girma et al., 2019).

The continuous development of a city and the depletion of green areas contribute to the increased focus on biodiversity and ecological health status. As highlighted by (Nor et al., 2017) The rampant urbanization that is engaged now raises a decline in urban green spaces, and that inevitably reduces connectivity in and between these ecosystems. It also contributes to destroying the balanced network of biodiversity in urban areas and decreasing the amount of stability and efficiency of urban ecosystems. Knowledge of how urban green space decline is related to biodiversity reduction is essential for overcoming the problem caused by urbanization and maintaining sustainable urban development. Urban green spaces are an irreplaceable element for balanced ecosystems, a home for wildlife, a climate regulator, and also support for human well-being. One of the outcomes of evaluating the effect of the degradation of urban green space on biodiversity will be informed land use management, conservation plans, and policy legislations aimed at neutralizing unbridled urbanization on ecosystems and biodiversity.

VGI refers to location-based information voluntarily contributed by individuals across various digital platforms (Goodchild, 2007), including but not limited to social media, mobile applications, and collaborative projects. VGI encompasses a diverse array of user-generated content specific to locations, including geotagged images, check-ins, reviews, and other contributions originating from particular areas (Haklay, 2013; Kitchin et al., 2018). As individuals engage with their environments, they organically gather and exchange these data points. Typically, VGI is produced by citizens utilizing location-aware devices like smartphones to capture geographic data in forms such as geo-tagged text, photos, or videos (Honarparvar et al., 2019). Subsequently, this information is shared across various online social media platforms like Twitter, thereby contributing to the collective pool of VGI.

At the core of VGI lies the principle of democratizing geographic information, wherein users actively participate in data creation and management rather than being passive recipients of government datasets (Goodchild, 2007). The emergence of new digital platforms has facilitated collaborative contributions to geographic information, fostering the collective construction of digital representations and conceptualizations of geography (Calcagni et al., 2019). These platforms serve as mediums for user interactions, allowing individuals to collaboratively shape and envision geographical elements within the digital domain. VGI is a multifaced data resource in many domains' applications such as land administration (Ahmad & Khiyal, 2023), disaster response and community development (Ahmad, 2023a, 2024b), governance and public service delivery (Ahmad, 2023c, 2024c) and many more.

For various reasons, the utilization of VGI in assessing green spaces has experienced a notable surge. Firstly, the proliferation of mobile devices and internet accessibility has led to an unprecedented accumulation and distribution of location-based data (H. Huang et al., 2021). This surge allows citizens to actively participate in the generation of geographical data, thereby furnishing valuable insights into their preferences, needs, and experiences (Goodchild, 2007). Secondly, traditional methods of data collection are both costly and time-intensive. In contrast, VGI offers a real-time, cost-efficient alternative, empowering public sector entities to access dependable information swiftly for decision-making processes (Ahmad et al., 2022).

An exemplary instance of VGI is OpenStreetMap (OSM), established in 2004, which endeavors to construct and sustain an open, editable global database and map (Grinberger et al., 2022). Noteworthy for its expansive volunteer community and extensive spatial data repository, OSM epitomizes the concept of VGI. OpenStreetMap is a user-generated, editable, and free-to-use map data project that follows the peer production model similar to Wikipedia (Weber & Haklay, 2008). OSM is also valuable resources to measure sustainable development (Ahmad, 2023b; Ahmad & Ali, 2023) and to get insights of market trends (Ahmad, 2024a) beside its usage in many more application areas such as disaster management, education, health.

User-generated geographic information such as OSM can provide valuable insights into urban green space use and preferences as noted by (Heikinheimo et al., 2020). Moreover, OSM can be used for comparing indicators of urban green space provision, access, and spatial distribution within cities (Texier et al., 2018). OSM is a valuable source for estimating the value of green space data for business development (Ahmad, 2024d). Even OpenStreetMap data can be employed to generate customized pedestrian routes with green areas, social places, and quieter streets, based on user preferences (Novack et al., 2018). To this effect, the primary objectives of this chapter are as follows.

i. To investigate the implications of shrinking urban green spaces with the help of OpenStreetMap data.
ii. To recommend policy-makers, urban planners, and other nature conservation actors' ways to mitigate the risk associated with the loss of urban green space.

The chapter is organized as follows: Section 2 encompasses the literature review. Section 3 outlines the methods for obtaining OpenStreetMap data and identifying urban green spaces within the study area; Moreover, this section covers analysis techniques used to assess urban green space depletion. Section 4 presents the results illustrating the application of the methods discussed in a specific urban area; Section 5 outlines the implications of the findings and provides recommendations for urban green space management; and the last section summarizes the key findings of the chapter and discusses avenues for future research.

LITERATURE REVIEW

Urbanization is a process when the number of inhabitants increases in urban areas (Tisdale, 1942), it also includes people moving into industries and cities (Anderson, 1959). It is the swarming of people in massive crowding into large, compact, and homogeneous areas, a spatial manifestation of the urban process (Friedmann, 1966). With the rapid growth of the urban populations and surrounding regions, the urban process has become inexorable (Antrop, 2004). This large-scale historical process occurs in such a way that it stems from population dynamics due to food and disease control improvements in these areas resulting from technological progressions (Fox, 2012). Taking the form of the rise of built-up areas and population concentration in cities and towns is manifested by the increase in the number of people living in cities, with social, economic, and environmental consequences (Dociu, 2012; Ren, 2015). One can describe the process of 'Urbanization' as the change of human activities and the production factors from the countryside to the cities, which results in settlement, increased consumption, and environmental contamination. It's a transformational process, which in both negative and positive ways changes the health and well-being of people around. Globally, cities are always making major changes in social, economic, and environmental domains (Eckert & Kohler, 2014; Godfrey & Julien, 2005; Jungalwalla, 1968).

Socially, it reconstitutes societies, being inclusionary but sometimes proving to be a source of resistance to diversity through the exacerbation of poverty and crime. Economically, through agglomeration of functions triggers the process of economic development, but on the other side, it may strain infrastructure and worsen the wealth inequality situation. Concerning the environment, urbanization means the destruction of habitat, pollution, and resources scarce, in turn destroying biodiversity and contributing to global warming. The impression of infrastructure problems has arisen when the cities are expanding which is also influencing transportation, housing, and basic services. Moreover, urbanization affects public health in two dimensions: it provides better health care provision to citizens, but on the negative side, it is almost always accompanied by increased pollution, noise, and overall living stress. It is necessary to develop the solutions through the combination of comprehensive and sustainable approaches that are focused on urbanity, environment, and equability (Deng et al., 2015; Dociu, 2012; Nguyen & Nguyen, 2018; Zhao, 2023).

Green Spaces are the areas where nature and society come together, they can be the parks, gardens, forests, meadows, and many other natural landscapes within the urban or rural environments. Such areas can be marked by vegetation, like plants, trees, shrubs, and grasses. Green spaces can differ in size and can be of different types from small recreation areas to large natural reserves. They create the conditions needed for relaxation, recreation, and, the simple enjoyment of being in the midst of nature. Urban greenspaces also perform other important functions, such as supporting biodiversity, regulating urban climates, mitigating air and water pollution as well as improving overall environmental quality (Coolen & Meesters, 2012; Hoyle et al., 2019).

Urban greenspaces provide daily contact with nature, enhancing human well-being and biodiversity conservation (Fuller et al., 2007). Greenspaces are central to ameliorating urban living conditions (Dallimer et al., 2011). Increased greenspace exposure is associated with decreased salivary cortisol, heart rate, and blood pressure, and reduced risk of preterm birth, type II diabetes, and cardiovascular mortality (Twohig-Bennett & Jones, 2018). Subsequently, greenspaces demonstrate the possibility of reducing obesity and improving health, as greenspace shows positive associations with health indicators (Lachowycz & Jones, 2011). More importantly, the association of greenspace exposure with reduced mortality,

CVD morbidity, cardiometabolic factors, mental health issues, low birth weight, physical activity levels, sleep quality, and urban crime rates are also established by Yang et al. (2021).

Thus, it can be said that green spaces are critical in defense of a plethora of troubles that arise due to urbanization. First, they contribute to strengthening the quality of town and city life by promoting recreation, relaxation, and social interaction areas. This in turn raises the level of mental and overall health in urban residents. Meanwhile, greenspaces also help to alleviate the urban heat island effect through the provision of shades and cooling which in turn helps to reduce air conditioning energy consumption and health problems related to heat. Next, these spaces create an environment that supports the existence of many species by offering them places in which to live and maintain ecosystem balance. Furthermore, green spaces act as a natural water dam by retaining rainwater and in turn reducing the threats of flooding and soil erosion. Also, these spaces can improve the air quality by taking up pollutants and emitting oxygen, thereby, reducing the impacts of air pollution which are caused by rapid urbanization. Moreover, greenspaces create a positive economic impact by raising property values and encouraging investments thereby contributing to economic development and revitalization endeavors in urban areas (Kothencz et al., 2017; Lee & Maheswaran, 2011; Li et al., 2022; Madureira et al., 2015; Ugolini et al., 2022; Wood et al., 2018; Zhou & Rana, 2012).

The ecological and biodiversity effects of greenspaces on urban areas are significant. On the one hand, they sustain different kinds of plants and animals by serving as their habitats and sanctuaries and hence, they enhance biodiversity and ecological stability in the local area. Greenspaces not only act as transportation routes for wildlife through urban environments but also keep genetic variability and population connection as they are. Besides, greenspaces augment habitat diversification and as a result, they provide many kinds of micro-environments in which a high level of biodiversity occurs and gets maintained. Therefore, they protect native plants and work as gene banks for native species which cascades to preserving regional diversity. Greenspaces also conspicuously contribute to ecosystem functions by supporting pollinators including bees and butterflies which are needed by many plants to reproduce and for the wellbeing of the whole ecosystem. As well as being used as natural laboratories for environmental research and education, greenspaces also help urban residents understand the significance of biodiversity by making them aware of the same.

Geographical data also known as spatial data supports the mapping and marking of green areas, which are the areas for size, location, and distribution within the confines of urban regions. Such information is crucial for detailed knowledge about the scope and availability of green infrastructure and also to identify those parts without a green space that need some attention. Additionally, spatial data provides the ability to analyze open spaces/urban green cover, which in turn helps to evaluate the quality and composition of these areas and their surroundings. Green networks, whose spatial integrity can be perceived through the use of data such as vegetation density, fragmentation, and connectivity, can provide valuable information about the ecological integrity and functional connectivity of the networks. Moreover, spatial data can evaluate ecosystem services that are delivered by greenspace in terms of carbon sequestration, air and water purification, and habitat provision. Through the estimation and mapping of ecosystem services, spatial data can be used as a guide to the long-term planning processes of urban development, conservation, and sustainability. Furthermore, spatial information can assist the monitoring and evaluation processes by making it possible to check the trends of green space dynamics such as urban growth, land use changes, and habitat loss (Chen et al., 2022; Dong et al., 2022; Feltynowski, 2023; C. Huang et al., 2021; Y. Huang et al., 2021; Łaszkiewicz et al., 2020; Texier et al., 2018).

OSM (OpenStreetMap) is free and useful information for vegetation mapping. OSM is an open-source initiative that enables the general public to participate and edit geographical details like where parks, forests, gardens, and other green facilities are located. Individuals can enter information regarding the location, size, and elemental components of greenspaces, including the zones of trails, playgrounds, and picnic areas. This community-sourced approach results in a diverse and living dataset that can be used for many purposes including making green space maps, studying, and planning what to do in the future. Researchers, policymakers, and community organizations can use the OSM's data to examine the distribution and accessibility of greenspaces, trend analysis of the changes in green spaces, and identify regions that may require extra attention for conservation or development. Furthermore, OSM data can be combined with other sources of information such as satellite images and government datasets for building maps even more detailed and comprehensive at scale ranging from local to regional and global (Ahmad, 2024d; Ludwig et al., 2021; Teeuwen et al., 2024).

METHODOLOGY

To address the objectives of this chapter, this section underscored the main study area, data sources, data extraction techniques, and methods employed to analyze the extracted data.

Study Area

The study area for this chapter is the Punjab province (Figure 1), the most densely populated province of Pakistan. Spreading over an area of about 205,344 square kilometers, the entire Punjab is defined by different geographical structures- from cultivated plains to semi-arid areas. There are metropolitan centers of Lahore, Faisalabad, and Rawalpindi in the province and other little towns and villages in the rural areas. Punjab plays a pivotal role in the country's economy, contributing significantly to agriculture, industry, and commerce. Punjab was selected as the study location due to its strategic targeting, high population, and quick urbanization. The research focuses on Punjab in an attempt to provide information, which will illuminate the link between the decline of urban green spaces and biodiversity, particularly in a case of a highly populated and urbanizing region of Pakistan.

Data

OSM is a worldwide project that is designed to gather, store, and share geospatial data for free. OSM data contains information about different geographical elements like roads, buildings, and land use from voluntarily donated data across the globe. OSM has been selected for many reasons, including the overall completeness and accessibility of the data as well as the provision of fine-grained details of urban green spaces and biodiversity features. The data in OSM is also frequently updated and the users can also collect data using the API for analysis.

For this chapter, three keys correlated with land use, nature, and leisure are selected to extract data related to green spaces. These keys are selected because they are the most appropriate in terms of the nature of urban green spaces as well as measuring biodiversity. The details of the keys and values selected for this study are tabulated in Table 1.

Figure 1. Study area

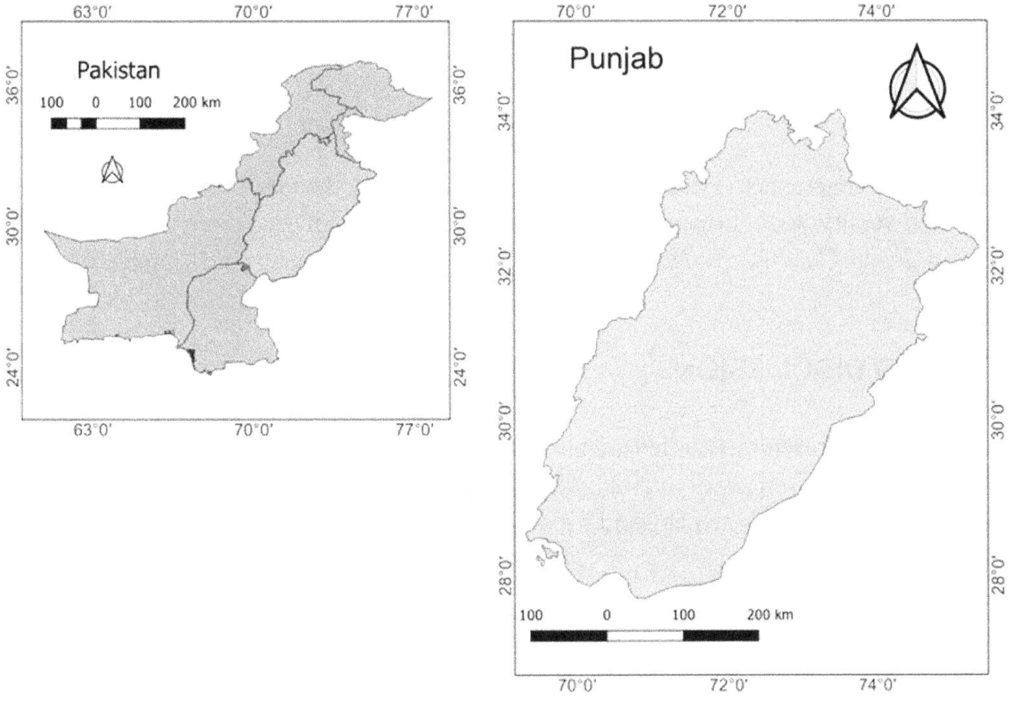

Table 1. Keys and values related to green spaces

Key	Values
Landuse	Grass Forest Recreation Ground
Leisure	Park Garden Golf Course Water Park
Natural	Wood Wetland Grassland

From the Table 1, it can be seen that total nine values selected against three keys. Although there are many other values available to be included, but to make the things simple, only three most relevant values are selected against each key. The OSM data is downloaded through the Overpass API (https://overpass-turbo.eu/) by querying tags related to urban green spaces. The Overpass API is a powerful tool for extracting specific data from the OSM database. It allows users to query OSM data using a specialized query language called Overpass QL. With Overpass API, users can retrieve various types of geographic data, such as points of interest, roads, buildings, and more, based on specific criteria and geographical regions. Green spaces are identified using OSM tags "land use", "natural", and "leisure". The key: value pairs are distinct, and hence they can be detected and extracted through the process. The data has been extracted in JSON format and the "PK-PB" code has been used to extract data for the Punjab Pakistan bounding area. Data from 2019 to 2023 has been extracted through Overpass API.

Methods

The purpose of this chapter is to evaluate green space depletion in the urban environment and its consequences for biodiversity. Assessing urban green space depletion involves temporal analysis of the OSM data. Python in Google Colab (https://colab.research.google.com/)) has been used to conduct analysis work.

RESULTS AND DISCUSSION

The following section presents a detailed analysis of the green spaces in the Punjab Province of Pakistan over the five-year period from 2019 to 2023. The general trends of all three keys are shown in Figure 2, whereas a detailed analysis of each key and its values is presented below.

Leisure Key

The summary statistics of leisure activities (Table 2) reveal notable trends over the five years. The line chart below (Figure 3) illustrates the trend of key leisure activities over the study period. We observe

Figure 2. Trends of land use, natural, and leisure keys over time

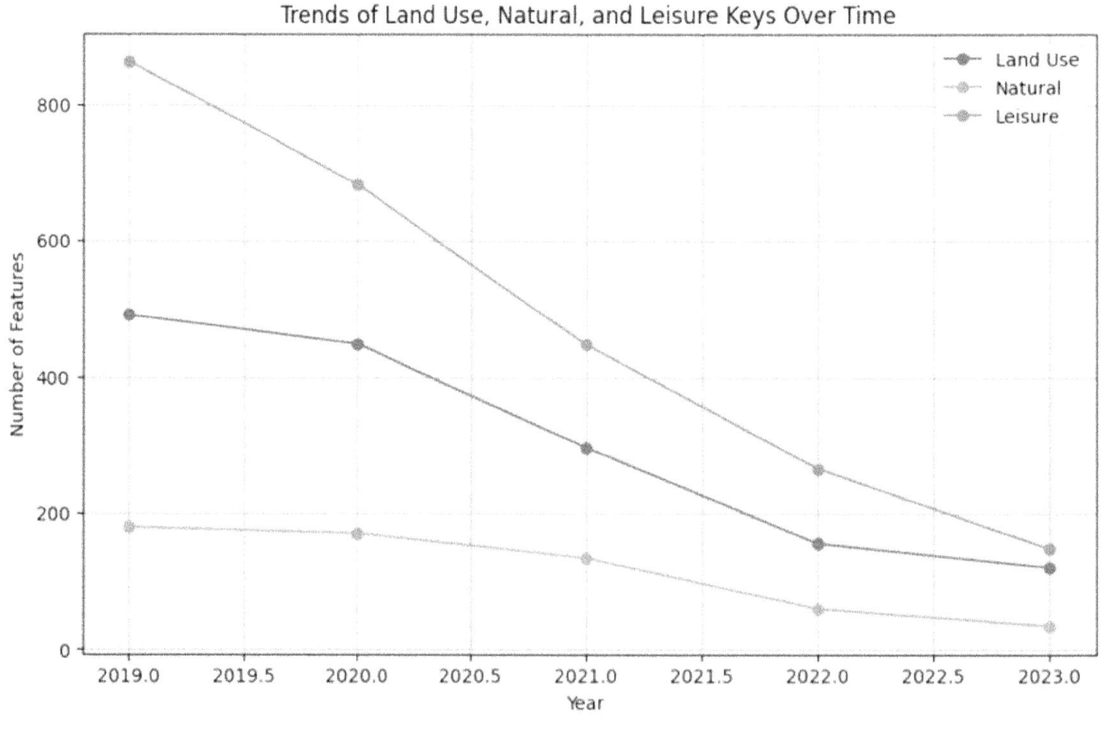

Table 2. Summary statistics of leisure key

Values	2019	2020	2021	2022	2023
Park	833	655	428	250	135
Garden	17	16	12	11	9
Golf Course	11	10	7	4	4
Water Park	2	2	1	1	0
Total	**863**	**683**	**448**	**266**	**148**

a general downward trend in the counts of parks, gardens, golf courses, and water parks from 2019 to 2023. This decline may indicate various factors such as changes in recreational preferences, urban development, or environmental considerations.

Among the leisure activities, parks exhibit the highest count in each year, followed by gardens, golf courses, and water parks. However, there is a significant decrease in the counts of all leisure activities from 2019 to 2023. For instance, the count of parks decreased from 833 in 2019 to 135 in 2023, indicating a substantial reduction in park areas over the years. Similarly, the counts of gardens, golf courses, and water parks also show declining trends.

Natural Key

The summary statistics of natural key (Table 3) provide insights into the distribution and changes in woodlands, wetlands, and grasslands values over the study period. The line chart (Figure 4) below il-

Figure 3. Trends of leisure key over time

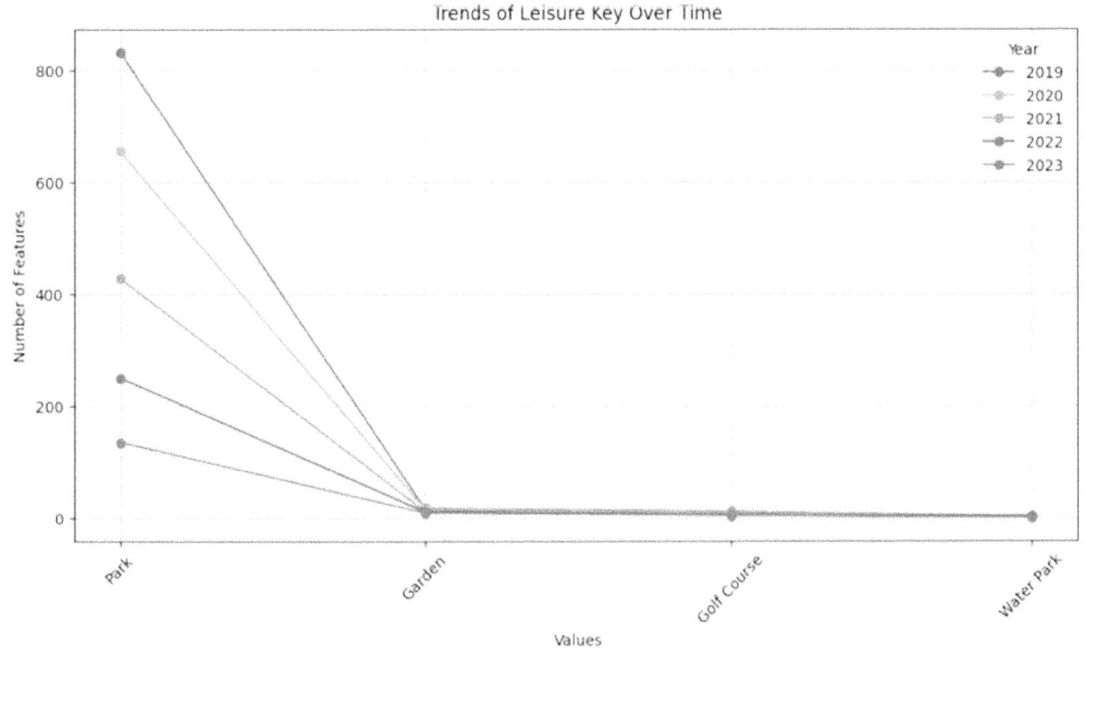

lustrates the trend of key natural features from 2019 to 2023. We observe a general decline in the counts of woodlands and wetlands, while the count of grasslands shows fluctuating patterns.

Woodlands exhibit the highest count among natural features in each year, followed by wetlands and grasslands. However, there is a noticeable decrease in the counts of woodlands and wetlands over the five-year period. For example, the count of woodlands decreased from 102 in 2019 to 28 in 2023, indicating a significant reduction in woodland areas. Similarly, the counts of wetlands and grasslands also show decreasing trends over the years.

Landuse Key

The summary statistics of land use (Table 4) provide insights into the distribution and changes in grasslands, forests, and recreation grounds over the study period. The line chart (Figure 5) below illustrates the trend of key land use categories from 2019 to 2023. We observe a general decline in the counts of grasslands, forests, and recreation grounds over the five years.

Table 3. Summary statistics of natural key

Values	2019	2020	2021	2022	2023
Wood	102	101	87	48	28
Wetland	50	45	40	7	5
Grassland	28	25	8	5	22
Total	**180**	**171**	**135**	**60**	**35**

Figure 4. Trends of natural key over time

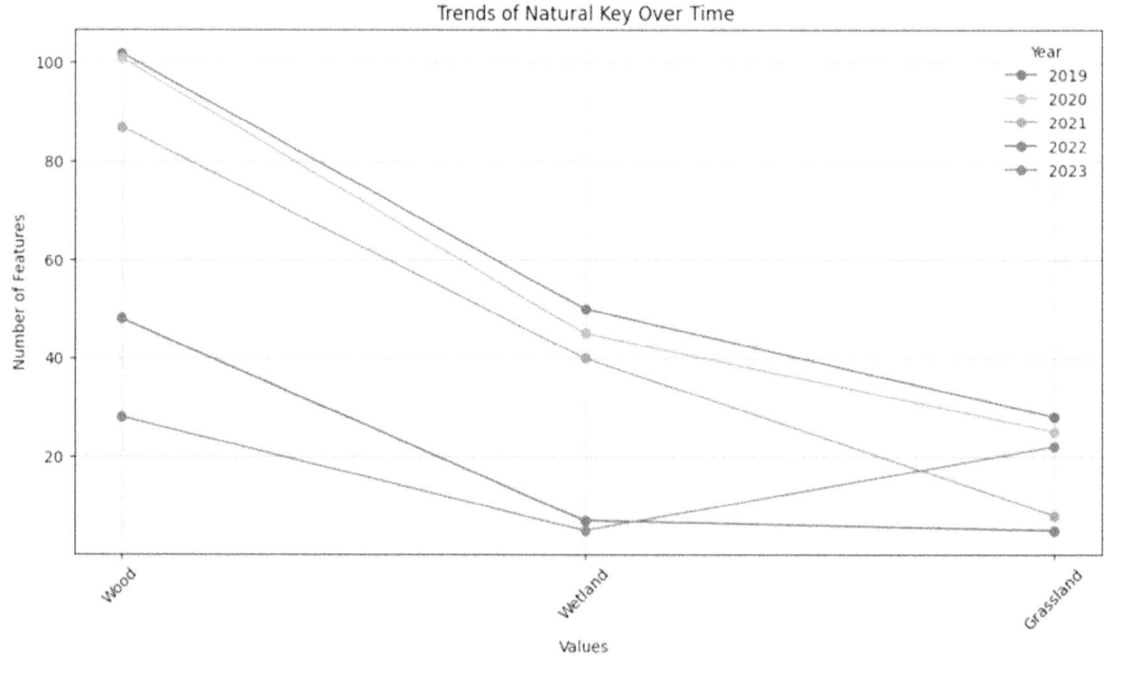

Grasslands exhibit the highest count among land use categories each year, followed by forests and recreation grounds. However, there is a significant decrease in the counts of all land use categories from 2019 to 2023. For instance, the count of grasslands decreased from 290 in 2019 to 60 in 2023, indicating a substantial reduction in grassland areas. Similarly, the counts of forests and recreation grounds also show declining trends over the years.

The summary statistics and trend analysis provide valuable insights into the changes and trends in leisure activities, natural features, and land use over the five-year period. These findings are essential for urban planning, environmental management, and decision-making processes to ensure sustainable development and conservation efforts.

Table 4. Summary statistics of Landuse key

Values	2019	2020	2021	2022	2023
Grass	290	251	168	80	60
Forest	181	181	117	68	59
Recreation Ground	20	17	12	8	2
Total	**491**	**449**	**297**	**156**	**121**

Figure 5. Trends of Landuse key over time

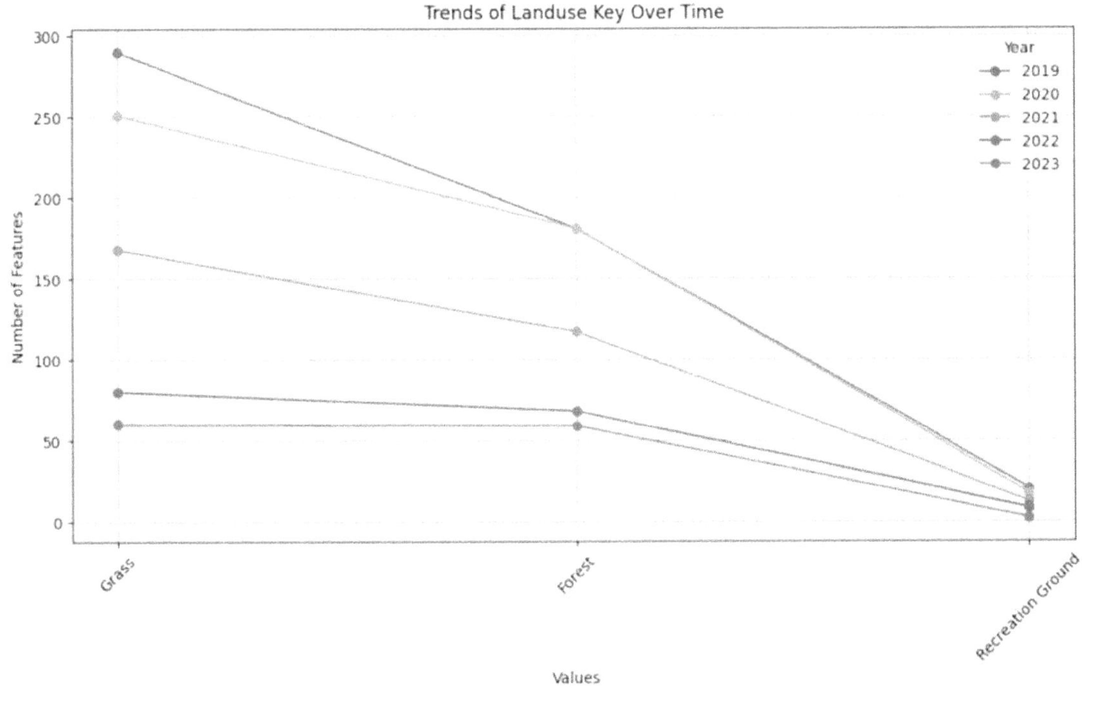

IMPLICATIONS AND RECOMMENDATIONS

Implications for Biodiversity Conservation

The findings from the analysis of leisure activities, natural features, and land use have significant implications for biodiversity conservation that are elaborated below.

Loss of Habitat Diversity

Wong et al. (2023) argued urban planners and landscape architects should aim to create green areas with regular shapes to reduce edge effects. Additionally, they should prioritize preserving semi-natural vegetation and incorporating freshwater bodies to bolster and maintain bird diversity within cities. Similarly, Park et al. (2020) asserted future design and planning of green spaces should take into account the characteristics of species' original habitats. They also argued that creating green spaces that mimic agricultural land, forests, and wetlands can help preserve the integrity of former habitats, thereby promoting biodiversity and ecological function. Forgoing in view, the downward trajectories ranged from forests to grasslands, woodlands, wetlands, and recreation areas, and suggest trends in loss of habitat diversity in the study areas. The decline of these natural areas could be accompanied by the loss of biodiversity as water bodies, forests, and grasslands provide habitats to a great number of plants and animals.

Fragmentation of Ecosystems

Habitat loss and fragmentation cause substantial alterations to landscapes, which in turn impact plant diversity and introduce uncertainty regarding the response of natural areas to future global changes. This uncertainty holds significant implications, especially considering that biodiversity loss often entails the elimination of crucial ecosystem functions (Zambrano et al., 2019). There decrease in parklands and recreational grounds will lead to the formation of ecological islands that are separated, which will eventually decrease connectivity between natural areas. Fragmentation results in a number of disadvantages including limiting species' capacity to migrate, regulating the germination as well as ensuring sustained populations.

Impact on Wildlife

Loss of natural habitats, like forests and wetlands, will likely bring bad consequences for animal species that depend on these areas for food, shelter, and breeding sites. Extinction of species can disturb ecosystem functions and cause the system to become less robust to environmental disturbances. For example, the study conducted by X. Yang et al. (2020) in Nanjing, China revealed that park area, habitat diversity, and proximity to the city center emerged as significant positive factors influencing bird species richness within Nanjing's urban parks.

Threats to Ecosystem Services

The ecosystem services, that natural habitats offer, in addition to filtering water and neutralizing carbon dioxide, help in pollination. Declines in these habitats can compromise the provision of ecosystem services, affecting human well-being and livelihoods. For example, Wiederkehr et al. (2020) observed that urbanization exerted a 32% greater influence on function compared to structure, primarily due to heightened algal proliferation in urban streams. These alterations in ecological processes, specifically ecosystem functioning, likely result in a decline of ecosystem services, with detrimental implications for ecology, society, and the economy.

Potential for Species Extinctions

Natural habitat loss and corrosion will trigger more and more species' extinctions, especially in the context of generalized and endemic ones. This leads to various consequences on the internal dynamics and adaptability of the ecosystem due to the loss of biodiversity. Similarly, Kehoe et al. (2021) examined the potential for cascading extinctions resulting from habitat loss, pollution, species invasions, climate change, and overexploitation. The authors suggested that habitat loss and pollution, in particular, pose the greatest risk for such extinctions due to their ability to alter community structure, the physical environment, and community robustness.

Recommendations for Policy Makers

To address the challenges posed by declining biodiversity and habitat loss, policymakers can consider the following recommendations:

Strengthen Habitat Protection

Implement and enforce regulations to protect and restore natural habitats, including woodlands, wetlands, grasslands, and forests. Create protected areas, conservation corridors, and green areas for the protection of biodiversity and to foster a network of viable connections between habitats.

Promote Sustainable Land Use Planning

Incorporate biodiversity preservation into the implementation of land use policies, including development. Give preference to the preservation of living spaces and ecological corridors, at the same time encouraging sustainable techniques for managing the land to reduce the amount and extent of habitat loss as a consequence of fragmentation and degradation.

Invest in Restoration and Reforestation

Support initiatives for habitat restoration and reforestation to enhance ecosystem resilience and biodiversity recovery. To develop a sense of ownership and responsibility towards our natural resources, involve the community in restoration projects, and encourage their participation and engagement.

Enhance Urban Green Infrastructure

Incorporation of green infrastructure, comprised of parks, green roofs, and urban forests, into urban planning and design is an effective strategy. Develop urban greenery, ecological networks, and wildlife habitats in cities, keeping in mind air pollution control and people's sensorial wellness.

Monitor and Assess Biodiversity

Establish monitoring programs that measure variations in biodiversity and habitat health in the long run. Use scientific research and data-driven approaches to assess the effectiveness of conservation measures and adapt policies accordingly.

Promote Public Awareness and Education

Raise awareness about the importance of biodiversity conservation and the role of individuals, communities, and businesses in preserving natural habitats. Provide educational programs and outreach initiatives to foster environmental stewardship and sustainable behavior.

With these measures in place, policymakers will definitely be proportional in the conservation and restoration of biodiversity as they will allow for the continuous development of ecosystems that will be available for future generations. The partnership among the officials of different government agencies, concerned stakeholders, and the public is necessary to tackle the innumerable and complex issues of biodiversity conservation.

CONCLUSION

In this study, we conducted a comprehensive analysis of green spaces in the Punjab Province of Pakistan from 2019 to 2023. OSM data has been used to conduct analysis on the green space trends using the three tags such as leisure, natural, and land use. The findings reveal significant trends and implications for biodiversity conservation and urban planning.

Firstly, our analysis identified a consistent decline in green spaces across all key categories over the five-year period. Leisure activities, including parks, gardens, golf courses, and water parks, experienced a notable reduction in counts, indicating potential challenges in recreational access and urban greenery. Similarly, natural features such as woodlands, wetlands, and grasslands showed declining trends, raising concerns about habitat loss and biodiversity depletion. Land use categories, including grasslands, forests, and recreation grounds, also exhibited decreasing counts, highlighting the ongoing pressure on natural landscapes due to urbanization and development.

The observed trends have significant implications for biodiversity conservation. Loss of habitat diversity, fragmentation of ecosystems, impact on wildlife, threats to ecosystem services, and potential species extinctions are among the key concerns arising from the decline in green spaces. These findings underscore the urgent need for proactive measures to address habitat loss, promote sustainable land use practices, and enhance biodiversity conservation efforts in the region.

In light of these implications, we offer several recommendations for policymakers to consider. Strengthening habitat protection, promoting sustainable land use planning, investing in restoration and reforestation, enhancing urban green infrastructure, monitoring and assessing biodiversity, and promoting public awareness and education are essential strategies to mitigate the negative impacts of green space depletion and safeguard biodiversity for future generations.

By implementing these recommendations, policymakers can contribute to the preservation and restoration of green spaces, fostering resilient ecosystems and healthy communities. Collaboration among government agencies, stakeholders, and the public is crucial to addressing the complex challenges of biodiversity conservation and urban development effectively.

REFERENCES

Ahmad, M. (2023a). Connecting People and Places: How Citizen Diplomacy and VGI Are Strengthening Disaster Response and Community Development. In S. Kavoğlu & E. Köksoy (Eds.), *Global Perspectives on the Emerging Trends in Public Diplomacy* (pp. 195–226). IGI Global. doi:10.4018/978-1-6684-9161-4.ch009

Ahmad, M. (2023b). Exploring the Role of OpenStreetMap in Mapping Religious Tourism in Pakistan for Sustainable Development. In V. João Pereira Domingues Martinho, J. Augusto Guerra da Rocha Nunes, M. Lúcia de Jesus Pato, & L. Andrade de Matos Castilho (Eds.), *Experiences, Advantages, and Economic Dimensions of Pilgrimage Routes* (pp. 23–40). IGI Global. doi:10.4018/978-1-6684-9923-8.ch002

Ahmad, M. (2023c). Leveraging Social Media Geographic Information for Smart Governance and Policy Making: Opportunities and Challenges. In C. Chavadi & D. Thangam (Eds.), *Global Perspectives on Social Media Usage Within Governments* (pp. 192–213). IGI Global. doi:10.4018/978-1-6684-7450-1.ch013

Ahmad, M. (2024a). Connecting the Dots: Harnessing OpenStreetMap for Big Data Analytics and Market Insights. In D. Darwish (Ed.), *Big Data Analytics Techniques for Market Intelligence* (pp. 329–347). IGI Global. doi:10.4018/979-8-3693-0413-6.ch013

Ahmad, M. (2024b). From Crisis to Opportunity: How VGI Is Transforming Disaster Response and Recovery Efforts. In S. Siyal (Ed.), *Using Crises and Disasters as Opportunities for Innovation and Improvement* (pp. 215–232). IGI Global., doi:10.4018/978-1-6684-9522-3.ch013

Ahmad, M. (2024c). The Role of Data Science and Volunteered Geographic Information in Enhancing Government Service Delivery: Opportunities and Challenges. In O. S. Ogunleye (Ed.), *Machine Learning and Data Science Techniques for Effective Government Service Delivery* (pp. 254–281). IGI Global.

Ahmad, M. (2024d). Unleashing Business Potential: Harnessing OpenStreetMap for Intelligent Growth and Sustainability. In S. Singh, S. S. Rajest, S. Hadoussa, A. J. Obaid, & R. Regin (Eds.), *Data-Driven Intelligent Business Sustainability* (pp. 177–198). IGI Global. doi:10.4018/979-8-3693-0049-7.ch013

Ahmad, M., & Ali, A. (2023). Mapping the Future of Sustainable Development Through Cloud-Based Solutions: A Case Study of OpenStreetMap. In Promoting Sustainable Management Through Technological Innovation (pp. 153–176). IGI Global. doi:10.4018/978-1-6684-9979-5.ch011

Ahmad, M., Khayal, M. S. H., & Tahir, A. (2022). Analysis of Factors Affecting Adoption of Volunteered Geographic Information in the Context of National Spatial Data Infrastructure. *ISPRS International Journal of Geo-Information*, *11*(2), 120. doi:10.3390/ijgi11020120

Ahmad, M., & Khiyal, M. S. H. (2023). Assessment of land administration in Pakistan and the potential role of volunteered geographic information. In Handbook of Research on Driving Socioeconomic Development With Big Data. IGI Global. doi:10.4018/978-1-6684-5959-1.ch014

Anderson, N. (1959). Urbanism and Urbanization. *American Journal of Sociology*, *65*(1), 68–73. doi:10.1086/222628

Antrop, M. (2004). Landscape change and the urbanization process in Europe. *Landscape and Urban Planning*, *67*(1–4), 9–26. doi:10.1016/S0169-2046(03)00026-4

Calcagni, F., Amorim Maia, A. T., Connolly, J. J. T., & Langemeyer, J. (2019). Digital co-construction of relational values: Understanding the role of social media for sustainability. *Sustainability Science*, *14*(5), 1309–1321. doi:10.1007/s11625-019-00672-1

Chen, J., Wang, C., Zhang, Y., & Li, D. (2022). Measuring spatial accessibility and supply-demand deviation of urban green space: A mobile phone signaling data perspective. *Frontiers in Public Health*, *10*, 1029551. doi:10.3389/fpubh.2022.1029551 PMID:36339177

Coolen, H., & Meesters, J. (2012). Private and public green spaces: Meaningful but different settings. *Journal of Housing and the Built Environment*, *27*(1), 49–67. doi:10.1007/s10901-011-9246-5

Dallimer, M., Tang, Z., Bibby, P. R., Brindley, P., Gaston, K. J., & Davies, Z. G. (2011). Temporal changes in greenspace in a highly urbanized region. *Biology Letters*, *7*(5), 763–766. doi:10.1098/rsbl.2011.0025 PMID:21429910

Deng, X., Huang, J., Rozelle, S., Zhang, J., & Li, Z. (2015). Impact of urbanization on cultivated land changes in China. *Land Use Policy*, *45*, 1–7. doi:10.1016/j.landusepol.2015.01.007

Dociu, M. (2012). The Socio-Economic Impact of Urbanization. *International Journal of Academic Research in Accounting, Finance and Management Sciences, 2*(1).

Dong, Q., Cai, J., Chen, S., He, P., & Chen, X. (2022). Spatiotemporal Analysis of Urban Green Spatial Vitality and the Corresponding Influencing Factors: A Case Study of Chengdu, China. *Land (Basel)*, *11*(10), 1820. doi:10.3390/land11101820

Eckert, S., & Kohler, S. (2014). Urbanization and health in developing countries: a systematic review. In World health & population, 15(1). doi:10.12927/whp.2014.23722

Feltynowski, M. (2023). Urban green spaces in land-use policy – types of data, sources of data and staff – the case of Poland. *Land Use Policy*, *127*, 106570. doi:10.1016/j.landusepol.2023.106570

Fox, S. (2012). Urbanization as a global historical process: Theory and evidence from sub-saharan africa. *Population and Development Review*, *38*(2), 285–310. doi:10.1111/j.1728-4457.2012.00493.x

Friedmann, J. (1966). Two Concepts of urbanization: A Comment. *Urban Affairs Quarterly*, *1*(4), 78–84. doi:10.1177/107808746600100406

Fuller, R. A., Irvine, K. N., Devine-Wright, P., Warren, P. H., & Gaston, K. J. (2007). Psychological benefits of greenspace increase with biodiversity. *Biology Letters*, *3*(4), 390–394. doi:10.1098/rsbl.2007.0149 PMID:17504734

Girma, Y., Terefe, H., Pauleit, S., & Kindu, M. (2019). Urban green spaces supply in rapidly urbanizing countries: The case of Sebeta Town, Ethiopia. *Remote Sensing Applications: Society and Environment*, *13*, 138–149. doi:10.1016/j.rsase.2018.10.019

Godfrey, R., & Julien, M. (2005). Urbanisation and health. In Clinical Medicine, Journal of the Royal College of Physicians of London, 5(2). doi:10.7861/clinmedicine.5-2-137

Goodchild, M. F. (2007). Citizens as sensors: The world of volunteered geography. In GeoJournal. doi:10.1007/s10708-007-9111-y

Grinberger, A. Y., Minghini, M., Juhász, L., Yeboah, G., & Mooney, P. (2022). OSM Science—The Academic Study of the OpenStreetMap Project, Data, Contributors, Community, and Applications. In ISPRS International Journal of Geo-Information, 11(4). doi:10.3390/ijgi11040230

Haklay, M. (2013). Citizen science and volunteered geographic information: Overview and typology of participation. In Crowdsourcing Geographic Knowledge: Volunteered Geographic Information (VGI) in Theory and Practice. Springer. doi:10.1007/978-94-007-4587-2_7

Heikinheimo, V., Tenkanen, H., Bergroth, C., Järv, O., Hiippala, T., & Toivonen, T. (2020). Understanding the use of urban green spaces from user-generated geographic information. *Landscape and Urban Planning*, *201*, 103845. doi:10.1016/j.landurbplan.2020.103845

Honarparvar, S., Forouzandeh Jonaghani, R., Alesheikh, A. A., & Atazadeh, B. (2019). Improvement of a location-aware recommender system using volunteered geographic information. *Geocarto International*, *34*(13), 1496–1513. doi:10.1080/10106049.2018.1493155

Hoyle, H., Jorgensen, A., & Hitchmough, J. D. (2019). What determines how we see nature? Perceptions of naturalness in designed urban green spaces. *People and Nature*, *1*(2), 167–180. doi:10.1002/pan3.19

Huang, C., Yang, J., Clinton, N., Yu, L., Huang, H., Dronova, I., & Jin, J. (2021). Mapping the maximum extents of urban green spaces in 1039 cities using dense satellite images. *Environmental Research Letters*, *16*(6), 064072. doi:10.1088/1748-9326/ac03dc

Huang, H., Yao, X. A., Krisp, J. M., & Jiang, B. (2021). Analytics of location-based big data for smart cities: Opportunities, challenges, and future directions. *Computers, Environment and Urban Systems*, *90*, 101712. doi:10.1016/j.compenvurbsys.2021.101712

Huang, Y., Lin, T., Zhang, G., Zhu, Y., Zeng, Z., & Ye, H. (2021). Spatial patterns of urban green space and its actual utilization status in China based on big data analysis. *Big Earth Data*, *5*(3), 391–409. doi:10.1080/20964471.2021.1939990

Jabbar, M., Yusoff, M. M., & Shafie, A. (2022). Assessing the role of urban green spaces for human well-being: a systematic review. In GeoJournal, 87(5). doi:10.1007/s10708-021-10474-7

Jungalwalla, N. (1968). Urbanization and health. *Israel Journal of Medical Sciences*, *4*(3). doi:10.4038/jccpsl.v14i1.2941 PMID:5698240

Kehoe, R., Frago, E., & Sanders, D. (2021). Cascading extinctions as a hidden driver of insect decline. In Ecological Entomology, 46(4). doi:10.1111/een.12985

Kitchin, R., Lauriault, T. P., & Wilson, M. W. (2018). Understanding Spatial Media. In Understanding Spatial Media. Sage. doi:10.4135/9781526425850.n1

Kothencz, G., Kolcsár, R., Cabrera-Barona, P., & Szilassi, P. (2017). Urban green space perception and its contribution to well-being. *International Journal of Environmental Research and Public Health, 14*(7), 766. doi:10.3390/ijerph14070766 PMID:28704969

Lachowycz, K., & Jones, A. P. (2011). Greenspace and obesity: A systematic review of the evidence. *Obesity Reviews, 12*(5). doi:10.1111/j.1467-789X.2010.00827.x PMID:21348919

Łaszkiewicz, E., Czembrowski, P., & Kronenberg, J. (2020). Creating a map of the social functions of urban green spaces in a city with poor availability of spatial data: A sociotope for lodz. *Land (Basel), 9*(6), 183. Advance online publication. doi:10.3390/land9060183

Le Texier, M., Schiel, K., & Caruso, G. (2018). The provision of urban green space and its accessibility: Spatial data effects in Brussels. *PLoS One, 13*(10), e0204684. doi:10.1371/journal.pone.0204684 PMID:30332449

Lee, A. C. K., & Maheswaran, R. (2011). The health benefits of urban green spaces: A review of the evidence. In Journal of Public Health (Vol. 33, Issue 2). doi:10.1093/pubmed/fdq068

Li, F., Li, M., & Feng, X. (2022). Ecological Benefit Analysis of Urban Green Space Based on Three-Dimensional Green Quantity. *Photonirvachak (Dehra Dun), 50*(11), 2267–2276. doi:10.1007/s12524-022-01603-x

Ludwig, C., Hecht, R., Lautenbach, S., Schorcht, M., & Zipf, A. (2021). Mapping Public Urban Green Spaces Based on OpenStreetMap and Sentinel-2 Imagery Using Belief Functions. *ISPRS International Journal of Geo-Information, 10*(4), 251. Advance online publication. doi:10.3390/ijgi10040251

Madureira, H., Nunes, F., Oliveira, J. V., Cormier, L., & Madureira, T. (2015). Urban residents' beliefs concerning green space benefits in four cities in France and Portugal. *Urban Forestry & Urban Greening, 14*(1), 56–64. Advance online publication. doi:10.1016/j.ufug.2014.11.008

Nguyen, H. M., & Nguyen, L. D. (2018). The relationship between urbanization and economic growth an empirical study on ASEAN countries. *International Journal of Social Economics, 45*(2), 316–339. Advance online publication. doi:10.1108/IJSE-12-2016-0358

Nor, A. N. M., Aziz, H. A., Nawawi, S. A., Jamil, R. M., Abas, M. A., Hambali, K. A., Yusoff, A. H., Ibrahim, N., Rafaai, N. H., Corstanje, R., Harris, J., Grafius, D., & Perotto-Baldivieso, H. L. (2021). Evolution of green space under rapid urban expansion in southeast Asian cities. *Sustainability (Basel), 13*(21), 12024. doi:10.3390/su132112024

Nor, A. N. M., Corstanje, R., Harris, J. A., Grafius, D. R., & Siriwardena, G. M. (2017). Ecological connectivity networks in rapidly expanding cities. *Heliyon, 3*(6), e00325. doi:10.1016/j.heliyon.2017.e00325 PMID:28706999

Novack, T., Wang, Z., & Zipf, A. (2018). A system for generating customized pleasant pedestrian routes based on openstreetmap data. *Sensors (Basel), 18*(11), 3794. doi:10.3390/s18113794 PMID:30404175

Park, Y., Kim, C. J., Choi, M. B., & Do, Y. (2020). A comparative study of carabid beetles in green spaces and former natural habitats. *Diversity*, *12*(12), 479. doi:10.3390/d12120479

Ren, G. Y. (2015). Urbanization as a major driver of urban climate change. In Advances in Climate Change Research, 6(1). doi:10.1016/j.accre.2015.08.003

Teeuwen, R., Milias, V., Bozzon, A., & Psyllidis, A. (2024). How well do NDVI and OpenStreetMap data capture people's visual perceptions of urban greenspace? *Landscape and Urban Planning*, *245*, 105009. doi:10.1016/j.landurbplan.2024.105009

Tisdale, H. (1942). The process of urbanization. *Social Forces*, *20*(3), 311. doi:10.2307/3005615

Twohig-Bennett, C., & Jones, A. (2018). The health benefits of the great outdoors: A systematic review and meta-analysis of greenspace exposure and health outcomes. *Environmental Research*, *166*, 628–637. doi:10.1016/j.envres.2018.06.030 PMID:29982151

Ugolini, F., Massetti, L., Calaza-Martínez, P., Cariñanos, P., Dobbs, C., Krajter Ostoić, S., Marin, A. M., Pearlmutter, D., Saaroni, H., Šaulienė, I., Vuletić, D., & Sanesi, G. (2022). Understanding the benefits of public urban green space: How do perceptions vary between professionals and users? *Landscape and Urban Planning*, *228*, 104575. doi:10.1016/j.landurbplan.2022.104575

Weber, P., & Haklay, M. (2008). OpenStreetMap: User-generated street maps. *IEEE Pervasive Computing*, *7*(4).

Wiederkehr, F., Wilkinson, C. L., Zeng, Y., Yeo, D. C. J., Ewers, R. M., & O'Gorman, E. J. (2020). Urbanisation affects ecosystem functioning more than structure in tropical streams. *Biological Conservation*, *249*, 108634. doi:10.1016/j.biocon.2020.108634

Wolch, J. R., Byrne, J., & Newell, J. P. (2014). Urban green space, public health, and environmental justice: The challenge of making cities "just green enough.". *Landscape and Urban Planning*, *125*, 234–244. doi:10.1016/j.landurbplan.2014.01.017

Wong, J. S. Y., Soh, M. C. K., Low, B. W., & Er, K. B. H. (2023). Tropical bird communities benefit from regular-shaped and naturalised urban green spaces with water bodies. *Landscape and Urban Planning*, *231*, 104644. doi:10.1016/j.landurbplan.2022.104644

Wood, E., Harsant, A., Dallimer, M., de Chavez, A. C., McEachan, R. R. C., & Hassall, C. (2018). Not all green space is created equal: Biodiversity predicts psychological restorative benefits from urban green space. *Frontiers in Psychology*, *9*(NOV), 2320. Advance online publication. doi:10.3389/fpsyg.2018.02320 PMID:30538653

Yang, B. Y., Zhao, T., Hu, L. X., Browning, M. H. E. M., Heinrich, J., Dharmage, S. C., Jalaludin, B., Knibbs, L. D., Liu, X. X., Luo, Y. N., James, P., Li, S., Huang, W. Z., Chen, G., Zeng, X. W., Hu, L. W., Yu, Y., & Dong, G. H. (2021). Greenspace and human health: An umbrella review. In Innovation, 2(4). doi:10.1016/j.xinn.2021.100164

Yang, X., Tan, X., Chen, C., & Wang, Y. (2020). The influence of urban park characteristics on bird diversity in Nanjing, China. *Avian Research*, *11*(1), 45. doi:10.1186/s40657-020-00234-5

Zambrano, J., Garzon-Lopez, C. X., Yeager, L., Fortunel, C., Cordeiro, N. J., & Beckman, N. G. (2019). The effects of habitat loss and fragmentation on plant functional traits and functional diversity: what do we know so far? In Oecologia, 191(3). doi:10.1007/s00442-019-04505-x

Zhang, W. J. (2008). A forecast analysis on world population and urbanization process. *Environment, Development and Sustainability*, *10*(6), 717–730. doi:10.1007/s10668-007-9081-8

Zhao, L. (2023). The Economic and Environmental Impacts of Urbanization in China. *Highlights in Business. Economics and Management*, *5*, 142–146. doi:10.54097/hbem.v5i.5041

Zhou, X., & Rana, M. M. P. (2012). Social benefits of urban green space: A conceptual framework of valuation and accessibility measurements. In Management of Environmental Quality, 23(2). doi:10.1108/14777831211204921

Chapter 10
Need for Biodiversity Conservation and Conservation Strategies

Alper Durmaz
https://orcid.org/0000-0001-6927-3283
Ali Nihat Gokyigit Botanical Garden Application and Research Center, Turkey

Emine İncilay Torunoğlu
https://orcid.org/0000-0003-4641-0067
Department of Medical Biochemistry, Faculty of Medicine, Necmettin Erbakan University, Turkey

Betül Aydın
https://orcid.org/0000-0002-9092-1350
Department of Molecular Biology and Genetics, Faculty of Science, Gazi University, Turkey

Erdi Can Aytar
https://orcid.org/0000-0001-6045-0183
Department of Horticulture, Faculty of Agriculture, Usak University, Turkey

ABSTRACT

Biodiversity conservation is imperative for the well-being of our planet, impacting ecosystem stability, productivity, and resilience against environmental stressors. This chapter emphasizes the significance of biodiversity in maintaining ecological balance and outlines key conservation strategies. Pollution and climate change pose significant threats to biodiversity, leading to disruptions in ecosystems and loss of species. Moreover, climate change-induced shifts in species abundance and distribution necessitate holistic conservation approaches. Plant ecology plays a pivotal role in conservation by studying species interactions and ecosystem dynamics. Conservation strategies must consider ecological tolerances and promote sustainable resource use in agriculture, forestry, and ecosystem restoration. Protected areas serve as vital biodiversity reserves, offering essential ecosystem services and supporting cultural and aesthetic values. Overall, effective conservation strategies are essential for preserving biodiversity and ensuring the long-term sustainability of our planet.

DOI: 10.4018/979-8-3693-3330-3.ch010

INTRODUCTION

Biodiversity stands out as a critical factor for our planet's well-being. The enrichment of our environment comes with an increase in the variety of species and genetic diversity. When the diversity of plant species per unit area rises, it creates more intricate environments, ensuring stability in the ecosystem and providing numerous benefits for both the environment and humans. Consequently, it is a well-established fact that plant communities in areas with high biodiversity exhibit enhanced stability and productivity. Different activities involving various animals, like grazing and mowing, have distinct effects on overall species richness or specific communities and species within a community. Plant communities exhibit changes based on their environmental preferences. Therefore, employing indicator values, as considered by Ellenberg, proves highly useful in interpreting the distribution of local flora and patterns of endemism.

In essence, to make accurate observations about the biodiversity of living groups in an area, especially in the field of Ecology and ecological indicators, it is crucial to understand how these observations should be interpreted. Plant ecology, as a scientific discipline, explores the interactions between plants and their environments. Research in this field aims to grasp the adaptations of plants to their growth environments, their roles and interactions within ecosystems, relationships between plants, and how plants respond to environmental factors. Plant ecology encompasses topics such as biological diversity, ecosystem dynamics, soil plants, and plant-community interactions.

Plant ecologists study ecosystems and various habitats to comprehend the factors influencing plant growth, including conditions for seed germination, growth, flowering, and reproduction. Their research contributes to the balanced functioning of ecosystems by examining the roles of plant species in nutrient and water cycles, as well as their interactions with other organisms.

Pollution and climate change are posing significant threats to biodiversity, resulting in substantial losses across various ecosystems. The presence of pollutants from sources like industrial activities and agricultural runoff introduces harmful substances into ecosystems, negatively impacting plant and animal life. Additionally, the changing climate adds further stress, altering habitats and disrupting the delicate balance of ecosystems.

The combined impact of pollution and climate change is evident in the loss of biodiversity, as species struggle to adapt to rapidly changing environmental conditions. This disruption affects the intricate relationships within ecosystems, causing a ripple effect across different trophic levels. Addressing these dual threats comprehensively is crucial to developing effective conservation strategies and mitigating the ongoing decline in biodiversity.

Wastes generated by human activities pose a significant threat to the proper functioning of soil, representing the majority of soil contaminants, especially in areas heavily impacted by agriculture and industry. Soil pollution in these regions can disturb the balance among plants, soil, and microorganisms, affecting soil composition, function, microbial activity, and physiological processes. This disturbance can have adverse effects on biological diversity and the overall functioning of ecosystems since the relationship between biological diversity and ecosystem order becomes more vulnerable due to environmental pollution and stress gradients.

It is crucial to understand how pollution gradients affect the richness and abundance of species, the evolutionary diversity within a community, and the variety of forms and functions present. Additionally, investigating the impact of these gradients on an ecosystem's ability to perform multiple functions, known as ecosystem multifunctionality, is essential. Despite ongoing efforts to combat climate change, the Earth has warmed by more than 1°C since the 1850–1900 period, and this warming trend is expected to continue until at least 2050. Achieving the necessary reduction in greenhouse gas emissions to limit global warming requires comprehensive strategies. Biodiversity, in all its aspects, must confront and adapt to the challenges posed by climate change. In response to climate change, species, communities, and ecosystems are undergoing shifts in abundance and distribution. There is an increase in species adapted to warmer conditions and a decline in those adapted to colder environments. Many species are expanding their ranges toward the poles in terrestrial, freshwater, and marine environments, while cold-adapted species experience contractions. Changes in phenology, such as earlier spring events and later autumn events, can lead to mismatches between dependent species, directly impacting ecosystem functions and services.

Addressing pollution and climate change challenges, especially in the context of biodiversity loss, requires a holistic ecological perspective. Studying the ecological tolerances of species within biodiversity and understanding how pollution and climate change parameters affect them can help determine the timing and extent of potential losses. Implementing effective conservation strategies and minimizing losses in biodiversity reserve areas will only be achievable through a comprehensive evaluation of all factors involved from a broad perspective.

Plant ecology plays a crucial role in both fundamental and applied scientific fields. For example, the principles of plant ecology aid in developing strategies for sustainable natural resource use and promoting ecosystem health in areas such as agriculture, forestry, natural area management, and ecosystem restoration. Efficiently applying knowledge is vital for promoting biodiversity across diverse areas. Particularly in habitats with well-known ecological requirements, conservation and utilization scenarios can be simulated to closely resemble natural conditions. These areas offer various benefits, providing essential services like primary production, provisioning services (food, freshwater, fiber, and bioenergy), and serving as habitats for animals and plants. They also contribute to climate regulation, carbon sequestration, erosion control, water regulation, air quality regulation, soil formation, pest control, and waste treatment.

Natural habitats and specific microclimatic zones offer numerous advantages, serving not only ecological functions but also cultural and aesthetic purposes. Many countries recognize the economic contributions of biodiversity and encourage activities like horse riding, bird watching, agriculture, and ecotourism in biodiverse areas. From an ecological standpoint, understanding the interactions between biodiversity and ecosystem function is crucial. The number of plant species is a straightforward and long-utilized parameter in assessing ecosystems. Changes in biodiversity often accompany shifts in the functional composition of vegetation, influencing numerous ecosystem functions and services. In conclusion, ecosystems with natural and undisturbed stability hold significant educational and scientific value due to their unique biological diversity, housing many rare plant, animal, and insect species.

Because of that, the conservation of biodiversity is crucial. Conservation of biodiversity especially contributes to the stability of ecosystems, provides economic value, encompasses medicinal resources, supports genetic diversity, holds cultural and aesthetic significance, contributes to climate regulation, enhances scientific understanding, and involves ethical responsibilities. Implementing effective conservation strategies is vital to ensure the continued benefits of biodiversity and bolster resilience against diverse threats. This is essential for future generations' well-being and our planet's overall sustainability.

UNDERSTANDING THE IMPORTANCE OF BIODIVERSITY CONSERVATION

The terrestrial expanse of our planet, comprising roughly three-quarters of its landmass, alongside two-thirds of its oceans, confronts unparalleled threats to biodiversity and ecological integrity. Despite concerted efforts through a succession of international agreements aimed at reversing these trends, neither the perils of extinction nor the underlying stressors such as climate change have shown signs of abatement, let alone reversal. With the impending rollout of the global biodiversity framework post-2020, there emerges a singular opportunity to reevaluate our stewardship of this planet and urgently reassess our priorities for the future. This necessitates meticulous consideration of where our future endeavours should be directed and what additional data, we require to manage these systems effectively (Zhang et al. 2023). While ecosystems rich in diversity like tropical forests and coral reefs have garnered significant attention, other ecosystems, despite their significance in terms of unique species and human cultures such as medicinal, cultural, food, and spiritual uses, often go unnoticed, neglected, or even marginalized. For instance, arid, dry, and semi-arid landscapes constitute nearly one-third of the Earth's land surface and are home to 18.5% of the world's population (Maestre et al., 2021).

Establishing conservation priorities depends on the values and interests of stakeholders involved in decision-making processes. Many conservationists agree that global rarity and threat status are pivotal criteria for prioritizing species. Additionally, there is advocacy for evaluating the phylogenetic diversity of protected species, as demonstrated by initiatives like the EDGE (Evolutionarily Distinct and Globally Endangered) program. Species that are endemic or restricted to specific geographical regions often receive precedence in conservation efforts. For example, in South Africa, the Cape Floristic Region, recognized as a global priority due to its abundance of endemic taxa, prioritizes endemic species in systematic conservation plans. Similarly, in Gabon, endemic species are focal points in national planning efforts, alongside other criteria. Endemic species also play a crucial role in identifying biodiversity hotspots, which are areas where at least 0.5% of the world's plant species are endemic (Lim et al., 2023).

Recent discussions regarding the determination of national priorities for species conservation propose a structured approach. Initially, taxonomic units are selected for evaluation, typically focusing on species. Subsequently, the distribution of these species is assessed. Finally, the proportion of the species' distribution within the area is calculated. Threshold values for species distribution ratios within an area may be categorized as high, medium, or low, aiding in the determination of national responsibility. However, until recently, precise threshold values for these categories were lacking. The Important Biodiversity Area (IUCN, 2016) methodology extends these concepts by suggesting specific thresholds for the percentage of the global population present in an area. This is assessed using proxies such as adult individuals or measures like distribution/range ratio, adjusted based on the threat status of species as outlined in the IUCN Red List of Threatened Species and other pertinent factors (Schmeller et al., 2014; Lim et al., 2023).

In conclusion, the conservation of biological diversity is paramount, given the urgent necessity posed by the unique threats confronting the terrestrial and marine ecosystems of our planet. Despite concerted international efforts, the risks of extinction and ecological degradation persist. However, the imminent implementation of the global biodiversity framework post-2020 presents a significant opportunity to reassess our stewardship of Earth's conservation and redefine our conservation priorities. This necessitates a meticulous examination of where our efforts should be directed and whether additional data are required for effective management.

Determining conservation priorities entails considering various factors, including global rarity, threat status, and the phylogenetic diversity of species. Endemic species play a crucial role, particularly in identifying biodiversity hotspots and guiding conservation strategies. Recent discussions advocate for a structured approach to determining national priorities, with an emphasis on evaluating taxonomic units, species distribution, and species proportions within an area.

Additionally, the Important Biodiversity Area methodology proposes specific threshold values based on the global population's presence in an area, taking into account factors such as species threat status. Despite remaining challenges, such as the precise definition of threshold values, the integration of these methodologies into conservation planning can enhance the efficacy of conservation efforts and contribute to preserving the biological diversity of our planet for future generations.

HUMAN ACTIVITIES AND BIODIVERSITY LOSS

Human activities release greenhouse gases into the atmosphere, notably carbon dioxide and methane, which results in a rise in atmospheric temperature due to the greenhouse effect. Consequently, this contributes to climate change and its associated environmental impacts. The repercussions of climate change have profound effects on the environment, biosphere, and biodiversity. Factors such as elevated temperatures, sea-level elevation, and an increased frequency of extreme weather events can disrupt ecosystems and have adverse impacts on biodiversity. This can lead to alterations in the distribution and habitats of plant and animal species, thereby changing ecosystems (Tong et al., 2023). Furthermore, the implications of climate change on respiratory health are significant. Elements like rising temperatures, air pollution, and shifts in pollen seasons can worsen the frequency and severity of asthma and other allergic respiratory ailments. Notably, the combination of air pollution and allergens can induce more severe symptoms in susceptible individuals (Pawankar, 2019).

Human economic activities, particularly the release of greenhouse gases and aerosols into the atmosphere, drive rapid changes in atmospheric composition and alterations in land cover and land use. This stands as one of the primary catalysts of climate change. Socio-economic development, manifested in behaviors, policies, actions, and underlying ideologies such as religious, cultural, or political beliefs, serves as the driving force behind climate change. Certain socio-economic activities are directly linked to climate change. The underlying causes of climate change are, in part, societal trends that may be influenced by long-term policy measures. These trends encompass demographic, economic, socio-political, scientific and technological, cultural, and religious factors, and can be seen as the foundational driving forces behind any human-induced alterations in natural and socio-economic systems. Factors like energy consumption, transportation, land use, trade, and tourism play pivotal roles in determining the impact of climate change on biodiversity. In Europe, energy consumption emerges as one of the most influential factors contributing to the rise in greenhouse gas (GHG) emissions. These emissions stem from the combustion of fossil fuels (oil, coal, natural gas) across sectors such as energy production, transportation, industry, and housing (Omann et al., 2009).

Drawing from data provided by the Intergovernmental Panel on Climate Change (IPCC) and the Convention on Biological Diversity (CBD), potential repercussions of climate change on ecosystems and biodiversity have been documented. Alterations in rainfall patterns and temperature regimes can prompt shifts in the boundaries of ecosystems, leading to expansions in some areas and contractions in others. Many species may find their habitats shifting poleward from their current locations. The timing and pace of such migrations will vary across species. Factors like droughts or outbreaks of diseases can decimate small populations, and the recovery of these populations may prove challenging. The breeding timing of species can impact the duration of the growing season in certain regions. Climate change can also hasten the spread of diseases and pests. Furthermore, the rising sea levels pose a significant threat to habitats such as coastal wetlands, mangroves, and coral reefs (CBD, 2023; Suarez et al., 2002).

In a recent investigation led by Habibullah and colleagues (2022), the study aimed to explore the ramifications of climate change on biodiversity loss across 115 countries worldwide. Biodiversity loss was assessed through the compilation of threatened species data encompassing amphibians, birds, fish, mammals, mollusks, plants, and reptiles. Various climate change parameters including temperature, precipitation, and occurrences of natural disasters were examined, with governance indicators and economic development levels serving as control variables. Robust regression techniques, including ordinary least squares and M-estimation with robust standard errors, were employed for the analysis. The findings from this research reveal a notable correlation between temperature, precipitation, natural disaster frequency, and biodiversity loss attributable to climate change. Moreover, higher levels of economic development were associated with increased biodiversity loss. Conversely, the study identified that favorable governance practices, such as effective control of corruption, regulatory quality, and adherence to the rule of law, were linked to a reduction in biodiversity loss. Thus, the study underscores the potential efficacy of embracing good governance practices, promoting environmental conservation efforts, and regulating greenhouse gas emissions as viable strategies to mitigate biodiversity loss (Habibullah et al., 2022).

Pollution stands as a primary catalyst for biodiversity decline, wreaking havoc on ecosystems and exerting detrimental effects on living organisms. The deleterious impacts of air, water, and soil pollution extend to the health and reproductive capabilities of plants and animals, precipitating reductions in population sizes and species richness. Chemical contaminants like pesticides, fertilizers, and industrial effluents disrupt ecosystem equilibrium and contribute to the dwindling numbers of numerous species. Concurrently, climate change emerges as an escalating menace to biodiversity, with its potential to reshape the distribution and abundance of myriad species. Escalating temperatures, shifting precipitation patterns, and intensified extreme weather phenomena impede the ability of plants and animals to thrive and propagate within their native habitats. Consequently, this engenders alterations in species distributions, modifications in community compositions, and heightened susceptibility to diseases and other stressors (Bellard et al., 2012).

In a study by Dulsat-Masvidal and colleagues (2023), it was reported that chemical pollution is overlooked for surface waters in natural areas. In this study, 411 water samples from 140 Important Bird and Biodiversity Areas (IBAs) in Spain were analyzed to evaluate the presence and distribution of 59 organic micropollutants (OMPs) and their impact on environmentally significant areas. According to the research findings, lifestyle compounds, pharmaceuticals, and organophosphate esters (OPEs) were the most common chemical families. However, pesticides and perfluoroalkyl substances (PFASs) were detected in approximately 25% of the samples. The average concentrations of the detected substances ranged from 0.1 to 301 ng/L. Based on spatial data, agricultural areas were identified as the most significant source of all OMPs in natural areas. Lifestyle compounds and PFASs were also identified as significant sources of pharmaceuticals in surface waters. The study identified 15 OMPs at levels that could pose a high risk to freshwater IBA ecosystems, with insecticide chlorpyrifos, antidepressant venlafaxine, and perfluorooctanesulfonic acid (PFOS) being prominent compounds. This study, the first to quantify water pollution in IBAs, highlights the emerging threat of OMPs to freshwater ecosystems critical for biodiversity conservation (Dulsat-Masvidal et al., 2023).

POLLUTION AND CLIMATE CHANGE: DUAL THREATS TO BIODIVERSITY

Biodiversity is the measure of life diversity on Earth, providing resilience and stability to ecosystems (Linhares et al., 2023). It encompasses the diversity of plants, animals, and microorganisms, including genetic diversity, species diversity, and populations. Additionally, it includes habitats, ecosystems, and natural areas. This diversity enriches the natural environment and provides many benefits to humans and other organisms, such as food, clothing, and shelter. However, factors such as overexploitation of natural resources, habitat destruction, pollution, and climate change negatively impact biodiversity (Prakash & Verma, 2022).

Habitat destruction and fragmentation are major causes of biodiversity loss. Human activities such as deforestation, urbanization, and agriculture have led to the destruction and degradation of natural habitats, resulting in displacement and extinction of many species. This habitat loss also disrupts the balance of ecological communities and can lead to the decline of specialized species dependent on specific habitat types. Another significant cause of biodiversity loss is the overexploitation of natural resources, including activities such as hunting, fishing, and logging. Unsustainable harvesting of plants and animals for commercial purposes can lead to the decline and extinction of many species, disrupt ecosystems, and reduce genetic diversity within populations. For example, overfishing has led to the depletion of many marine species, threatening the stability of marine food webs and ecosystems (Leddin, 2024).

Human activities have been known to cause dramatic declines in biodiversity (Linhares et al., 2023). Increasing human activities such as agriculture and forestry, land use changes, and hunting threaten the diversity of natural life (Young et al., 2005). Conflicts between biodiversity conservation and human activities are increasing, despite growing conservation efforts, and are a leading concern of the 21st century (Johnson et al., 2017; Tilman et al., 2017). To reverse this trend, significant changes in individual behaviors are needed. Individuals' daily activities, no matter how small, directly or indirectly affect biodiversity (Balmford et al., 2021; Nielsen et al., 2021).

In their study, Li et al. (2020) aimed to investigate the impact of human-induced global change on river biodiversity. They used environmental DNA samples from a large river basin in China to profile biodiversity from bacteria to invertebrates. They then analyzed the dependencies of these species, particularly with regard to C/N/P cycling-related ecosystem functions. This study addresses different measures of biodiversity, such as taxonomic, phylogenetic, and functional diversity, and examines how human activities affect this diversity and associated ecosystem functions. Ultimately, the study reports that human land use impacts biodiversity and ecosystem functions in river ecosystems (Li et al., 2020).

EFFECTIVE CONSERVATION STRATEGIES

For many years, the global population has been striving to find effective responses to the biodiversity crisis affecting wildlife, implementing both radical and impactful measures. Regional conservation efforts, which include protected areas and other local conservation measures, have been expanding and diversifying gradually. However, the effectiveness of regional or local conservation strategies in mitigating biodiversity loss is diminished due to insufficient consideration of ecological approaches and implementation deficiencies by local or national authorities.

Despite the efforts made by national governments over the past 20 years to expand protected areas, there has been limited success in enhancing biodiversity and broadening the scope of ecosystem activities and strategies. Many conservation strategies focus particularly on various and specific habitats, such as productive fisheries and the rehabilitation of terrestrial and marine carbon cycles. In the coming years, for local conservation strategies to effectively preserve biodiversity, humanity must more effectively contribute to globally established biodiversity goals. Measures aimed at preventing species extinctions and conserving the most intact ecosystems must be supported by rigorous monitoring and enforcement. Additionally, collaboration with local communities, civil society organizations, and private environmental entities, all of which play significant roles in biodiversity conservation, is essential.

The long-term success of habitat-based conservation depends on various factors, including securing adequate funding for proposed actions, planning to prevent and, if possible, reverse the effects of climate change, and integrating biodiversity conservation more robustly into land, water, and sea management policies under international biodiversity agreements and conservation plans. (Maxwell et al., 2020)

Protected Areas and Habitat Preservation

Globally, many protected areas with different management strategies complement each other by covering various amounts of land and hosting diverse biodiversity. This emphasizes the shared responsibility for conservation efforts across humanity.

Scientific studies have mainly focused on evaluating how close protected areas are to meeting the global conservation target of safeguarding at least 20% of terrestrial areas by the near future. In general, it's evident that different management approaches complement each other at both regional and sub-continental scales, leading to a notable increase in both the number and size of protected areas. For example, a closer look at unreported protected areas reveals that those with the strictest protection tend to be larger and richer in biodiversity, while those falling under Category VI are smaller but often situated nearer to human activities. It's widely recognized that the distribution of IUCN categories across the global protected area network isn't uniform, with regional differences being influenced by a variety of historical, geographical, social, and economic factors. In North America, the presence of extensive wild areas facilitates the establishment of protected areas falling under Categories I-III, whereas in Europe, high population density and fragmented landscapes explain the prevalence of Categories IV-VI. The prevalence of strictly protected areas in Southeast Asia is linked to traditional and cultural factors, while in Central America, protected areas falling under Categories V-VI are influenced by the overlap of indigenous territories and biodiversity hotspots.(Leberger et al., 2020)

The effectiveness of various conservation efforts and strategies in safeguarding biodiversity needs thorough global scrutiny. It's crucial to recognize a shared responsibility and synergy among different conservation categories in preserving vital habitats for plants, birds, mammals, amphibians, and all living organisms. Overall, there's little variation in the stability of protected areas across conservation categories worldwide. However, it's important to acknowledge that stability and conservation levels vary across different ecoregions, continents, and taxonomic groups. This highlights the absence of a universal hierarchy among protected areas in biodiversity conservation. Instead, it underscores the importance of assessing their relative contributions for each taxonomic group within specific geographical contexts.

Expanding protected areas worldwide is vital to tackle global environmental changes effectively. However, understanding how different conservation strategies complement each other remains crucial. Conservation policies involve a complex mix of ecological, social, economic, political, and cultural factors. Therefore, it's essential to grasp how various factors influence the choice of management strategies across different geographic contexts and their responsiveness to different goals. While protected areas dedicated to strict nature conservation should lead conservation efforts, those managed as alternatives, integrated with human activities, also hold promise for biodiversity conservation. Such areas can help preserve fragmented or human-impacted landscapes and bolster ecological resilience. Nonetheless, given the lack of long-term evidence for their effectiveness, it's essential to emphasize full support for the management strategies underpinning these protected areas. Regions where conservation strategies are consistently applied tend to have a higher proportion of protected areas covering natural lands or achieve greater success in conservation compared to others (Vimal et al., 2021).

Habitat Restoration and Rehabilitation

Habitat restoration and rehabilitation refer to deliberate actions aimed at repairing, enhancing, or reconstructing ecosystems that have been degraded, damaged, or destroyed. These efforts play a crucial role in biodiversity conservation by revitalizing natural habitats and promoting the recovery of native plant and animal species. Restoration activities may include removing invasive species, replanting native vegetation, restoring wetlands or forests, and creating wildlife corridors to reconnect fragmented habitats. By restoring and rehabilitating habitats, ecosystems become more resilient to environmental stresses, such as those caused by human activities or natural disasters. In the context of renewable energy projects like offshore wind, onshore wind, solar, and bioenergy, habitat restoration and rehabilitation are paramount. During the construction phase of these projects, activities such as land clearing and infrastructure development can disrupt natural habitats, leading to habitat loss and fragmentation. To mitigate these impacts, it is essential to integrate habitat restoration and rehabilitation measures into project planning and implementation.

For example, in offshore wind projects, where marine ecosystems are affected, habitat restoration efforts may involve creating artificial reefs or establishing marine protected areas to compensate for habitat loss. Similarly, in onshore wind and solar projects, degraded habitats can be restored by replanting native vegetation and implementing sustainable land management practices. By incorporating habitat restoration and rehabilitation into renewable energy projects, it is possible to minimize their ecological footprint and contribute to the overall conservation of biodiversity. The importance of habitat restoration and rehabilitation in mitigating the impacts of renewable energy projects on biodiversity are mutual. By integrating these practices into project planning and implementation, it is possible to minimize habitat loss, fragmentation, and degradation, thus promoting the recovery and conservation of natural ecosystems.

Species-Specific Conservation Programs

Conserving species is crucial for protecting biodiversity. It involves assessing the endangered status of specific species and developing strategies to ensure their survival, like preventing habitat loss, establishing protected areas, and controlling illegal activities like hunting and trade. These efforts play a vital role in safeguarding habitats for many species and directly address threats they face. However, the success of species conservation programs relies on broader biodiversity conservation strategies. For instance, preventing habitat loss is essential for preserving critical habitats, and combating illegal activities helps maintain populations of endangered species. Preserving genetic diversity is also crucial for the long-term health of species. Species conservation is a key aspect of biodiversity conservation. When combined with other strategies such as habitat restoration and community education, it contributes significantly to the sustainable preservation of biodiversity.

Species-specific conservation efforts rely on a range of international agreements and programs aimed at safeguarding endangered species and preserving biodiversity. These initiatives serve as crucial frameworks for coordinating global efforts to protect and conserve threatened species. Here are some of the key international agreements and programs in this regard:

CITES (Convention on International Trade in Endangered Species of Wild Fauna and Flora): CITES regulates and monitors international trade in endangered species to ensure their conservation. It aims to prevent exploitation and unsustainable trade of endangered species through international cooperation (Andersson et al., 2021).

IUCN Red List of Threatened Species: Compiled by the International Union for Conservation of Nature (IUCN), the Red List categorizes and assesses the conservation status of species worldwide. It serves as a vital tool for identifying species at risk and prioritizing conservation efforts (Crosti et al., 2020).

Convention on Biological Diversity (CBD): The CBD is an international treaty dedicated to conserving biodiversity, promoting sustainable use of natural resources, and ensuring equitable sharing of benefits. It emphasizes the conservation of species and their habitats through policy frameworks and action plans (Essl et al., 2020).

Species Survival Commission (SSC): Operating under the umbrella of the IUCN, the SSC brings together various expert groups to address the conservation needs of endangered species. It coordinates research and conservation actions to prevent species extinction. (Sigwart et al., 2023)

Global Environment Facility (GEF): As an international financial institution, the GEF provides funding for biodiversity conservation projects worldwide. It supports initiatives aimed at protecting species and promoting their sustainable management (Kumar, 2022).

RAMSAR Convention: This convention focuses on the conservation of wetlands and their biodiversity. It promotes international cooperation to ensure the wise use and conservation of wetland ecosystems, which are vital habitats for many species (Stroud et al., 2022).

Marine Turtle Conservation Action Plan: Designed to protect marine turtles and restore their populations, this international plan aims to conserve their habitats and combat illegal harvesting and trade.

Great Ape Survival Partnership (GRASP): GRASP is a partnership dedicated to conserving great ape species in Africa and Asia. It works to protect ape habitats, mitigate threats, and promote sustainable management practices. These international agreements and programs play a crucial role in coordinating efforts to conserve species and their habitats on a global scale. They underscore the importance of collaborative action and shared responsibility in addressing biodiversity conservation challenges (Mitani et al., 2024).

Sustainable Land Use Practices

The immediate impacts of altering land use involve the loss of habitats and changes in the environment. Yet, some effects are complex and not easily noticeable or quantifiable. Additionally, changes in land use can interact with other global factors. For example, the construction of roads can facilitate the spread of invasive species and provide easier access for direct exploitation. While interactions between land-use changes and climate change are significant, they have not received sufficient attention.

IPBES's global report underscores that the principal driver behind the decline in nature and biodiversity is irresponsible and unsustainable alteration of land use. Specifically, human activities such as hunting, fishing, logging, and harvesting are estimated to contribute to over half of the human impacts on terrestrial and freshwater ecosystems. The expansion of human influence has been observed to encompass a considerable portion of the terrestrial biosphere. Throughout this process, human activities like agricultural expansion, cultivation areas, and plantations emerge as primary causes of global land-use change.

Land use change negatively affects biological diversity, especially due to intensified agriculture and forestry, which are linked to a decrease in species richness. Similarly, urbanization leads to a reduction in bird and plant density. There's a growing need for collaboration between scientists and local authorities to better grasp the connection between land use change and biodiversity. Geographical challenges limit the applicability of findings and the reliability of future projections, making it hard to devise biodiversity conservation strategies. Investigating various regional land-use mechanisms is crucial for establishing a global perspective. Popular watershed management practices, often focused on forests, may overshadow the importance of other diverse habitats like tropical and subtropical savannas and grasslands. Attention should be given to studying how land use change interacts with other global factors such as climate change.

The importance of understanding the effects of past land use and societal preferences is clear. Using different criteria can lead to inaccuracies in assessing the impacts of habitat degradation caused by land use changes. The dominance of studies from North America and Europe in land use and habitat conservation suggests a neglect of regional differences in land use pressures over time. Failing to create a balanced database that considers all these challenges on a global scale and includes every relevant variable may result in our global understanding of the impacts of land use change on species being incomplete. This limited understanding could hinder the implementation of biodiversity conservation strategies, especially in areas with intense land use. Addressing these issues would improve the reliability of projections about global changes in biodiversity and ensure the success of conservation efforts aimed at protecting species threatened by habitat loss and degradation (Davison et al., 2021).

International Cooperation and Policy Frameworks

International agreements serve as a crucial framework for the conservation and sustainable use of biodiversity. Here are some key international agreements aimed at preserving biological diversity: Convention on Biological Diversity (CBD) Adopted in 1992, this treaty aims to conserve biological diversity, promote its sustainable use, and ensure the fair and equitable sharing of benefits derived from genetic resources.

Kyoto Protocol, this 1997 protocol focuses on combating climate change by reducing greenhouse gas emissions and regulating carbon trading, thereby aiming to mitigate the impact of climate change on biodiversity (Kolawole & Iyiola, 2023).

Paris Agreement, adopted in 2015, this agreement aims to limit global warming to 1.5 °C above pre-industrial levels and addresses mitigation, adaptation, and finance to tackle climate change, recognizing its significant impact on biodiversity (Smith et al., 2022).

Montreal Protocol, established in 1987, this protocol aims to protect the ozone layer by regulating the production and consumption of ozone-depleting substances, which contribute to biodiversity loss (Streck, 2023).

Madrid Protocol, this agreement, adopted in 1992, aims to restrict the production and consumption of substances that deplete the ozone layer, thus protecting biodiversity from harmful UV radiation (Haward, 2021).

UNESCO World Heritage Convention, enacted in 1972, this convention seeks to preserve and protect unique cultural and natural sites worldwide, contributing to the conservation of biodiversity (Gorenflo & Romaine, 2021).

Aichi Biodiversity Targets, these targets, adopted by CBD parties in 2010, aim to sustainably conserve and enhance biodiversity by 2020, guiding international efforts in biodiversity conservation (Nguyen & Vuong, 2021).

Agenda 2030 and Sustainable Development Goals (SDGs), Adopted by the United Nations in 2015, the SDGs aim to promote sustainable development and include specific goals related to biodiversity conservation, recognizing its integral role in achieving sustainability (Zhang et al., 2022).

Rotterdam Convention, this treaty, adopted to regulate hazardous chemicals in international trade, aims to control and reduce the adverse effects of chemicals on biodiversity conservation (Wang, 2023).

Cartagena Protocol, as part of the CBD, this protocol regulates the safe transfer, handling, and use of genetically modified organisms (GMOs) to ensure biodiversity conservation (Zhao, 2020).

CBD Protocols, Various protocols under the CBD, such as the Protocol on Biosafety, Forest Biodiversity Protocol, and Protocol on Protected Areas, focus on specific aspects of biodiversity conservation (Kiehn et al., 2021).

These international agreements and initiatives play a vital role in tackling the challenges of conserving biodiversity and ensuring its sustainable use, recognizing how closely connected biodiversity is to human well-being and environmental health. However, effectively putting these agreements into action and achieving their goals requires global cooperation and unwavering dedication. It's also crucial to allocate more resources and efforts at local, national, and international levels. Protecting biodiversity is crucial for the future of our planet and humanity, calling for all stakeholders to come together around a common vision.

The Convention on Biological Diversity provides a global framework for conserving biodiversity, promoting its sustainable use, and ensuring fair and equitable sharing of benefits from genetic resources. By regulating international trade in endangered wildlife and plants, these agreements serve as essential tools for biodiversity conservation. Additionally, by encouraging the preservation and sustainable use of wetlands, they contribute to safeguarding aquatic ecosystems. Similarly, all agreements establish international standards and foster collaboration to address threats to biodiversity and protect natural habitats. By providing a critical framework for biodiversity conservation, these agreements enable the international community to work together effectively to preserve the rich diversity of life on our planet.

Harnessing Plant Ecology for Biodiversity Conservation

The relentless degradation of terrestrial natural habitats due to human activities presents an ominous challenge to plant species worldwide, pushing many to the brink of extinction. This peril is often instigated by the widespread clearance of native vegetation to make way for agriculture, triggering a domino effect of erosion, salinization, and the encroachment of invasive species. Adding to the complexity, climate change has emerged as a potent new adversary, intensifying the existential threats faced by plant life. Alarmingly, over half of the world's plant species call 34 global biodiversity hotspots home, areas that once sprawled across 15.7% of the Earth's landmass but have since dwindled to a mere 2.3%. This rapid decline puts countless endemic species in a precarious position, edging them closer to the precipice of extinction (Reed et al., 2011).

The inception of the International Union for Conservation of Nature (IUCN) Red List of Threatened Plants in 1998 marked a pivotal moment in biodiversity conservation, cataloging over 8,000 species as of 2010. While in situ conservation remains a cornerstone, it proves insufficient to address the multifaceted challenges of preserving endangered species. In scenarios where traditional methods like seed banking falter, alternative strategies become imperative. Preserving genetic diversity emerges as paramount, with micropropagation offering a versatile means to propagate diverse genotypes from regional gene pools (Rogers, 2003). Coordinated efforts led by international bodies are indispensable for integrated conservation, combining in situ and ex situ methodologies. Despite financial constraints, botanical gardens are assuming an increasingly vital role in safeguarding endangered species through collection and maintenance initiatives.

Plant conservation progresses through in situ and ex situ conservation methods, utilizing botanical gardens and seed banks, necessitating additional approaches for species most at risk of extinction. Complementary strategies, such as ex situ seed storage combined with in vitro and cryopreservation storage, can play a crucial role in conserving the biological diversity of many habitats facing extinction (Cruz-Cruz et al., 2013).

Preserving plant biodiversity is essential for maintaining ecosystem health and resilience. Conservation efforts can be categorized into in situ and ex situ methods. In situ conservation involves protecting plant species within their natural habitats, aiming to maintain their ecological roles and interactions. Conversely, ex situ conservation involves safeguarding plant species outside their native environments, often in controlled settings such as botanical gardens, seed banks, or gene banks. While in situ conservation is crucial for preserving genetic diversity and ecosystem functions, it may not always be sufficient due to habitat degradation and loss. Ex situ methods serve as valuable complements, providing a safety net for species facing imminent threats in the wild. These methods allow for the storage of seeds, tissues, or whole plants, ensuring genetic diversity is preserved for future restoration efforts. Both in situ and ex situ approaches offer unique advantages and challenges. In situ conservation supports species within their natural habitats, allowing for the maintenance of complex ecological relationships. However, it requires effective habitat management and protection, which can be challenging in the face of human activities and environmental changes. On the other hand, ex situ conservation provides a means to safeguard species that are at risk of extinction in the wild. It offers opportunities for research, education, and reintroduction programs. However, ex situ methods may pose risks such as genetic erosion, loss of adaptive traits, and limited representation of natural populations. Ultimately, a combination of in situ and ex situ strategies is often necessary to comprehensively address the conservation needs of plant species. By integrating these approaches and adapting them to specific species and ecosystems, conservationists can enhance the resilience of plant biodiversity in the face of mounting threats. (Ramsay et al.,2000; Cruz-Cruz et al., 2013).

Plant conservation strategies play a critical role in preserving biological diversity. Especially with the threat to terrestrial natural ecosystems due to human impact, the conservation and sustainability of plant species are of paramount importance. These strategies aim to preserve the diversity and health of plant species and ecosystems, rescue endangered species, and pass them on to future generations. *In vitro* and *ex situ* conservation methods ensure the preservation of plant species outside their natural habitats. These methods play an important role in preserving biological diversity and contributing to its sustainability. Therefore, effective implementation of plant conservation strategies is vital for the long-term preservation of natural ecosystems and plant species.

During the *in vitro* propagation process, commonly used supporting material such as agar provides support from culture induction to rooting. However, the quality of plant roots grown in agar-based media is often low, which can adversely affect planting success. Particularly in woody plant species, plantlets grown in agar-based media may rapidly wither after transplantation. This is due to the decrease or absence of epicuticular and cuticular waxes, which can cause rapid water loss in plants post-transplantation. Additionally, stomata that do not open under dark or low relative humidity conditions can also contribute to this situation. This can lead to rapid water loss in plants post-transplantation, resulting in low survival rates (Sarasan, 2010).

Ex vitro symbiotic seed germination, on the other hand, is a method used to germinate orchid seeds using substrates pre-inoculated with compatible fungi. This method is advantageous for the mass production of orchid species as it does not require laboratory facilities or axenic conditions. In recent years, studies on orchid conservation have provided significant support for these conservation strategies (Deniz et al. 2022; Aytar and Özdener Kömpe, 2023; Aytar et al., 2023).

ETHICAL AND CULTURAL DIMENSIONS OF BIODIVERSITY CONSERVATION

Examining the Ethical Responsibilities Associated With Biodiversity Preservation and the Conservation of Species Diversity

Biodiversity, which refers to the variety of life forms on Earth, is crucial for maintaining the stability and resilience of ecosystems. Conservation efforts should focus not only on studying and preserving individual species, but also on addressing the ethical complexities inherent in this endeavour. Biodiversity conservation is based on ethical principles derived from various philosophical traditions. The four main areas of ethics most relevant to the ethical questions raised by work in ecology and biodiversity management are (traditional) normative ethical theory, research ethics, animal ethics and environmental ethics. Normative ethical theory, which can be summarised as doing the right thing and being good within the human community, includes utilitarianism, deontological ethics and virtue ethics (Minteer and Collins, 2005). Utilitarianism emphasises species and ecosystems as providers of resources or services for people (Loreau, 2014). Deontological ethics emphasises fulfilling one's moral responsibilities, arguing that biodiversity has intrinsic value for humans, independent of its utility (Karlsson and Björnberg, 2021). Virtue ethics, on the other hand, focuses on the internalisation and display of character traits that are conducive to harmonious coexistence with nature.

Every species represents a distinct evolutionary origin and ecological purpose, contributing to the diversity and richness of ecosystems. From a utilitarian perspective, species diversity improves the resilience of ecosystems, providing buffers against environmental perturbations and supporting essential ecosystem services. However, beyond its instrumental value, each species possesses intrinsic worth and deserves protection, regardless of its usefulness to humans. Conservation efforts should aim to protect species diversity not only for human benefit but also out of respect for the intrinsic value of each living organism (Taylor et al., 2020).

Conservation interventions frequently involve challenging ethical trade-offs between competing priorities, such as preserving biodiversity, promoting economic development, and ensuring social justice (Brittain et al., 2020). For example, managing protected areas may require relocating indigenous communities or limiting access to essential resources, which raises concerns about fairness and equality. Ethical concerns about the manipulation of species and ecosystems are raised by the emergence of technologies like gene editing and de-extinction. Decision-making frameworks that prioritize transparency, inclusivity, and deliberative democracy can be beneficial in navigating these complex ethical landscapes (Biasetti and De Mori, 2021).

Preserving biodiversity is not only a scientific endeavour, but also a moral obligation that is founded on ethical principles and obligations. By navigating ethical dilemmas, upholding principles of stewardship and justice, and fostering public engagement, we can strive towards a future where biodiversity thrives, ensuring the well-being of ecosystems and humanity alike. Embracing our ethical responsibilities towards biodiversity conservation is essential for shaping a sustainable and harmonious relationship with the natural world.

Discussing the Cultural Significance of Biodiversity and its Role in Shaping Human Societies and Traditions

The diverse human societies have a profound cultural significance for biodiversity, which is the rich tapestry of life on Earth. Biodiversity shapes human cultures in myriad ways, influencing practices, customs, and worldviews, embedded in folklore, traditions and spiritual beliefs. Attitudes towards biodiversity and conservation practices are shaped by cultural values and beliefs. Indigenous cultures frequently have strong spiritual ties to the land and its people, which encourages them to practice stewardship that preserves biodiversity for future generations (Freitas et al., 2020). Western societies have had a long history of exhibiting anthropocentric attitudes, which have focused on nature as primarily a resource that can be used for human benefit. To promote inclusive and effective conservation strategies that resonate with local communities, it is crucial to recognize and respect diverse cultural perspectives (Nan et al., 2021).

Indigenous cultures often view nature and biodiversity as sacred, representing spiritual connections to the land and its inhabitants (Turner et al., 2022). Creation myths frequently feature biodiversity, with animals, plants, and landscapes holding divine significance (Nongbri, 2006). Through generations of observation and interaction with their environments, indigenous and traditional communities have acquired a wealth of knowledge about biodiversity. Insights into plant medicines, sustainable resource management practices, and ecosystem dynamics are included in this traditional ecological knowledge. Traditional knowledge systems are founded on biodiversity, which guides indigenous societies in their relationships with the natural world and informs cultural practices (Langton and Rhea, 2005; Pushpangadan et al., 2018).

Biodiversity has an impact on cultural practices and traditions, including foodways, agriculture, art, and storytelling. The diversity of local ecosystems is often reflected in traditional cuisines by incorporating a variety of wild plants, animals, and fungi. Moreover, biodiversity is a source of inspiration for artistic expression through rituals, ceremonies, and artistic representations, which encourages a deep appreciation for nature (Pretty et al., 2009; Ogawa et al., 2021).

Communities' identities and cultural heritage are shaped by biodiversity, which connects them to specific landscapes and ecosystems. Strong cultural ties are maintained by indigenous peoples to their ancestral land, which is characterized by unique biodiversity hotspots. Indigenous communities gain a sense of belonging and continuity through the storage of traditional knowledge, languages, and customs in these cultural landscapes (Parviainen, 2015; Valkó et al., 2018).

In summary, biodiversity not only contributes to ecological stability but also serves as the cornerstone of human cultures and traditions. We can enhance our appreciation for the interconnectedness of life and promote greater respect for the natural world by acknowledging the cultural significance of biodiversity. The preservation of biodiversity is not solely about ecological preservation but also about protecting the cultural heritage and identities of various human societies.

CONCLUSION

This chapter emphasizes the significance of biological diversity by elucidating its numerous benefits, which affect the stability of various ecosystems and human quality of life. Particularly, it highlights the contribution of biological diversity to the balanced functioning of ecosystems, focusing on the diversity of plant species and their ecological interactions. However, human impacts such as pollution and climate change pose threats to this valuable diversity. To address these threats, comprehensive conservation strategies need to be developed. Additionally, it is noted that scientific research plays a crucial role in supporting conservation efforts by contributing valuable knowledge about biological diversity and ecosystem health. In conclusion, the preservation of biological diversity is not only essential for maintaining ecosystem balance and human well-being but also critical for the welfare of future generations and the sustainability of our planet.

REFERENCES

Andersson, A. A., Tilley, H. B., Lau, W., Dudgeon, D., Bonebrake, T. C., & Dingle, C. (2021). CITES and beyond: Illuminating 20 years of global, legal wildlife trade. *Global Ecology and Conservation*, *26*, e01455. doi:10.1016/j.gecco.2021.e01455

Aytar, E. C., Harzli, I., & Özdener Kömpe, Y. (2023). Phytochemical Analysis of Anacamptis coriophora Plant Cultivated Using Ex Vitro Symbiotic Propagation. *Chemistry & Biodiversity*, *20*(12), e202301218. doi:10.1002/cbdv.202301218 PMID:37870554

Aytar, E. C., & Özdener Kömpe, Y. (2023). Effect of different substrates on in vitro symbiotic seed germination for soilless production of Anacamptis laxiflora orchid. *Acta Botanica Croatica, 82*(2), 0-0.

Balmford, A., Bradbury, R. B., Bauer, J. M., Broad, S., Burgess, G., Burgman, M., Byerly, H., Clayton, S., Espelosin, D., Ferraro, P. J., Fisher, B., Garnett, E. E., Jones, J. P. G., Marteau, T. M., Otieno, M., Polasky, S., Ricketts, T. H., Sandbrook, C., Sullivan-Wiley, K., & Nielsen, K. S. (2021). Making more effective use of human behavioural science in conservation interventions. *Biological Conservation*, *261*, 109256. doi:10.1016/j.biocon.2021.109256

Bellard, C., Bertelsmeier, C., Leadley, P., Thuiller, W., & Courchamp, F. (2012). Impacts of climate change on the future of biodiversity. *Ecology Letters*, *15*(4), 365–377. doi:10.1111/j.1461-0248.2011.01736.x PMID:22257223

Biasetti, P., & De Mori, B. (2021). The ethical matrix as a tool for decision-making process in conservation. *Frontiers in Environmental Science*, *9*, 584636. doi:10.3389/fenvs.2021.584636

Brittain, S., Ibbett, H., de Lange, E., Dorward, L., Hoyte, S., Marino, A., Milner-Gulland, E. J., Newth, J., Rakotonarivo, S., Veríssimo, D., & Lewis, J. (2020). Ethical considerations when conservation research involves people. *Conservation Biology*, *34*(4), 925–933. doi:10.1111/cobi.13464 PMID:31953971

CBD. (2003). *Interlinkages Between Biological Diversity and Climate Change, CBD Technical Series 10*. Secretariat of the Convention of Biological Diversity.

Crosti, R., Arcangeli, A., Campagnuolo, S., Castriota, L., Falautano, M., Maggio, T., & Andaloro, F. (2020). Assessing worth of marine protected areas for the protection of threatened biodiversity using IUCN Red List and Red List Index. A pilot study in six mediterranean areas. *Ecological Indicators*, *119*, 106765. doi:10.1016/j.ecolind.2020.106765

Cruz-Cruz, C. A., González-Arnao, M. T., & Engelmann, F. (2013). Biotechnology and conservation of plant biodiversity. *Resources*, *2*(2), 73–95. doi:10.3390/resources2020073

Davison, C. W., Rahbek, C., & Morueta-Holme, N. (2021). Land-use change and biodiversity: Challenges for assembling evidence on the greatest threat to nature. *Global Change Biology*, *27*(21), 5414–5429. doi:10.1111/gcb.15846 PMID:34392585

Deniz, İ. G., Kömpe, Y. Ö., Harzli, I., Aytar, E. C., Mutlu, V. A., & Uysal, D. İ. (2022). From seed to flowering tuberous orchid using ex vitro symbiotic seed germination: A breakthrough study with Anacamptis sancta. *Rhizosphere*, *24*, 100597. doi:10.1016/j.rhisph.2022.100597

Dulsat-Masvidal, M., Ciudad, C., Infante, O., Mateo, R., & Lacorte, S. (2023). Water pollution threats in important bird and biodiversity areas from Spain. *Journal of Hazardous Materials*, *448*, 130938. doi:10.1016/j.jhazmat.2023.130938 PMID:36860036

Essl, F., Latombe, G., Lenzner, B., Wilson, J. R. U., Genovesi, P., Pagad, S., Seebens, H., & Smith, K. (2020). The Convention on Biological Diversity (CBD)'s Post-2020 target on invasive alien species – what should it include and how should it be monitored? *NeoBiota*, *62*, 99–121. doi:10.3897/neobiota.62.53972

Freitas, C. T., Lopes, P. F., Campos-Silva, J. V., Noble, M. M., Dyball, R., & Peres, C. A. (2020). Co-management of culturally important species: A tool to promote biodiversity conservation and human well-being. *People and Nature*, *2*(1), 61–81. doi:10.1002/pan3.10064

Gorenflo, L. J., & Romaine, S. (2021). Linguistic diversity and conservation opportunities at UNESCO World Heritage Sites in Africa. *Conservation Biology*, *35*(5), 1426–1436. doi:10.1111/cobi.13693 PMID:33448452

Habibullah, M. S., Din, B. H., Tan, S. H., & Zahid, H. (2022). Impact of climate change on biodiversity loss: Global evidence. *Environmental Science and Pollution Research International*, *29*(1), 1073–1086. doi:10.1007/s11356-021-15702-8 PMID:34341937

Haward, M. (2021). Biodiversity in Areas Beyond National Jurisdiction (BBNJ): The Commission for the Conservation of Antarctic Marine Living Resources (CCAMLR) and the United Nations BBNJ agreement. *The Polar Journal*, *11*(2), 303–316. doi:10.1080/2154896X.2021.1984658

Johnson, C. N., Balmford, A., Brook, B. W., Buettel, J. C., Galetti, M., Guangchun, L., & Wilmshurst, J. M. (2017). Biodiversity losses and conservation responses in the Anthropocene. *Science*, *356*(6335), 270–275. doi:10.1126/science.aam9317 PMID:28428393

Karlsson, M., & Edvardsson Björnberg, K. (2021). Ethics and biodiversity offsetting. *Conservation Biology*, *35*(2), 578–586. doi:10.1111/cobi.13603 PMID:32757221

Kiehn, M., Fischer, F., & Smith, P. P. (2021). The Nagoya protocol and access and benefit sharing regulations of the Convention on Biological Diversity (CBD) and its impacts on botanic gardens' collections and research. *Perspectives in Agriculture, Veterinary Science, Nutrition and Natural Resources*, *16*(34), PAVSNNR202116034. doi:10.1079/PAVSNNR202116034

Kolawole, A. S., & Iyiola, A. O. (2023). *Environmental Pollution: Threats, Impact on Biodiversity, and Protection Strategies*. doi:10.1007/978-981-19-6974-4_14

Kumar, A. (2022). 4. Global Environment Facility (GEF). *Yearbook of International Environmental Law*, *31*(1), 296–301. doi:10.1093/yiel/yvab055

Langton, M., & Rhea, Z. M. (2005). Traditional indigenous biodiversity-related knowledge. *Australian Academic and Research Libraries*, *36*(2), 45–69. doi:10.1080/00048623.2005.10721248

Leberger, R., Rosa, I. M. D., Guerra, C. A., Wolf, F., & Pereira, H. M. (2020). Global patterns of forest loss across IUCN categories of protected areas. *Biological Conservation*, *241*, 108299. doi:10.1016/j.biocon.2019.108299

Leddin, D. (2024). The impact of climate change, pollution and biodiversity loss on digestive health and disease. *Gastro Hep Advances*. doi:10.1016/j.gastha.2024.01.018

Li, F., Altermatt, F., Yang, J., An, S., Li, A., & Zhang, X. (2020). Human activities' fingerprint on multitrophic biodiversity and ecosystem functions across a major river catchment in China. *Global Change Biology*, *26*(12), 6867–6879. doi:10.1111/gcb.15357 PMID:32936984

Lim, D. Y., Starnes, T., & Plumptre, A. J. (2023). Global priorities for biodiversity conservation in the United Kingdom. *Biological Conservation*, *277*, 109798. doi:10.1016/j.biocon.2022.109798

Linhares, Y., Kaganski, A., Agyare, C., Kurnaz, I. A., Neergheen, V., Kolodziejczyk, B., Kędra, M., Wahajuddin, M., El-Youssf, L., dela Cruz, T. E., Baran, Y., Pešić, M., Shrestha, U., Bakiu, R., Allard, P. M., Rybtsov, S., Pieri, M., Siciliano, V., & Flores Bueso, Y. (2023). Biodiversity: The overlooked source of human health. *Trends in Molecular Medicine*, *29*(3), 173–187. doi:10.1016/j.molmed.2022.12.002 PMID:36585352

Loreau, M. (2014). Reconciling utilitarian and non-utilitarian approaches to biodiversity conservation. *Ethics in Science and Environmental Politics*, *14*(1), 27–32. doi:10.3354/esep00149

Maestre, F. T., Benito, B. M., Berdugo, M., Concostrina-Zubiri, L., Delgado-Baquerizo, M., Eldridge, D. J., Guirado, E., Gross, N., Kéfi, S., Le Bagousse-Pinguet, Y., Ochoa-Hueso, R., & Soliveres, S. (2021). Biogeography of global drylands. *The New Phytologist*, *231*(2), 540–558. doi:10.1111/nph.17395 PMID:33864276

Maxwell, S. L., Cazalis, V., Dudley, N., Hoffmann, M., Rodrigues, A. S. L., Stolton, S., Visconti, P., Woodley, S., Kingston, N., Lewis, E., Maron, M., Strassburg, B. B. N., Wenger, A., Jonas, H. D., Venter, O., & Watson, J. E. M. (2020). Area-based conservation in the twenty-first century. *Nature 2020 586:7828, 586*(7828), 217–227. doi:10.1038/s41586-020-2773-z

Minteer, B. A., & Collins, J. P. (2005). Ecological ethics: Building a new tool kit for ecologists and biodiversity managers. *Conservation Biology*, *19*(6), 1803–1812. doi:10.1111/j.1523-1739.2005.00281.x

Mitani, J. C., Abwe, E., Campbell, G., Giles-Vernick, T., Goldberg, T., McLennan, M. R., Preuschoft, S., Supriatna, J., & Marshall, A. J. (2024). Future coexistence with great apes will require major changes to policy and practice. *Nature Human Behaviour*, *2024*(4), 1–12. doi:10.1038/s41562-024-01830-x PMID:38374442

Nan, M. A., Lun, Y. A. N. G., Qingwen, M. I. N., Keyu, B., & Wenhua, L. (2021). The significance of traditional culture for agricultural biodiversity—Experiences from GIAHS. *Journal of Resources and Ecology*, *12*(4), 453–461. doi:10.5814/j.issn.1674-764x.2021.04.003

Nguyen, M. H., & Vuong, Q. H. (2021). Evaluation of the Aichi Biodiversity Targets: The international collaboration trilemma in interdisciplinary research. *Pacific Conservation Biology*, *28*(6), 517–531. doi:10.1071/PC21026

Nielsen, K. S., Marteau, T. M., Bauer, J. M., Bradbury, R. B., Broad, S., Burgess, G., Burgman, M., Byerly, H., Clayton, S., Espelosin, D., Ferraro, P. J., Fisher, B., Garnett, E. E., Jones, J. P. G., Otieno, M., Polasky, S., Ricketts, T. H., Trevelyan, R., van der Linden, S., & Balmford, A. (2021). Biodiversity conservation as a promising frontier for behavioural science. *Nature Human Behaviour, 5*(5), 550–556. doi:10.1038/s41562-021-01109-5

Nongbri, T. (2006). *Culture and biodiversity: Myths, legends and the conservation of nature in the hills of North-East India.* Indian Anthropologist, 1-21.

Ogawa, M., Soga, M., & Yoshida, T. (2021). Participation of diverse actors and usage of traditional and local knowledge in local biodiversity strategies and action plans of Japanese municipalities. *Ecology and Society*, *26*(3), art26. doi:10.5751/ES-12612-260326

Omann, I., Stocker, A., & Jäger, J. (2009). Climate change as a threat to biodiversity: An application of the DPSIR approach. *Ecological Economics*, *69*(1), 24–31. doi:10.1016/j.ecolecon.2009.01.003

Parviainen, J. (2015). Cultural heritage and biodiversity in the present forest management of the boreal zone in Scandinavia. *Journal of Forest Research*, *20*(5), 445–452. doi:10.1007/s10310-015-0499-9

Pawankar, R. (2019). Climate change, air pollution, and biodiversity in Asia Pacific: Impact on allergic diseases. *Asia Pacific Allergy*, *9*(2), e11. doi:10.5415/apallergy.2019.9.e11 PMID:31089453

Prakash, S., & Verma, A. (2022). *Anthropogenic Activities and Biodiversity Threats*. https://papers.ssrn.com/abstract=4048276

Pretty, J., Adams, B., Berkes, F., De Athayde, S. F., Dudley, N., Hunn, E., & Pilgrim, S. (2009). The intersections of biological diversity and cultural diversity: Towards integration. *Conservation & Society*, *7*(2), 100–112. doi:10.4103/0972-4923.58642

Pushpangadan, P., George, V., Ijinu, T. P., & Chithra, M. A. (2018). Biodiversity, bioprospecting, traditional knowledge. Sustainable development and value added products: A review. *Journal of Traditional Medicine & Clinical Naturopathy*, *7*(01), 1–7. doi:10.4172/2573-4555.1000256

Ramsay, M. M., Jackson, A. D., & Porley, R. D. (2000). A pilot study for the ex situ conservation of UK bryophytes. EuroGard.

Reed, B. M., Sarasan, V., Kane, M., Bunn, E., & Pence, V. C. (2011). Biodiversity conservation and conservation biotechnology tools. *In Vitro Cellular & Developmental Biology. Plant*, *47*(1), 1–4. doi:10.1007/s11627-010-9337-0

Rogers, S. M. (2003). Tissue culture and wetland establishment of the freshwater monocots Carex, Juncus, Scirpus, and Typha. *In Vitro Cellular & Developmental Biology. Plant*, *39*(1), 1–5. doi:10.1079/IVP2002358

Sarasan, V. (2010). Importance of in vitro technology to future conservation programmes worldwide. *Kew Bulletin*, *65*(4), 549–554. doi:10.1007/s12225-011-9250-7

Sigwart, J. D., Pollom, R., & Böhm, M. (2023). The IUCN Species Survival Commission launches a new Red List Authority to assess marine invertebrates. *Oryx*, *57*(4), 418–419. doi:10.1017/S003060532300042X

Smith, P., Arneth, A., Barnes, D. K. A., Ichii, K., Marquet, P. A., Popp, A., Pörtner, H. O., Rogers, A. D., Scholes, R. J., Strassburg, B., Wu, J., & Ngo, H. (2022). How do we best synergize climate mitigation actions to co-benefit biodiversity? *Global Change Biology*, *28*(8), 2555–2577. doi:10.1111/gcb.16056 PMID:34951743

Streck, C. (2023). Synergies between the Kunming-Montreal Global Biodiversity Framework and the Paris Agreement: The role of policy milestones, monitoring frameworks and safeguards. *Climate Policy*, *23*(6), 800–811. doi:10.1080/14693062.2023.2230940

Stroud, D. A., Davidson, N. C., Finlayson, C. M., & Gardner, R. C. (2022). Development of the text of the Ramsar Convention: 1965–1971. *Marine and Freshwater Research*, *73*(10), 1107–1126. doi:10.1071/MF21312

Suárez, A., Watson, R. T., & Dokken, D. J. (2002). *Climate change and biodiversity*. Research Gate.

Taylor, B., Chapron, G., Kopnina, H., Orlikowska, E., Gray, J., & Piccolo, J. J. (2020). The need for eco-centrism in biodiversity conservation. *Conservation Biology*, *34*(5), 1089–1096. doi:10.1111/cobi.13541 PMID:32424955

Tilman, D., Clark, M., Williams, D. R., Kimmel, K., Polasky, S., & Packer, C. (2017). Future threats to biodiversity and pathways to their prevention. *Nature, 546*(7656), 73–81. doi:10.1038/nature22900

Tong, S., Beggs, P. J., Davies, J. M., Jiang, F., Kinney, P. L., Liu, S., Yin, Y., & Ebi, K. L. (2023). Compound impacts of climate change, urbanization and biodiversity loss on allergic disease. *International Journal of Epidemiology*, *52*(3), 655–663. doi:10.1093/ije/dyac197 PMID:36228124

Turner, N. J., Cuerrier, A., & Joseph, L. (2022). Well grounded: Indigenous Peoples' knowledge, ethnobiology and sustainability. *People and Nature*, *4*(3), 627–651. doi:10.1002/pan3.10321

Valkó, O., Tóth, K., Kelemen, A., Miglécz, T., Radócz, S., Sonkoly, J., Tóthmérész, B., Török, P., & Deák, B. (2018). Cultural heritage and biodiversity conservation–plant introduction and practical restoration on ancient burial mounds. *Nature Conservation*, *24*, 65–80. doi:10.3897/natureconservation.24.20019

Vimal, R., Navarro, L. M., Jones, Y., Wolf, F., Le Moguédec, G., & Réjou-Méchain, M. (2021). The global distribution of protected areas management strategies and their complementarity for biodiversity conservation. *Biological Conservation*, *256*, 109014. doi:10.1016/j.biocon.2021.109014

Wang, S. (2023). International law-making process of combating plastic pollution: Status Quo, debates and prospects. *Marine Policy*, *147*, 105376. doi:10.1016/j.marpol.2022.105376

Young, J., Watt, A., Nowicki, P., Alard, D., Clitherow, J., Henle, K., Johnson, R., Laczko, E., McCracken, D., Matouch, S., Niemela, J., & Richards, C. (2005). Towards sustainable land use: Identifying and managing the conflicts between human activities and biodiversity conservation in Europe. *Biodiversity and Conservation*, *14*(7), 16411661. doi:10.1007/s10531-004-0536-z

Zhang, S., Zhou, Y., Yu, R., Xu, X., Xu, M., Li, G., Wang, W., & Yang, Y. (2022). China's biodiversity conservation in the process of implementing the sustainable development goals (SDGs). *Journal of Cleaner Production*, *338*, 130595. doi:10.1016/j.jclepro.2022.130595

Zhang, Y., Tariq, A., Hughes, A. C., Hong, D., Wei, F., Sun, H., Sardans, J., Peñuelas, J., Perry, G., Qiao, J., Kurban, A., Jia, X., Raimondo, D., Pan, B., Yang, W., Zhang, D., Li, W., Ahmed, Z., Beierkuhnlein, C., & Ma, K. (2023). Challenges and solutions to biodiversity conservation in arid lands. *The Science of the Total Environment*, *857*, 159695. doi:10.1016/j.scitotenv.2022.159695 PMID:36302433

Zhao, J. (2020). The role of international organizations in preventing conflicts between the SPS Agreement and the Cartagena Protocol on Biosafety. *Review of European, Comparative & International Environmental Law*, *29*(2), 271–281. doi:10.1111/reel.12326

Chapter 11
Strategies for Conservation in the Face of Biodiversity Decline

Asad Azeem
https://orcid.org/0000-0001-6550-3571
Department of Plant Breeding and Genetics, University of Layyah, Layyah, Paksiatan

Sami Ul-Allah
Department of Plant Breeding and Genetics, Bahauddin Zakariya University, Multan, Pakistan

Sana Khan
Department of Plant Breeding and Genetics, University of Agriculture, Faisalabad, Pakistan

Sonia Azeem
Department of Zoology, Government College University, Faisalabad, Layyah Campus, Layyah, Pakistan

Huma Shahid
Department of Environmental Science, Government College University, Faisalabad, Pakistan

Muhammad Zia Ul Haq
https://orcid.org/0000-0001-9914-3367
Department of Agronomy, University of Agriculture, Faisalabad, Pakistan

Waqas Ahmad
Department of Horticulture, University of Layyah, Layyah, Pakistan

ABSTRACT

In this chapter, the authors have analyzed the important role of different strategies such as satellite-based monitoring, remote sensing approach, as well as geographic information systems (GIS) in biodiversity conservation. Through the exploration of the multi-dimensional role of satellites, the research papers emphasize LiDAR and multispectral imaging as well as the effective integration of GIS. These techniques demonstrate how they can be used to tackle today's biodiversity problems. Spatial mapping and informed decision-making can be used to solve the complex problems created by the loss of biodiversity at the core level. And lastly, this chapter stresses the importance of all the techniques in protecting ecosystems, by using integrated approaches which combine the advantages of all the techniques and preserve biodiversity for a long time.

DOI: 10.4018/979-8-3693-3330-3.ch011

INTRODUCTION

Biodiversity is the changeability found in living things from all sources, such as aquatic, marine, and other terrestrial ecosystems, and the ecological complexes that comprise them. This includes variety within species as well as among species and within ecosystems (Hald-Mortensen, 2023). The incredible diversity of life on Earth is its most remarkable trait. The fact that there is biodiversity raises several critical scientific queries, such as how many species there are in the world. What causes and how does the Earth's biodiversity differ so much across its surface? How do various species coexist and evolve with their abiotic conditions within an ecosystem? What effects does this coexistence have on the evolution of species, and how does it alter over time? Despite recent advancements in science, these questions continue to be complicated (Lomolino et al., 2017). However, the functions and services of biodiversity provide the foundation of human well-being and sustainable development. These services include crop pollination, food and genetic resource provisioning, disease control, primary production, nutrient cycling, air and water purification, mitigation of greenhouse gas emissions and climate change, and educational, cultural, and spiritual benefits (Mi et al., 2021). However, biodiversity is being threatened by human-caused global changes in land use and climate, which are affecting both land and marine biodiversity from the poles to the tropics(Watson et al., 2019). What is biodiversity's response and adaptation to the Anthropocene's drastic changes? To effectively advise biodiversity protection, how can scientists assess and forecast the dangers of global extinctions and local diversity loss? These are the science of biodiversity's most pressing problems (Mi et al., 2021).

Global biodiversity loss threatens the resilience of land and marine-based ecosystems, which could impact human health and well-being and the underlying economy. To address the issue, policymakers and investors must comprehensively know the factors contributing to biodiversity loss (Hald-Mortensen, 2023). According to the 2019 IPBES Worldwide Measurement Report on Biodiversity and Ecosystem Services, the damage of biodiversity is caused by pollution, alterations in land and marine use, climate change, direct exploitation of living things, and the introduction of alien species. Agricultural expansion is responsible for 85% of the species at risk, starting with the primary driver, land-use change, specifically deforestation for agriculture (Hald-Mortensen, 2023). The European Union and its member states have pledged to stop biodiversity deficit at both the continental and worldwide levels over the past few decades, and they have established specific objectives to achieve this. Legal frameworks like the Birds and Habitats Directives (currently referred to as the Nature Directives) and the Marine Strategy Framework Directive, as well as programs like the EU Biodiversity Strategy and financial instruments like the LIFE program, promote attaining these aims (Hermoso et al., 2022). However, this significant effort has not stopped the loss of biodiversity on a continental scale. Although the Biodiversity Strategy for 2020 aims to improve the conservation status of 50% of species and 100% of habitats, only 15% of habitats and 27% of species recorded in the Birds Directive and Habitats Directive, respectively, are deemed to be at no predictable hazard of extinction. Climate change is increasingly impacting ecosystem services, biodiversity, and society. Thus, the EU is facing more environmental challenges than just meeting its targets for biodiversity conservation (Hermoso et al., 2022).

The data summarizing the state of conservation worldwide lend credibility to the idea that conservation is a crisis discipline, which has often been applied to the field (Gavin et al., 2018). For instance, the number and range of roughly one-third of vertebrate species are declining, and 42% of the invertebrate species listed on the IUCN Red List face extinction (Ceballos et al., 2017). However, conservation funding may need to rise by an order of magnitude to solve these issues and reach the biodiversity targets the Convention on Biological Diversity set forth (McCarthy et al., 2012). The survival of biodiversity depends on protected areas (Watson et al.), and globally coordinated efforts to expand PAs could significantly improve species safeguarding (Montesino Pouzols et al., 2014). The potential importance of other aspects of diversity in expressing threatened biodiversity heritage or maintaining functional and locally stable ecosystems that provide human assistances is well recognized, but our understanding of how PAs development could improve these aspects of variety is substantially smaller (Mazel et al., 2014). Although techniques for utilizing phylogenetic variety in conservation planning were created more than twenty years before and are essential to conservation, conservation efforts still need to consider ancestral relationships between species (Pollock et al., 2017). The functional diversity has historically only been used to estimate functional diversity across large spatial scales in conservation planning due to the limitations imposed by the quickly growing datasets of functional characters, habitat types or broad vegetation and dendrogram-based functional diversity metrics (Thuiller et al., 2015). Our study demonstrates the potential for significant increases in biodiversity protection while emphasizing the necessity of clearly establishing a connection between targeted conservation goals and biodiversity measurements. We offer a framework and numerical methods for these aims for protecting biodiversity in its many forms. The objective of this chapter is to highlight the strategies for conservation in the face of biodiversity decline.

UNDERSTANDING BIODIVERSITY DECLINE

Causes and Drivers of Biodiversity Loss

What causes the widespread loss of natural resources? Direct drivers affect the environment directly, whereas indirect drivers change the pace, direction, level, and strength of direct drivers. Direct drivers are distinguished from indirect drivers by their direct physical effects on the environment (Brondizio et al., 2019). Direct drivers can be either human-caused (pollution, deforestation, hunting, anthropogenic climate modification or natural (volcanic eruptions, earthquakes, weather events), or a combination of both (e.g., zoonotic diseases exacerbated by anthropogenic climate alteration). All indirect factors are political, economic, cultural, demographic, institutional, social, and human-caused; social ideals and narratives serve as their ultimate foundation (Díaz & Malhi, 2022). The loss of biodiversity is also a result of climate change, which has an impact on almost all ecosystems. The major disruption of ecosystems could be caused by rising temperatures, the acidity of the ocean, and extreme occurrences like changed ocean currents (Hald-Mortensen, 2023). The disturbance in the marine food chain due to overfishing creates conditions that lead to the extinction of species by endangering their chances of survival and causing biodiversity loss. The loss and fragmentation of natural habitats are consequences of urbanization, which can be observed by urban development and expansion. The result of this expansion is a loss of biodiversity, which is exacerbated by rising pollution levels and the induction of non-native species. The induction of invasive species by human activity is a significant danger to biodiversity. Invasive species can severely disrupt ecosystems by outcompeting native species, changing the environment, and dispersing disease (Hald-Mortensen, 2023). Although there might be other factors, these are the main causes of biodiversity loss, and they highlight the necessity for international cooperation in attempts to lessen the harm and protect biodiversity worldwide. These drivers include the spread of exotic species, urbanization, maritime exploitation, agricultural growth, and climate change. The knowledge base for action is created by evaluating the drivers contributing to biodiversity loss Hald-Mortensen, 2023).

Global and Local Trends in Biodiversity Decline

According to the best thorough worldwide systematic evaluation to date (Jaureguiberry et al., 2022; Purvis et al., 2019) the greatest common direct effects of nature's debility globally change in sea and land usage (such as coastal development, the development of agricultural frontiers, and deforestation,), as well as direct exploitation (such as selective logging, fishing, and hunting), with invasive species (introduced accidentally or voluntarily by humans or, much less frequently, arriving on their own) and pollution (such as light and noise pollution, seawater acidification due to augmented levels of CO_2 in the atmosphere, and) having less significance. The overall important ranking of direct causes varies by region and realm. For instance, the primary cause of the reduction in marine biodiversity is direct exploitation, primarily from fishing. Numerous studies with greater precision and geographic restrictions demonstrate that the importance of invasive species is higher on oceanic islands than on continents and that regions with Arctic and Mediterranean climates are more susceptible to the influences of climate alteration. While it hasn't been the primary cause of changes in nature on a worldwide scale thus far, climate change is expected to have an increasing impact for the remainder of the twenty-first century (Change, 2018; Díaz & Malhi, 2022). These findings are mostly supported by two recent global studies that focused more narrowly and precisely on the dispersal arrays of a variety of marine creatures as well as terrestrial amphibians, birds, and mammals (Harfoot et al., 2021; O'Hara et al., 2021). How are landscapes and plots of nature changing at even finer scales? This is significant because most local to regional sizes are where conservation and management initiatives are typically carried out, as well as the majority of the advantages that biodiversity confers on society (De Palma et al., 2018; Gonzalez et al., 2016). How much is global species migration contributing to the increase in taxonomic and functional biodiversity of local ecosystems, even when native species and global biodiversity indicators are declining? Many factors, including variations in the spatiotemporal design, constraints in the research techniques and primary datasets, and ecological affects, contribute to this disagreement (De Palma et al., 2018; Gonzalez et al., 2016). The balance between local damages and new entrances of native or non-native species is sometimes obscured by a continuous or almost constant total number of species, which is the final factor. Weak worldwide trends may be disguising regional variations, such as rising temperatures in temperate and boreal regions due to climate change, and falling temperatures in tropical regions due to land-use change (Dornelas et al., 2014). Before drawing certain results, more data and a better integration of approaches are required. The discovery that the local species composition and relative abundance are changing due to an acceleration of local colonization and extinction appears to be the cause of these trends, or rather their lack thereof, is less contentious (Dornelas et al., 2019).

SIGNIFICANCE AND PRINCIPLES OF BIODIVERSITY CONSERVATION

Importance of Conservation Planning

Effective worldwide conservation is more important than ever due to the increasing pressures on biodiversity, especially those brought on by climate change (Scheffers et al., 2016). The reassembling of communities will occur due to changing conditions that affect the comparative appropriateness for species, the phenology and fitness of species, the location and accessibility of suitable habitat for species, the strength and direction of species interactions and more (Bellard et al., 2012). Many of the dangers to biodiversity are still present, and conservation efforts are now insufficient to stop the reduction in biodiversity (Maxwell et al., 2016; Visconti et al., 2016). Systematic conservation planning (currently referred to as "conservation planning") offers a visible context for planning for climate alteration adaptation and employs specific objectives to allocate conservation resources efficiently (Reside et al., 2018). The conservation planning approach's capacity to evaluate and consider various objectives, future scenarios, and the uncertainties surrounding these and other components is one of its strong points (Kujala et al., 2013). Although it is generally accepted that conservation planning must consider climate change, its implementation still has significant obstacles. These include the following: advice on integrating knowledge into on-the-ground action, financing for the execution of climate alteration adaptation plans, ecosystems and an inadequate understanding of how climate modification affects target species, and processes (Hilbert et al., 2014). Systematic conservation planning (SCP) aims to facilitate decision-making about the distribution of funds for the preservation of biodiversity (McIntosh et al., 2017). he Great Barrier Reef Marine Park's rezoning and several South African conservation schemes are two of the most well-known uses of SCP (Pressey et al., 2013). SCP has heavily impacted the conservation planning practices of huge conservation groups in terms of World Wildlife. National and local governments, as well as The Nature Conservancy and Fund (Driver et al., 2012; Kareiva et al., 2014). Nowadays, SCP is considered one of the most rigorous and convincing paradigms for conservation planning because of its specialized objectives and ecological principles (Watson et al., 2011). The foundation of SCP is that guidance based on evidence encourages responsibility and openness when handling complex and morally charged decisions. Thus, based on this reasoning, we are interested in how much SCP is genuinely helping decision-makers and how, when, and where it is most likely to impact conservation (McIntosh et al., 2017).

Principles of Effective Biodiversity Conservation

Loss of biodiversity jeopardizes both human populations' well-being and the long-term sustainability of ecological services. Global policy efforts like the Convention on Biological Diversity have yet to stop biodiversity loss (Gavin et al., 2018). Effective biodiversity conservation is based on a comprehensive strategy that maintains the diversity and unpredictability of life on Earth. The improved adaptability can be ensured by conservation initiatives that support genetic diversity, such as maintaining varied seed banks for plant species or creating natural corridors for wildlife migration. Thus, a crucial component of biodiversity conservation is understanding and preserving genetic variety, which is necessary for species' survival and long-term viability (Hald-Mortensen, 2023). Biodiversity promotes our material well-being by producing valuables like, timber, fuel, medicine and food as well as enhancing the natural environment's aesthetic value. It also plays a crucial role in the diversity of habitats, ecological processes and biotic communities, that make up the biosphere (Rawat & Agarwal, 2015). Biodiversity is the basis of life. It offers the air, drink, and food, organisms' necessity to survive. Wetlands eliminate pollutants from water, trees and plants absorb carbon dioxide to slow down global warming, and fungi and bacteria break down organic matter to enrich the soil. The maintenance of the ecosystem services provided by biodiversity is achieved by preserving and protecting soil, conserving and cleaning water, preserving hydrological cycles, regulating biochemical cycles, absorbing and dissolving pollutants and waste materials through decomposition, and determining and regulating the global climate (Rawat & Agarwal, 2015). The environmental right of every living thing on Earth is unique and deserves respect, regardless of its value to humans. Regardless of whether it is valuable to humans or not, every organism has an intrinsic right to survive. The natural world values human legacy since humans are a part of it (Rawat & Agarwal, 2015).

Balancing Conservation With Sustainable Development

The vast and varied diversity of plants, animals, and microorganisms must be preserved to sustain ecological balance, which is essential for the reciprocal survival and existence of all living things (Verma, 2018). Humans should take proper care to maintain ecological balance and preserve biodiversity in all of its forms since they are the ones who benefit from environmental balance and biodiversity and because their anthropogenic activities are primarily to blame for ecological imbalance and biodiversity loss (Verma, 2018). Future generations will undoubtedly benefit from eco-friendly and sustainable initiatives regarding safety, equitable and sustainable development, and excellent health. The United Nations CBD aims to guarantee biodiversity conservation, responsible use of genetic resources, and benefit sharing. The Strategic Plan (SP) for Biodiversity 2011–2020 was developed to move closer to the goal of humanity, "Living in Harmony with Nature" by 2050. When that vision comes to pass, biodiversity will be respected, preserved, restored, and used responsibly to maintain ecosystem services and a healthy planet vital to all human beings. The SP applies to other UN Conventions focusing on nature and has 20 global (Aichi) aims (Locke et al., 2019). These sustainable practices and conservation responses are meant to be used in tandem, providing a logical foundation for both international collaboration and coordinated national efforts towards significant efforts to save the Earth's ecosystem. Similar circumstances translate into comparable domestic response options and duties for nations. Developed countries can support initiatives in other regions, mainly when their trade footprints result in biodiversity loss in other nations (Locke et al., 2019).

CONSERVATION STRATEGIES OF BIODIVERSITY CONSERVATION

Following are the Biodiversity Conservation Strategies:

PROTECTED AREAS AND CONSERVATION RESERVES

Role of Protected Areas in Biodiversity Conservation

The degree to which protected areas can stop extinctions varies. The concept of "irreplaceability" refers to how much a site could contribute to conservation objectives or, on the other hand, how many opportunities would be lost if the site disappeared. The concept of "irreplaceability" refers to how much a site could contribute to conservation objectives or, on the other hand, how many opportunities would be lost if the site disappeared (Le Saout et al., 2013). Since protected areas are essential for preventing habitat loss, they are commonly acknowledged as the most significant conservation technique (Suding, 2011). There is abundant indication that PAs will be crucial for species conservation in the future, and they have helped species expand their range (Thomas & Gillingham, 2015). It is valuable for species to track climate space upslope because PAs are extremely found in high-elevation, difficult topography places. It is expected that ectothermic plants and animals, which make up the majority of upper-elevation specialist species, will lose their climatic niche from the tops of mountains, while mountainous regions will probably serve as homes for emerging groups (Shoo et al., 2011). More than 200,000 protected areas exist worldwide, many of which were established to preserve specific species of significance (Deguignet et al., 2014). However, beyond the long-term survival of the target species, their creation and effective management may have other advantages, such as the preservation of entire pools of biodiversity and the supply of ecosystem services like carbon sequestration (Viña & Liu, 2017). The establishment of nature reserves in China has accelerated, especially in the previous ten years. Approximately fifteen percent of the country's land surface was covered by 2729 nature reserves as of the end of 2014 (Xu et al., 2017). Being one of the 17 "mega-diversity" countries of the world, the need to preserve and reestablish the habitat of specific charismatic and endangered species, like the giant panda, the elephant, and the tige, has motivated the establishment of many of these reserves. The latter's status as a global symbol of biodiversity conservation led to the conservation of the species being given national priority (Viña & Liu, 2017).

Challenges and Success Stories

Addressing the CBD report's emphasis on Indigenous-led projects and protected/conserved lands, there were some significant constraints. For instance, certain protected/conserved lands belonging to Indigenous Peoples are either not designated as ICCAs due to the decision of the Indigenous organizations in charge of them or for other reasons, or they are designated as such without the agreement of the Indigenous Peoples (Jonas et al., 2017). Secondly, regions governed by both local and Indigenous groups are included in ICCAs (Tran et al., 2020). The historical and contemporary contexts, aspirations, and challenges faced by Indigenous Peoples differ significantly from those of local communities. These differences include their close relationship with the land and natural resources of their territory, their unique socio-political system, their language, beliefs, values, and culture and their perception of themselves as distinct from the leading sectors of their national society (Tran et al., 2020). IPCAs have demonstrated advantages for ecological conservation. However, they are also subject to the same national and international risks that affect other kinds of globally protected and conserved regions (Lotze et al., 2018; Schulze et al., 2018). Global resource extraction, climate change, invasive species, tourism and recreation, and other causes pose a threat to both terrestrial and marine ecosystems as well as PAs on a scale that is considerably larger than any one territory or jurisdiction can handle (Lotze et al., 2018; Schulze et al., 2018). Both non-Indigenous and Indigenous-led conservation projects, especially those in protected and conserved areas, must make commitments and make real efforts to solve these conservation concerns. The World Conservation Monitoring Center reports that there are about 1671 recognized species of birds, mammals, reptiles, and amphibians in Malaysia. At least 15,500 species of vascular plants, 23.2% of which are indigenous, may be found in Malaysia. Concurrently, 4.1% of Malaysia is classified as protected by IUCN classifications I–V. The several measures implemented to oversee protected areas align with the nation's dedication to conserving biodiversity, and when combined, they surpass the global biodiversity objective of ten percent (Hamzah, 2011). Sabah's ability to halt the loss of forest cover and implement reforestation initiatives has resulted in the restoration of wildlife habitats, such as those found in Danum Valley and Lower Kinabatangan. Consequently, the ecotourism sector in the State has gained momentum and is well recognized as an exemplar for sustainable ecotourism growth. As a result, many other Malaysian states are encouraging and expanding ecotourism in protected areas in the same manner as Sabah. As a result, many other Malaysian states are encouraging and expanding ecotourism in protected areas in the same manner as Sabah (Hamzah, 2011).

Designing and Managing Effective Conservation Reserves

It's not simple to preserve nature's legacy, even if humanity faces an urgent biodiversity crisis (Peh, 2011). The crucial instrument for achieving conservation goals for biodiversity preservation is the creation of reserve networks. Conservation goals may come from research supported by data or from recommendations made by policymakers. The amount of the natural landscape that must be preserved is up for debate, although reserve networks should ideally be sufficiently big to provide sufficient ecological representation and persistence (Françoso et al., 2015). The conservation targets set forth by the CBD (also known as the UNEP) and ratified by 168 nations, including Brazil, are the most broadly applicable. The recently approved revised and updated strategy plan for global biodiversity conservation for 2011–2020 in Aichi, Japan, includes the setting of conservation targets of 17% for terrestrial and inland water ecosystems and 10% for marine and coastal habitats. Brazilian PAs established by national, state, or local governments fall into one of twelve categories, which are divided into two groups: Passages for Sustainable Use (SU) and Integral Protection (IP) (Françoso et al., 2015). The latter seeks to encourage nature conservation and the balanced use of natural resources, while the former's main objective is the protection of natural resources. Many nations (particularly those that have ratified the CBD) devote significant resources to the identification, establishment, and upkeep of Protected Areas (Watson et al.), as this is the most pragmatic framework for the preservation of biodiversity. However, federal IPs, particularly those that need to be expropriated, such as parks, REBIOs, and EEs, must receive the majority of funding for the establishment of PAs in Cerrado. National parks are frequently huge, making them more effective in promoting ecological tourism, which benefits local people and society (Françoso et al., 2015). Natural ecosystems are disappearing and arable land is still growing almost everywhere in the world outside of New Zealand, Australia, and Europe. The agricultural matrix among regions of natural habitat is exploited by numerous wild species and provides critical resources for particular, even if future conversion to farming can be prevented (Mendenhall et al., 2014). Accordingly, in the upcoming decades, conservation regulations may include standard components like AES that promote farming methods that are less detrimental to wildlife (Batáry et al., 2015).

COMMUNITY-BASED CONSERVATION

Engaging Local Communities in Conservation Efforts

Community ecologists have long been curious about how species coexist in varied ecosystems where many species have similar resource requirements. According to classical theory, stable coexistence happens when fitness disparities are balanced by stabilizing niche differences, which leads to species more severely limiting themselves than their heterospecific competitors (Mi et al., 2021) The processes of dispersion limitation, abiotic filtering, resource competition, and trophic interactions with mutualists and natural enemies which govern a species' spread and abundance. Understanding the fundamental mechanisms of species coexistence is crucial in light of the fast degradation and destruction of natural ecosystems, as it can help prevent the local extinction of some species and guide ecosystem restoration efforts. Moreover, a community's ability to adapt to environmental changes is frequently attributed to its degree of redundancy (Mi et al., 2021). It would be easier to understand the resilience of systems and to restore them in order to maintain ecosystem function if species and guilds were better understood. It is becoming increasingly clear that without local communities participating in the creation, administration, and oversight of conservation programs, no legal requirements can be effective. Numerous government and non-governmental groups are launching measures to do this (Rawat & Agarwal, 2015). For instance, village residents' participation in the protection and rehabilitation of degraded forest areas close to their settlements is highly valued in the Joint Forest Management concept. Effective conservation efforts depend on the confidence and participation of local communities (Dobhal et al., 2011).

Examples of Successful Community-Led Conservation Projects

Conservation that is headed by locals as opposed to externally directed activities is known as community-led conservation. Since local natural resources have historically been essential to livelihoods, community-led conservation is a complicated and long-standing issue (Rytkönen & Hotakainen, 2020). Community activism on environmental issues has become more intense in recent years. The civic initiatives include the well-known transition town movement, local currencies, co-housing plans, and the creation of communal land trusts (Dunkley & Franklin, 2017). In the UK, a localist narrative supports these programs and prompts governing bodies to promote the advantages of community-driven projects (Catney et al., 2014). The need for biodiversity conservation transcends national boundaries. It is an important worldwide issue. In 1972, the preservation of biological diversity and the responsible utilization of its constituent parts gained prominence) (Rawat & Agarwal, 2015). According to (Doughty, 2016) local knowledge and involvement in the battle against climate change have a lot of promise. Local populations are knowledgeable about the weather, microclimates, and flora in their area. Frequently, they have prior experience adjusting to shifting climate conditions. For instance, the CBD has strengthened international cooperation in preserving biodiversity and promoting sustainable development. The Strategic Plan for Biodiversity 2011–2020, which includes the Aichi Biodiversity Targets, was adopted as a result of the CBD (Armitage et al., 2020). At the international level, the Strategic Plan and the Aichi goals continue to be important tools for enforcing socially and environmentally fair biodiversity conservation. The ratifying States have implemented a range of policies and legislation to facilitate the implementation of the CBD and these agreements, albeit to differing degrees of success (Armitage et al., 2020).

RESTORATION ECOLOGY

Population, community, and ecosystem ecology have emerged as divisions of ecological study that concentrate on distinct levels of organization, while restoration ecology is a very recent and all-encompassing field. Restoration ecology, defined as the study of the connections between organisms and their environment in the context of ecological restoration, encompasses all areas of ecological research and encompasses the study of genes to whole landscapes. The connection between restoration ecology and the broader concept of ecology is not accidental; restoration ecology may be regarded as a distinct field within ecological study, delineated by its specific setting. The primary objective of restoration ecology is to analyze and measure the deviations from a typical ecosystem condition, encompassing all forms of spatial and temporal fluctuations (Gerwing et al., 2022). It seeks to comprehend the factors that cause and control these deviations, and subsequently forecast the methods by which the system can be returned to a less disrupted state or the original state where it was before. Restoration ecology incorporates multiple interconnected fields like as hydrology, geomorphology, oceanography, and other relevant social science areas.

Rehabilitating Degraded Ecosystems

Restorationists strive to transition a damaged system's composition, structure, and dynamics to an ecological state that falls within an acceptable limit, in comparison to a less modified and perhaps more sustainable system. In this simplified paradigm, features relate to the structural components of an ecosystem. An illustration of this is the fact that floodplains play a crucial role in river ecosystems, since their relationship with both water and land is a fundamental characteristic of their structure(Palmer et al., 2017). Similarly, the composition of a forest encompasses several categories of tree sizes and characteristics of the canopy. One element of its arrangement is the spatial dispersion of trees throughout the terrain. Processes, also known as functions, include a diverse array of dynamic characteristics, including stream flow, fire patterns primary production, dispersion and migration, biogeochemical cycles, and population dynamics. Processes exhibit temporal and spatial variations, and ideally result in the restoration of a self-sustaining dynamic system that requires less human involvement compared to the restoration process itself (Beechie et al., 2010). The ecological restoration framework strives to restore the qualities of a whole system, including, food webs, species assemblages and functional attributes, to resemble those of reference systems. However, the process of restoration may span many decades (Kabir et al., 2015) . An unexpected consequence refers to a result that is not intentionally aimed at achieving complete ecosystem recovery and adherence to a suitable reference system. Efforts to counteract environmental damage, which do not involve ecological restoration, encompass the utilization of hardscapes or nonnative species to mitigate excessive soil erosion and run-off. Additionally, engineered systems that are incapable of sustaining themselves due to their design or location in a heavily altered landscape context, such as strip mining operations, chemically polluted brownfield sites, or severely eroded sites, are also employed (Mori & Isbell, 2024). Additional instances of projects with restricted goals encompass endeavors aimed at maximizing a solitary ecosystem service, such as fortifying a steep incline through the implementation of a monoculture of nonnative, deeply-rooted trees, or restoring a previously forested area that has undergone mining activities to a nonnative grassland.

Achieving Balance Between Conservation and Restoration

Conservation refers to the responsible and sustainable management and utilization of biodiversity and natural resources in order to avoid their decline or deterioration. It entails the careful equilibrium between resource use and their safeguarding and administration, achieved via active human participation. It encompasses both restoration and preservation activities. Due to the fact that many protected areas permit sustainable consumption of natural resources and industrial operations, the word 'protection' may sometimes and inadvertently be employed interchangeably with 'conservation'. Conservation biology assists professionals in evaluating the effects of human activities on ecosystems, reducing the decline of biodiversity, and maintaining the overall health of ecosystems (Mori & Isbell, 2024). It often prioritizes the preservation and safeguarding of resources, in line with the wider goals of conservation. The principles guiding objectives for conservation of nature and ecological restoration have typically centered on the prevention of ecological change or the restoration of ecosystems to a desired historical state. Attaining a healthy equilibrium between the conservation and restoration of natural resources is crucial for the effective management of the environment and the enduring viability of ecosystems.

Technological Innovations in Conservation

Our century has been characterized by technological innovation, which has greatly advanced humankind and profoundly changed numerous aspects of society Still, in spite of appeals to adopt new technologies while looking for answers, the advantages of recent technology advancements in biodiversity protection have not yet been fully acknowledged. Conservation projects, on the other hand, often use "old technologies" in their techniques. Technology offers essential tools for enhancing data collection and monitoring species, habitats, and threats, which helps inform management choices. The majority of these gadgets have one thing in common: their technology, or at least the ones that supported them, were first created for other uses (like military or medical) before being "adopted" by the conservation community. This may also help to explain the skepticism that current technology-based conservation techniques have toward: (1) monitoring animals and the surroundings they live in, given the widespread public's easy access to a variety of sophisticated monitoring instruments (GPS tracking devices or drones) and (2) Genetics for conservation (Czech, 2003).

Satellite-Based Monitoring

Satellites are crucially significant in biodiversity conservation as they have a multitude of functions (Lausch et al., 2024). The satellites tracked from the different stations on Earth serve as a vantage for Earth observation and therefore they contribute to extensive habitat mapping, monitoring and control of land use. More significantly, with special imaging and monitoring systems such as high-resolution imagery and synthetic aperture radar in the satellites, the precision and feasibility of the monitoring activities are highly elevated (Jiao, 2024).

Integration With Geographic Information Systems (GIS)

GIS technology along with satellite and remote sensing data enhances the impacts of the conservation activities (Aslam et al., 2024). GIS helps to produce more detailed maps and spatial analyses that give a clear picture of ecological structures. Therefore, well-informed decision-making for conservation is boosted by geospatial data visualization representation, enabling site selection and enhancement. Successful integration examples include landscape corridor mapping, land-use pattern analysis, and conservation-efficiency approaches through a spatially based approach. Therefore, the technologies such as; satellites, remote sensing, and GIS technologies together highlight their collective capabilities in addressing the complicated issues of biodiversity conservation (Pandey & Pandey, 2023).

Conservation Drones and AI Applications

Unprecedented environmental issues, including pollution, deforestation, biodiversity loss, and climate change, are present in the modern period. In spite of these obstacles, artificial intelligence has become a technical wonder with great potential for environmental preservation. With AI's capacity to analyze massive data sets, spot trends, and make deft judgments instantly, we have an unmatched chance to completely transform how we approach sustainable growth and environmental preservation(Chalmers et al., 2019; Raihan, 2023). The rapid progress in artificial intelligence (AI) has resulted in its extensive application in a multitude of fields, such as social networks, e-commerce, agriculture, environmental sustainability, education, healthcare, data manipulation, urban design, public safety, transportation, and environmental conservation (Rayhan, 2023). The development and application of artificial intelligence (AI) in any discipline require the utilization of large and high-quality datasets, Internet-of-things (IoT) network components, technological advances including high-resolution cameras, satellite technology, sensors, drones, and unmanned aerial vehicles (UAVs), as well as computational space and storage (Bondi et al., 2018).

Integrating Technology for Monitoring and Management

Data from monitoring is used to assess the efficacy of current laws and programs, to guide conservation policy, and to record the state and trends of biodiversity(Stephenson, 2020). Developing enough capacity for biodiversity monitoring presents a variety of difficulties, depending on the environment, species, and monitoring method in issue. A multi-pronged strategy with various programs for monitoring (such as comprehensive and intense sample regimes and earth observations) is a well-known technique which can be adopted in places where it can be challenging to offer enough biodiversity monitoring coverage (Schmeller et al., n.d.). This strategy also includes the essential capacity development initiatives. At the same time, statistical techniques must be developed to address any potential drawbacks that these new approaches may present. Despite having a diverse range of species in their flora and fauna, many tropical and subtropical nations are still falling behind in biodiversity monitoring due to resource constraints (Kühl et al., 2020). The scientific community of countries with greater experience in monitoring programs can support the creation of monitoring programs in species-rich environments, the identification of suitable monitoring sites, and the resolution of taxonomic obstacles through technology transfer.

ASSESSING BIODIVERSITY THREATS

Identifying Key Threats to Biodiversity

There is no indication that the process of reducing biodiversity and causing corresponding changes in the environment will slow down, as it is happening at a rate never seen in human history. Human activity has drastically impacted and altered almost every ecosystem on Earth, and it is constantly being converted for agricultural and other purposes. The number and geographic distribution of many plant and animal populations have reduced. Extinction of species is a normal phase of Earth's history; however, due to human activities, the pace of extinction has accelerated by at least hundred times relative to the natural rate. Numerous factors contribute to biodiversity loss (Rawat & Agarwal, 2015). The loss and transformation of habitats by changing land uses, overuse of biological resources, pollution, climate change, and invasive species are some of the reasons contributing to the rapid decline of biodiversity. Several environmental or human-caused variables frequently interact and intensify one another (Rawat & Agarwal, 2015). Overall, habitat degradation and change are the primary factors directly contributing to biodiversity loss worldwide. Destroying a habitat makes it impossible for anyone environment to sustain the species that calls it home. As a result of the displacement or destruction of existing creatures in the ecosystem, biodiversity is decreased (Agarwal et al., 2011). It occurs when the rate at which members of a specific species are being taken exceeds the rate at which the population being harvested can naturally reproduce. This can be accomplished by trading, hunting, fishing, obtaining food, etc. Many species, including trees, animals used for meat production, marine fish, and invertebrates, are still seriously threatened by overexploitation. An ecosystem will change as a result of newly imported species. Biological pollution is the standard term used to describe such introduced organisms. According to (Dumont, 2012) global losses of pristine biodiversity until the middle of the twenty-first century will be primarily dependent on the global rate of human birth. Therefore, maintaining human population control is essential to the preservation of biodiversity.

Tools and Methods for Threat Assessment

The past 200 years have seen a dramatic shift in the identification and monitoring of biodiversity. It is a given that the rate at which species are going extinct will surpass the rate at which biodiversity is being inventoried and characterized, given the current identification and monitoring methods. A paradigm shifts in the way we quantify and document biodiversity is needed, combining approaches at the landscape level (top-down) and species level (bottom-up) (Roy et al., 2012). It would be beneficial to identify regions that may be rich in biological diversity and to immediately begin a thorough inventory of biodiversity in such areas. Recent advancements in Geographic Information Systems and Remote Sensing technologies have allowed us to stratify vegetation types according to ecological gradients and environmental factors, as well as to categorize and map species congregation geographically. It's critical to have accurate and standardized protocols for biodiversity data collection, monitoring, and distribution. The modern instruments employed to identify and track biological diversity can be broadly divided into three groups: (i) characterizing biodiversity at the landscape scale (using multispectral or hyperspectral satellite data or aerial photography for mapping and modeling); field inventorying with the latest instruments and methods (such as GPS (Global Positioning System) tagging for field measurements and documentation using Personal Digital Assistants (PDAs); and (iii) accurate documentation and distribution through the use of information technology(Roy et al., 2012). The latest developments in the domains of remote sensing and geographic information systems, mobile computing, GPS technology, wireless connectivity, and the internet have made it possible to accurately and consistently document biological diversity. This revisit capability aids in identifying biodiversity hotspots and areas that require more research (Roy et al., 2012). According to its Red List categories and criteria 6,20, the IUCN has evaluated the risk of extinction for 61,000 animal species, nearly all of which are bird and mammal species. Each species has been assigned one of the following statuses. The extinction risk (v) of 0 indicates "least concern," while "near threatened" (v=1), "vulnerable" (v=2), "endangered" (v=3), "critically endangered" (v=4), and "extinct" or "extinct in the wild" (v=5) are the other classifications. The IUCN has become our standard nomenclature; if a species is classified as vulnerable, endangered, or critically endangered, it is deemed threatened. The percentage of all species that are threatened with extinction and the mean extinction risk value for all the species in a country are two additional metrics we utilized to assess the extinction risks posed by the mammal and bird species in that location (Tilman et al., 2017).

CONCLUSION

Taking everything into consideration, although, satellite-based monitoring, remote sensing techniques, and Geographic Information Systems (GIS) seem to be three different concepts, but the combination of these techniques shows the effectiveness they can achieve to support biodiversity conservation. Satellites that can observe the planet from a bird-eye view are one of the main contributors to habitat mapping, as well as to the monitoring of land use. Besides that, various types of remote sensing techniques allow us to investigate ecosystems in a more detailed way such as using LiDAR and multispectral imaging. The augmentation with GIS greatly helps metaphorical comprehension of ecological patterns and thus, facilitates informed decisions. Productive application of the technologies, like wildlife corridor mapping and land-use analysis, vividly demonstrate the force behind these tools. This complementary partnership lays a solid base for the resolution of the complex situations of biodiversity conservation, giving us a new way forward for protecting this planet from the extinction of diverse ecosystems.

REFERENCES

Agarwal, N., Singh, G., & Singh, H. (2011). Present status of Icthyofaunal diversity of Garhwal Himalayan river Bhilangna and its tributaries with reference to changing environment. *Environment Conservation Journal*, *12*(3), 101–108. doi:10.36953/ECJ.2011.120319

Armitage, D., Mbatha, P., Muhl, E. K., Rice, W., & Sowman, M. (2020). Governance principles for community-centered conservation in the post-2020 global biodiversity framework. *Conservation Science and Practice*, *2*(2), e160. doi:10.1111/csp2.160

Aslam, R. W., Shu, H., Naz, I., Quddoos, A., Yaseen, A., Gulshad, K., & Alarifi, S. S. (2024). Machine Learning-Based Wetland Vulnerability Assessment in the Sindh Province Ramsar Site Using Remote Sensing Data. *Remote Sensing (Basel)*, *16*(5), 928. doi:10.3390/rs16050928

Batáry, P., Dicks, L. V., Kleijn, D., & Sutherland, W. J. (2015). The role of agri-environment schemes in conservation and environmental management. *Conservation Biology*, *29*(4), 1006–1016. doi:10.1111/cobi.12536 PMID:25997591

Beechie, T. J., Sear, D. A., Olden, J. D., Pess, G. R., Buffington, J. M., Moir, H., Roni, P., & Pollock, M. M. (2010). Process-based principles for restoring river ecosystems. *Bioscience*, *60*(3), 209–222. doi:10.1525/bio.2010.60.3.7

Bellard, C., Bertelsmeier, C., Leadley, P., Thuiller, W., & Courchamp, F. (2012). Impacts of climate change on the future of biodiversity. *Ecology Letters*, *15*(4), 365–377. doi:10.1111/j.1461-0248.2011.01736.x PMID:22257223

Bondi, E., Fang, F., Hamilton, M., Kar, D., Dmello, D., Choi, J., Hannaford, R., Iyer, A., Joppa, L., Tambe, M., & Nevatia, R. (2018). Spot poachers in action: Augmenting conservation drones with automatic detection in near real time. *32nd AAAI Conference on Artificial Intelligence, AAAI 2018*, (pp. 7741–7746). IEEE. 10.1609/aaai.v32i1.11414

BrondizioE.DíazS. M.SetteleJ.NgoH.GuezeM.Aumeeruddy-ThomasY.BaiX.GeschkeA.MolnárZ.Niamir A. (2019). Assessing a planet in transformation: rationale and approach of the IPBES Global Assessment on Biodiversity and Ecosystem Services. doi:10.5281/zenodo.3831852

Catney, P., MacGregor, S., Dobson, A., Hall, S. M., Royston, S., Robinson, Z., Ormerod, M., & Ross, S. (2014). Big society, little justice? Community renewable energy and the politics of localism. *Local Environment*, *19*(7), 715–730. doi:10.1080/13549839.2013.792044

Ceballos, G., Ehrlich, P. R., & Dirzo, R. (2017). Biological annihilation via the ongoing sixth mass extinction signaled by vertebrate population losses and declines. *Proceedings of the National Academy of Sciences of the United States of America*, *114*(30), E6089–E6096. doi:10.1073/pnas.1704949114 PMID:28696295

ChalmersC.FergusP.WichS.MontanezA. C. (2019). *Conservation AI: Live Stream Analysis for the Detection of Endangered Species Using Convolutional Neural Networks and Drone Technology*. http://arxiv.org/abs/1910.07360

Change, P. C. (2018). *Global warming of 1.5° C*. World Meteorological Organization.

Czech, B. (2003). Technological Progress and Biodiversity Conservation: A Dollar Spent, a Dollar Burned. *Conservation Biology*, *17*(5), 1455–1457. doi:10.1046/j.1523-1739.2003.02370.x

De Palma, A., Sanchez-Ortiz, K., Martin, P. A., Chadwick, A., Gilbert, G., Bates, A. E., Börger, L., Contu, S., Hill, S. L., & Purvis, A. (2018). Challenges with inferring how land-use affects terrestrial biodiversity: study design, time, space and synthesis. In *Advances in ecological research* (Vol. 58, pp. 163–199). Elsevier. doi:10.1016/bs.aecr.2017.12.004

Deguignet, M., Juffe-Bignoli, D., Harrison, J., MacSharry, B., Burgess, N., & Kingston, N. (2014). *United Nations list of protected areas*. UNEP-WCMC.

Díaz, S., & Malhi, Y. (2022). Biodiversity: Concepts, patterns, trends, and perspectives. *Annual Review of Environment and Resources*, *47*(1), 31–63. doi:10.1146/annurev-environ-120120-054300

Dobhal, R., Kumar, A., & Rawat, S. (2011). Conservation and Management of Bioresources in Uttarakhand, India. *Community based bio-diversity conservation in the Himalay, The Energy and Resource Institute (TERI)*.

Dornelas, M., Gotelli, N. J., McGill, B., Shimadzu, H., Moyes, F., Sievers, C., & Magurran, A. E. (2014). Assemblage time series reveal biodiversity change but not systematic loss. *Science*, *344*(6181), 296–299. doi:10.1126/science.1248484 PMID:24744374

Doughty, C. A. (2016). Building climate change resilience through local cooperation: A Peruvian Andes case study. *Regional Environmental Change*, *16*(8), 2187–2197. doi:10.1007/s10113-015-0882-2

Driver, A., Sink, K. J., Nel, J. L., Holness, S., Van Niekerk, L., Daniels, F., Jonas, Z., Majiedt, P. A., Harris, L., & Maze, K. (2012). *National Biodiversity Assessment 2011: An assessment of South Africa's biodiversity and ecosystems*.

Dumont, E. (2012). Estimated impact of global population growth on future wilderness extent. *Earth System Dynamics Discussions*, *3*(1), 433–452. doi:10.5194/esdd-3-433-2012

Dunkley, R. A., & Franklin, A. (2017). Failing better: The stochastic art of evaluating community-led environmental action programs. *Evaluation and Program Planning*, *60*, 112–122. doi:10.1016/j.evalprogplan.2016.11.005 PMID:27842275

Françoso, R. D., Brandão, R., Nogueira, C. C., Salmona, Y. B., Machado, R. B., & Colli, G. R. (2015). Habitat loss and the effectiveness of protected areas in the Cerrado Biodiversity Hotspot. *Natureza & Conservação*, *13*(1), 35–40. doi:10.1016/j.ncon.2015.04.001

Gavin, M. C., McCarter, J., Berkes, F., Mead, A. T. P., Sterling, E. J., Tang, R., & Turner, N. J. (2018). Effective biodiversity conservation requires dynamic, pluralistic, partnership-based approaches. *Sustainability (Basel)*, *10*(6), 1846. doi:10.3390/su10061846

Gerwing, T. G., Hawkes, V. C., Gann, G. D., & Murphy, S. D. (2022). Restoration, reclamation, and rehabilitation: on the need for, and positing a definition of, ecological reclamation. *Restoration Ecology, 30*(7), 0–2. doi:10.1111/rec.13461

Gonzalez, A., Cardinale, B. J., Allington, G. R., Byrnes, J., Arthur Endsley, K., Brown, D. G., Hooper, D. U., Isbell, F., O'Connor, M. I., & Loreau, M. (2016). Estimating local biodiversity change: A critique of papers claiming no net loss of local diversity. *Ecology*, *97*(8), 1949–1960. doi:10.1890/15-1759.1 PMID:27859190

Hald-Mortensen, C. (2023). The Main Drivers of Biodiversity Loss: A Brief Overview. *Journal of Ecology and Natural Resources*, *7*(3), 000346. doi:10.23880/jenr-16000346

Harfoot, M. B., Johnston, A., Balmford, A., Burgess, N. D., Butchart, S. H., Dias, M. P., Hazin, C., Hilton-Taylor, C., Hoffmann, M., & Isaac, N. J. (2021). Using the IUCN Red List to map threats to terrestrial vertebrates at global scale. *Nature Ecology & Evolution, 5*(11), 1510-1519. doi:10.1038/s41559-021-01542-9

Hermoso, V., Carvalho, S., Giakoumi, S., Goldsborough, D., Katsanevakis, S., Leontiou, S., Markantonatou, V., Rumes, B., Vogiatzakis, I., & Yates, K. (2022). The EU Biodiversity Strategy for 2030: Opportunities and challenges on the path towards biodiversity recovery. *Environmental Science & Policy, 127*, 263–271. doi:10.1016/j.envsci.2021.10.028

Hilbert, D. W., Hill, R., Moran, C., Turton, S. M., Bohnet, I., Marshall, N. A., Pert, P. L., Stoeckl, N., Murphy, H. T., & Reside, A. E. (2014). *Climate change issues and impacts in the Wet Tropics NRM cluster region*. James Cook University.

Jaureguiberry, P., Titeux, N., Wiemers, M., Bowler, D. E., Coscieme, L., Golden, A. S., Guerra, C. A., Jacob, U., Takahashi, Y., Settele, J., Díaz, S., Molnár, Z., & Purvis, A. (2022). The direct drivers of recent global anthropogenic biodiversity loss. *Science Advances, 8*(45). doi:10.1126/sciadv.abm9982 PMID:36351024

Jiao, Z. (2024). The Application of Remote Sensing Techniques in Ecological Environment Monitoring. *Highlights in Science. Engineering and Technology, 81*, 449–455.

Jonas, H. D., Lee, E., Jonas, H. C., Matallana-Tobon, C., Wright, K. S., Nelson, F., & Enns, E. (2017). Will "other effective area-based conservation measures" increase recognition and support for ICCAs. *Parks, 23*(2), 63–78. doi:10.2305/IUCN.CH.2017.PARKS-23-2HDJ.en

Kabir, S. M. Z., Rabbi, F., & Chowdhury, M. B. (2015). Mine Closure Planning and Practice in Canada and Australia : A Comparative Review. *World Review of Business Research, 5*(3), 1–22.

Kareiva, P., Groves, C., & Marvier, M. (2014). The evolving linkage between conservation science and practice at T he N ature C onservancy. *Journal of Applied Ecology, 51*(5), 1137–1147. doi:10.1111/1365-2664.12259 PMID:25641980

Kühl, H. S., Bowler, D. E., Bösch, L., Bruelheide, H., Dauber, J., Eichenberg, D., Eisenhauer, N., Fernández, N., Guerra, C. A., Henle, K., Herbinger, I., Isaac, N. J. B., Jansen, F., König-Ries, B., Kühn, I., Nilsen, E. B., Pe'er, G., Richter, A., Schulte, R., & Bonn, A. (2020). Effective Biodiversity Monitoring Needs a Culture of Integration. *One Earth, 3*(4), 462–474. doi:10.1016/j.oneear.2020.09.010

Kujala, H., Burgman, M. A., & Moilanen, A. (2013). Treatment of uncertainty in conservation under climate change. *Conservation Letters, 6*(2), 73–85. doi:10.1111/j.1755-263X.2012.00299.x

Lausch, A., Selsam, P., Pause, M., & Bumberger, J. (2024). Monitoring vegetation-and geodiversity with remote sensing and traits. *Philosophical Transactions. Series A, Mathematical, Physical, and Engineering Sciences, 382*(2269), 20230058. doi:10.1098/rsta.2023.0058 PMID:38342219

Le Saout, S., Hoffmann, M., Shi, Y., Hughes, A., Bernard, C., Brooks, T. M., Bertzky, B., Butchart, S. H., Stuart, S. N., Badman, T., & Rodrigues, A. S. L. (2013). Protected areas and effective biodiversity conservation. *Science, 342*(6160), 803–805. doi:10.1126/science.1239268 PMID:24233709

Locke, H., Ellis, E. C., Venter, O., Schuster, R., Ma, K., Shen, X., Woodley, S., Kingston, N., Bhola, N., Strassburg, B. B., Paulsch, A., Williams, B., & Watson, J. E. M. (2019). Three global conditions for biodiversity conservation and sustainable use: An implementation framework. *National Science Review*, *6*(6), 1080–1082. doi:10.1093/nsr/nwz136 PMID:34691979

Lomolino, M., Riddle, B., & Whittaker, R. (2017). *Biogeography: Biological diversity across time and space*. Sinauer Associates.

Lotze, H. K., Guest, H., O'Leary, J., Tuda, A., & Wallace, D. (2018). Public perceptions of marine threats and protection from around the world. *Ocean and Coastal Management*, *152*, 14–22. doi:10.1016/j.ocecoaman.2017.11.004

Maxwell, S. L., Fuller, R. A., Brooks, T. M., & Watson, J. E. (2016). Biodiversity: The ravages of guns, nets and bulldozers. *Nature*, *536*(7615), 143–145. doi:10.1038/536143a PMID:27510207

Mazel, F., Guilhaumon, F., Mouquet, N., Devictor, V., Gravel, D., Renaud, J., Cianciaruso, M. V., Loyola, R., Diniz-Filho, J. A. F., Mouillot, D., & Thuiller, W. (2014). Multifaceted diversity–area relationships reveal global hotspots of mammalian species, trait and lineage diversity. *Global Ecology and Biogeography*, *23*(8), 836–847. doi:10.1111/geb.12158 PMID:25071413

McCarthy, D. P., Donald, P. F., Scharlemann, J. P., Buchanan, G. M., Balmford, A., Green, J. M., Bennun, L. A., Burgess, N. D., Fishpool, L. D., Garnett, S. T., Leonard, D. L., Maloney, R. F., Morling, P., Schaefer, H. M., Symes, A., Wiedenfeld, D. A., & Butchart, S. H. M. (2012). Financial costs of meeting global biodiversity conservation targets: Current spending and unmet needs. *Science*, *338*(6109), 946–949. doi:10.1126/science.1229803 PMID:23065904

McIntosh, E. J., Pressey, R. L., Lloyd, S., Smith, R. J., & Grenyer, R. (2017). The impact of systematic conservation planning. *Annual Review of Environment and Resources*, *42*(1), 677–697. doi:10.1146/annurev-environ-102016-060902

Mi, X., Feng, G., Hu, Y., Zhang, J., Chen, L., Corlett, R. T., Hughes, A. C., Pimm, S., Schmid, B., Shi, S., Svenning, J.-C., & Ma, K. (2021). The global significance of biodiversity science in China: An overview. *National Science Review*, *8*(7), nwab032. doi:10.1093/nsr/nwab032 PMID:34694304

Montesino Pouzols, F., Toivonen, T., Di Minin, E., Kukkala, A. S., Kullberg, P., Kuusterä, J., Lehtomäki, J., Tenkanen, H., Verburg, P. H., & Moilanen, A. (2014). Global protected area expansion is compromised by projected land-use and parochialism. *Nature*, *516*(7531), 383–386. doi:10.1038/nature14032 PMID:25494203

O'Hara, C. C., Frazier, M., & Halpern, B. S. (2021). At-risk marine biodiversity faces extensive, expanding, and intensifying human impacts. *Science*, *372*(6537), 84–87. doi:10.1126/science.abe6731 PMID:33795456

Palmer, M. A., Zedler, J. B., & Falk, D. A. (2017). Ecological theory and restoration ecology. Foundations of Restoration Ecology: Second Edition, 3–26. doi:10.5822/978-1-61091-698-1_1

Pandey, P. C., & Pandey, M. (2023). Highlighting the role of agriculture and geospatial technology in food security and sustainable development goals. *Sustainable Development (Bradford)*, *31*(5), 3175–3195. doi:10.1002/sd.2600

Peh, K. S. (2011). Crop failure signals biodiversity crisis. *Nature, 473*(7347), 284. doi:10.1038/473284d PMID:21593847

Pollock, L. J., Thuiller, W., & Jetz, W. (2017). Large conservation gains possible for global biodiversity facets. *Nature, 546*(7656), 141–144. doi:10.1038/nature22368 PMID:28538726

Pressey, R. L., Mills, M., Weeks, R., & Day, J. C. (2013). The plan of the day: Managing the dynamic transition from regional conservation designs to local conservation actions. *Biological Conservation, 166*, 155–169. doi:10.1016/j.biocon.2013.06.025

Purvis, A., Molnár, Z., Obura, D., Ichii, K., Willis, K., Chettri, N., Dulloo, M., Hendry, A., Gabrielyan, B., & Gutt, J. (2019). Status and trends-nature.

Raihan, A. (2023). *Original Research Article Artificial intelligence and machine learning applications in forest management and biodiversity conservation. 6*(2), 1–16. doi:10.24294/nrcr.v6i2.3825

Rawat, U., & Agarwal, N. (2015). Biodiversity: Concept, threats and conservation. *Environment Conservation Journal, 16*(3), 19–28. doi:10.36953/ECJ.2015.16303

RayhanA. (2023). AI and the Environment Toward Sustainable Development and Conservation. *Research Gate, July*. doi:10.13140/RG.2.2.12024.42245

Reside, A. E., Butt, N., & Adams, V. M. (2018). Adapting systematic conservation planning for climate change. *Biodiversity and Conservation, 27*(1), 1–29. doi:10.1007/s10531-017-1442-5

Roy, P., Roy, A., & Karnataka, H. (2012). Contemporary tools for identification, assessment and monitoring biodiversity. *Tropical Ecology, 53*(3), 261–272.

Scheffers, B. R., De Meester, L., Bridge, T. C., Hoffmann, A. A., Pandolfi, J. M., Corlett, R. T., Butchart, S. H., Pearce-Kelly, P., Kovacs, K. M., Dudgeon, D., Pacifici, M., Rondinini, C., Foden, W. B., Martin, T. G., Mora, C., Bickford, D., & Watson, J. E. M. (2016). The broad footprint of climate change from genes to biomes to people. *Science, 354*(6313), aaf7671. Advance online publication. doi:10.1126/science.aaf7671 PMID:27846577

Schulze, K., Knights, K., Coad, L., Geldmann, J., Leverington, F., Eassom, A., Marr, M., Butchart, S. H., Hockings, M., & Burgess, N. D. (2018). An assessment of threats to terrestrial protected areas. *Conservation Letters, 11*(3), e12435. doi:10.1111/conl.12435

Shoo, L. P., Storlie, C., Vanderwal, J., Little, J., & Williams, S. E. (2011). Targeted protection and restoration to conserve tropical biodiversity in a warming world. *Global Change Biology, 17*(1), 186–193. doi:10.1111/j.1365-2486.2010.02218.x

Stephenson, P. J. (2020). Technological advances in biodiversity monitoring: Applicability, opportunities and challenges. *Current Opinion in Environmental Sustainability, 45*, 36–41. doi:10.1016/j.cosust.2020.08.005

Suding, K. N. (2011). Toward an era of restoration in ecology: Successes, failures, and opportunities ahead. *Annual Review of Ecology, Evolution, and Systematics, 42*(1), 465–487. doi:10.1146/annurev-ecolsys-102710-145115

Thomas, C. D., & Gillingham, P. K. (2015). The performance of protected areas for biodiversity under climate change. *Biological Journal of the Linnean Society. Linnean Society of London, 115*(3), 718–730. doi:10.1111/bij.12510

Thuiller, W., Maiorano, L., Mazel, F., Guilhaumon, F., Ficetola, G. F., Lavergne, S., Renaud, J., Roquet, C., & Mouillot, D. (2015). Conserving the functional and phylogenetic trees of life of European tetrapods. *Philosophical Transactions of the Royal Society B: Biological Sciences, 370*(1662), 20140005. doi:10.1098/rstb.2014.0005

Tilman, D., Clark, M., Williams, D. R., Kimmel, K., Polasky, S., & Packer, C. (2017). Future threats to biodiversity and pathways to their prevention. *Nature, 546*(7656), 73–81. doi:10.1038/nature22900 PMID:28569796

Tran, T. C., Ban, N. C., & Bhattacharyya, J. (2020). A review of successes, challenges, and lessons from Indigenous protected and conserved areas. *Biological Conservation, 241*, 108271. doi:10.1016/j.biocon.2019.108271

Verma, A. K. (2018). Ecological balance: An indispensable need for human survival. *Journal of Experimental Zoology India, 21*(1), 407–409.

Viña, A., & Liu, J. (2017). Hidden roles of protected areas in the conservation of biodiversity and ecosystem services. *Ecosphere, 8*(6), e01864. doi:10.1002/ecs2.1864

Visconti, P., Bakkenes, M., Baisero, D., Brooks, T., Butchart, S. H., Joppa, L., Alkemade, R., Di Marco, M., Santini, L., Hoffmann, M., Maiorano, L., Pressey, R. L., Arponen, A., Boitani, L., Reside, A. E., van Vuuren, D. P., & Rondinini, C. (2016). Projecting global biodiversity indicators under future development scenarios. *Conservation Letters, 9*(1), 5–13. doi:10.1111/conl.12159

Watson, J. E., Grantham, H. S., Wilson, K. A., & Possingham, H. P. (2011). Systematic conservation planning: past, present and future. *Conservation biogeography*, 136-160.

Watson, R., Baste, I., Larigauderie, A., Leadley, P., Pascual, U., Baptiste, B., Demissew, S., Dziba, L., Erpul, G., & Fazel, A. (2019). *Summary for policymakers of the global assessment report on biodiversity and ecosystem services of the Intergovernmental Science-Policy Platform on Biodiversity and Ecosystem Services*. IPBES Secretariat.

Xu, W., Xiao, Y., Zhang, J., Yang, W., Zhang, L., Hull, V., Wang, Z., Zheng, H., Liu, J., Polasky, S., Jiang, L., Xiao, Y., Shi, X., Rao, E., Lu, F., Wang, X., Daily, G. C., & Ouyang, Z. (2017). Strengthening protected areas for biodiversity and ecosystem services in China. *Proceedings of the National Academy of Sciences of the United States of America, 114*(7), 1601–1606. doi:10.1073/pnas.1620503114 PMID:28137858

Chapter 12
Policymaking and Strategies for Socio-Ecological, Environmental, and Green Socio-Technological Innovation

José G. Vargas-Hernandez
Tecnológico Nacional de México, Mexico

Omar C. Vargas-González
https://orcid.org/0000-0002-6089-956X
Tecnológico Nacional de Mexico, Ciudad Guzmán, Mexico

ABSTRACT

This study aims to analyze the relationships between policymaking and strategies for socio-ecological, environmental, and green socio-technological innovation. The analysis departs from the assumption that the environment of a socio-ecological and socio-technological innovation can be fostered by policymaking and strategies aimed to provide the transformation of pure research on scientific knowledge into innovative commercial products and services. The method employed is analytic-reflective based on the theoretical and empirical literature review. It is concluded that policymaking and strategies accelerate the creation and development processes of socio-ecological, environmental, and green socio-technological innovation.

INTRODUCTION

The technological innovation trajectories are inseparable from the scale and scope economies, mass production, consumption, and the expansion of global markets. The global commitment to economic growth and social development is linked to technological and socio technological innovations unmeshed in complex systems in scale and scope, involving supranational institutions, national and local governments, transnational and multinational corporations, trading blocs, etc. Global economic development,

DOI: 10.4018/979-8-3693-3330-3.ch012

political revolution, technological innovation change and resource scarcity are some factors that have an impact on socio-ecological interactions at regional and local levels (Haber et al., 2006). Global economic growth has become an unsustainable pattern of socio-ecological and technological innovation development for solving some operational challenges such as investments and incentives, global supply chains and delivery service models, etc. Innovative service models are perceived as turning innovation into solutions for problems.

In the logic of mass production and technological innovation systems characteristic of marketspaces make possible to conceive uncoupling high technology and global scale innovations increasingly used in medicine, biotechnology, electronics, etc. (Oladapo et al., 2021). The economic and socio-ecological subsystems lead to an integrated renewable resources innovation model based on quantitative methods and information technologies to forecast processes and objects suitable for systemic crises. Technological innovation is stimulated by direct access information. Integrated models of socio-ecological economy and socio-humanitarian systems are based on quantitative methods and innovation technologies to forecast and control environmental conditions of uncertainty, stability, and risks. Theoretical models of systemic sustainability promote the emergence of a global living technological laboratory network innovative initiatives and the emergence of an auto-catalytic socio-eco-technical system focusing on individual projects of systemic sustainability that contribute to the collective creation for a thrivable planet. (Laszlo, et al., 2012). Integrated models based on quantitative methodologies and innovative technologies are used to forecast and control the dynamics of socio ecological-economic and humanitarian systems under conditions of complexity, uncertainty, and instability (Ramazanov, 2019).

BACKGROUND

Scientific knowledge on the innovation of socio-ecological economic development using methods for science and technology policies in aggregated economic models have been developed by Heets & Seminozhenko (2006), Klebanova & Kizima (2010), Kleiner (2013), Ayvazyan et al. (2012) and others. Socio-ecological innovation for economic development supports individual and group creativity to create and develop technologies to market more effectively (Marxt & Brunner, 2012). Research and development have integrated mathematical models, methods, and innovation technologies to make decisions, forecast and manage the dynamics of socio-ecological, economic, and socio-humanitarian systems under conditions of uncertainty, complexity, and risk. The integrated models used to approach socio-ecological-economic and humanitarian systems are supported by the emerging concept of sustainable development as the combination of the economic, social, and environmental dimensions framed by knowledge creation, technology innovation and other factors influencing the environment and the structure of economic growth (Heets &Seminozhenko, 2006; Ramazanov, 2018; Babenko et al., 2019; Solou, 2002).

Research and development of an integrated stochastic model of the dynamics of socio-ecological and economic, innovative, and humanitarian systems and processes for optimal forecast and control in innovation economy is necessary to function at the macro and micro levels in complicated conditions of crises, uncertainty, and instability. Integrated stochastic models of the dynamics of socio-ecological economic growth and development and socio-humanitarian systems, including other sectors such as production, R&D and education considering socio-ecological innovation based on mathematical methods and innovation technology in conditions of uncertainty, instability, and risk, must be created and developed.

Innovation studies can identify three sub-bodies of literature towards responsible research and innovation, mission-oriented policy of innovation and capacious innovation. Responsible research on innovation is a theoretical construct and a policy concept is linked to tradition of assessment technology (Schot & Rip 1997). Responsible research innovation is the transparent and interactive processes where the societal actors and innovators are mutually responsible with ethical acceptability, sustainability and social desirability of the innovation process, products and services with control over the research direction, development of technology and innovation to stakeholders (von Schomberg, 2012). Responsible research innovation advocates democratic control intervening in innovation processes.

This study is created based on the interrelationships of three main topics. It begins with and analysis on technological innovation followed by the socio-ecological, environmental, and green socio-technological innovation and finally it analyzes the policymaking and strategic implications to provide some conclusions.

TECHNOLOGICAL INNOVATION

Economic globalization brings changes in world-wide challenges to improve the economic growth at the cost of competition and degradation of natural resources and socio-ecosystems, and labor division supported by technological innovation and post-industrialization innovative technologies aimed to create new opportunities in changing production modes and consumption patterns (Dimitrova, 2013). The techno-economic is the dominant technological paradigm resulting from technological and organizational innovations that cause economic, social, organizational, and institutional impact (González Álvarez, 1997). The technological paradigm shift affects the traditional economic sectors requiring the modernization of production processes and the adoption of management techniques. The techno-economic standard is formed by a set of principles applied in the generation of technological and organizational innovations producing economic growth.

Modernity has increased the per capita access to energy, resources, and materials associated to complexity overhead available by a consistent innovation pattern of innovative technologies and gadgets affluence of consumer durables. Modernization refers to the development of specific applications of technological innovation that have an impact on efficiency and quality. Revolutionary innovations and scientific shamanism, remains as had pertained since the mid-18th century, the same as state peasants in their relationship to land. The technological innovations and management structures and practices after 1960s made the herders to change techniques and practices (Pelto, 1973). Orthodox innovation institutions are facilitative or constraining, static and inert with capacious innovation conceptualizations leading to institutional entrepreneurship operating with technological innovation and change.

The component of technological innovation versus traditional approaches as for example the Porter's Five Forces Model, organizational business themselves for competitive advantage. Technology innovation is a change factor responding to creative management action in the strategic role within the innovation process and not in a static system such as the 5 Forces model of Porter (Henderson, 1994). The contributions of Porter (1980) to the analysis of innovation related to processes and products in corporate strategy although he underestimates the ability of management to implement innovation strategies the capacity for change and technological discontinuities of the industrial structure.

The innovation modes are the science and technology model, the science, technology and innovation and the doing, using, and interacting models, they have differentiated knowledge bases of innovation. These models aimed to predict the state and the economy innovation and development result from the dynamics of capital, labor knowledge and innovative technology resources, consumption, investments, and natural resources.

Evolution provides the analytical instrument to encompass the organizational network with the conditions to assimilate technological innovations (Trist, 1976; OECD, 2001c). Change framework considers that social transformation and evolution benefits from the utilization of localized visions leading to local diversification of social and technological innovation for attracting the postcapitalist world through the real time and responsive networks to support the nods for the needed learning the innovations developed. The evolutive societal innovation interactions from a macro-perspective links the global economic, demographic, and environmental developments with energy metabolism and technological innovation changes (Krausmann & Fisher-Kowalski, 2013)

Open innovation diplomacy is related to the different ecosystems and organizations able to collaborate and bridges the separated domains of social, organizational, technological, and cultural (Carayannis & Campbell, 2011). Incumbent organizations can dominate the technological frontier, increasing their preeminence with their tenure. Sustainable society is built with knowledge and experience development in transformation and innovation processes becoming strategic variables implying economic, social, technical, technological, political, and cultural innovation.

Management of complex organizational systems develop and function considering the conditions of instability to ensure increasing economic growth supported by achievements of innovation technology and science. Management innovation strategies lead to processes that combine information technologies, quality, and innovation processes, etc. to achieve significant market positions. The managerial component of the strategic advantage of well-used technology is crucial because managing innovation technology effectively requires a clear understanding of the evolution, maturation, and diffusion of the technology itself throughout the global economy (Price, 1996).

Organizational technological change is paradoxically associated with organizational stability. Defenses imposed by the management inabilities to face situations of strategic nature decisions affect organizations in case of technological innovation discontinuities and rejection of change. The balance of an organization can be seriously affected many times by social impacts produced by the introduction of a technological innovation. Models of change management contribute to balance evolution and the process of adaptation become more sensitive internal and externally to management experience, financing, and innovative technological resources, making the organizational technological and social aspects in an innovation system.

The organizational size has positive effects in the technological industry is because the market tends to be more independent. Recent cites of new patents are from firms elaborating in current developments of leading age technological innovation fields.

Setting the focus to provide a platform for contextual technologically innovative designs framed by a meta-level objective for a thrivable planet to catalyze the emerging networking initiatives and activities through the system level.

Urban living labs are defined as the physical region where stakeholders form public-private-people partnerships among public agencies, businesses, universities, and users can collaborate to create, develop, validate and test new and innovative technologies, products, services, and systems in contexts (Juujärvi & Pesso, 2013). The strategy Research and Innovation for Smart Specialisation (RIS3) is an ex-ante condition for receiving support to achieve a shift in the structure to solve the problems of standardized regional innovation policies (Landabaso, 2014; Tödtling & Trippl, 2005).

SOCIO-ECOLOGICAL, ENVIRONMENTAL, AND GREEN SOCIO-TECHNOLOGICAL INNOVATION

Innovation is creating knowledge and technologies to be used in socio-ecological sustainability and the use of natural resources. Natural resources and commitments develop innovative socioeconomic, ecological, and technological contexts, leading to diversification and transitioning to alternative production forms.

The inclusion of natural environment in the quintuple helix for knowledge and innovation model is the component for knowledge innovation and production serving for the survival, preservation and vitalization of humankind, green technologies, sustainable development and socio-ecology become the societal knowledge and innovation production (Carayannis & Campbell, 2010). Environmental pressures result in technological innovations such as demands of market, competition, and government regulations and from internal pressures such as improvement of working quality of life and active planning that enable organizations to cope with changes. Active planning start from the normative level including the organizational, group and members values with its actors, and the community in which they work (de Melo, 1997b).

The dynamism of innovation environments based on the sociological perspective treats it as and organizational ecology leading to the technological innovation component compared to traditional approaches to small businesses seeking competitive advantages.

Every day, human future may exceed technoscientific innovation engaged on socioecological futures as cultural inheritances are being within particular localities engaged with future scenarios (Gibson, Head & Carr 2015).

The socio-ecological model is more implicit in the transformation towards sustainable society, deserving more attention in smart specialization policies which has different ends that the science and technology model (Coenen & Morgan, 2019). The socio-ecological model of innovation is the research and innovation strategies discerned in RIS3 (Jensen et al. 2007;Asheim et al. 2017; European Commission 2012). The smart specialization promoted by research and innovation strategies is predicated on an innovation model of conventional science and technology (European Commission 2012).

Smart urban innovation technology is a means to improve urban life leading to the risk of surveillance and loss of privacy are prevalent (Hollands, 2008; Kitchin, 2014). The increased automation and roboticization of services and other domains using artificial intelligence and smart innovation technologies increases the risks of innovation (Karvonen et al. 2019).

Ecosystems services of high nature added value delivery through public interventions aimed to empower communities aimed to foster technological innovation in multifunctional landscapes. This emerging situation attracts the systemic sustainability to demonstrate the socio-ecological innovations spanning the economic, social, technological, infrastructural, and agricultural domains (Laszlo, et al., 2012). Socio-ecological innovation for economic development is supported by creativity of individuals and groups to create and develop knowledge and technology (Marxt & Brunner, 2012).

The effectiveness of socio-ecological and technological innovation in transition to sustainable development implementation must consider parameters such as innovation regulatory policies the rational use of production resources, the political situations, functioning of the economy and market performance approach to analysis of renewable industrial purposes resources, goods, and services, investments, demographics (The European Sustainable Development Strategy, 2001).

The socio-ecological sustainable transitions approaches bring changes in the societal system in which the socio-political actors share a set of rules while the subordinated social groups share different sets of rules leading to distinguish the regimes among the technological innovation, socio-ecological, socio-cultural, socio-technical and policy regimes. The socio-technological system is a set of heterogenous and interlinked material and technical elements such as structures and artefacts that encompasses societal production, diffusion, and use of technology with societal functions such as communications, transports, nutrition, etc., represented through networking of actors and social groups with formal, cognitive, and normative rules guiding activities (Papachristos, Sofianos, & Adamides, 2013).

The sociotechnical concepts from both social and technological innovation perspectives provide the basis for formulating and implementing innovations making possible to design flexible organizational structures to facilitate communication more compatible with innovative management methods and content (de Melo, 1997). Socio-economic and socio-cultural factors are linked between them and affect socio-technological innovation and development, and therefore all these factors have a relevant significance in the determination of the socio-ecological niche including group values, norms, attitudes, labor, and land tenure allocations to community and household tasks, livelihood strategies, labor organization and marketing, food supply and demand, household food preferences and habits, etc.

The socio-ecological development poses biophysical limits to economic growth on a planet of finite resources unless technological innovations and systems overcome the biophysical costs at the expenses of socio-ecological spaces and enhance health and quality of life of people in local communities (Daly & Farley 2004). Large-scale transition to socio-ecological and technological innovation has unpredictable and inevitable uncertain consequences in economic growth, social justice, equality, equity, and environmental sustainability of communities, improvement of quality of life in a horizon moral enhancement (Beck & Beck-Gernsheim, 2002; Bauman, 2000). From the socio-ecological perspective, technology innovation has no power but with human agency fulfills functions associated with social and organizational structures (Geels, 2002) and other combinations of transition forms are required to achieve sustainability transitions.

Environmental sustainability strategies are the efficiency, consistency in eco-innovation, and sufficiency which can be measured to address the environmental sustainability crisis. The consistency strategy of environmental sustainability aims to the restructuring of economies though socio-ecological and technological innovations supported with resource productivity and enhanced efficiency measures for decoupling economic growth and environmental damage (Gouverneur & Netzer, 2014). All types beyond technological innovation involving actors and agents in the strategy setting and development

Greening economy and investments in eco-innovation and clean innovation technologies are aimed to reduce the destructive levels of temperature increase (Rockström et al. 2016). And the demand for green innovation and knowledge society solutions utilizing socio-ecological and economic resources, technology, and entrepreneurship as the drivers. Environmental regulations stimulate the green technological innovation activities. Environmental regulation affects green technological innovation (Luo et al. 2021). More intense and strict environmental regulations strengthen the effectiveness and efficiency of green finance on green technological innovation. As environmental regulations become more stringent, it motivates the firm to achieve green technological innovation with green finance investment (Luo et al. 2021).

Environmental regulations positively moderate the relationship between green finance and green technological innovation. Firms adapt their green technological innovation strategy to become more outcome oriented increasing the effect of green finance. Green finance has a positive effect in driving green technological innovation. An adaptive approach addresses the issues of managing and planning the technological innovation processes, technology transfer and training in innovation environments. Social and technological innovation in organizations leads to competitive advantage able to mobilize technological knowledge, skills, and experience to develop the innovation strategy to create and develop processes innovation and market leading products and services that allow to adapt and respond to the threats of the changing environment and become successful (Tidd et al, 1997).

Green technology innovation comprises the stages of research and development, start-up, patenting, upscaling, dealing with risk, commercialization, and adoption. The factors influencing these stages are the organizational setting, the technology-specific characteristics, the funding available, and the commercial market conditions (Pizzol & Andersen, 2022). Adoption of new green technology innovation requires support from training and preparation to avoid the harmful effects on organizational development in less developed and peripheral regions (González Álvares, 1997), ranging from the waste of technology underuse to alterations of organizational social fabric.

Technological innovation autonomy through adequate technical training of efficiency in human resources and infrastructure must be available to organizational and institutional actors (Dias, Rosenthal & de Melo, 2000). Technological infrastructure must be compatible with the local capacities for innovation development of efforts by national and local development agencies which results in less risks and lower costs (OECD, 1999a). Building the infrastructure and capital investment for scientific research and technological innovation lead to consulting and training people to operating successfully (OECD, 1999a, 2001b, 2001c).

The local socio-ecological factors contribute to the design of the socio-ecological niche concept in the context of biophysical variables at the local level that co-determine and constrain productivity. These factors in combination with interactions delineate the socio-ecological niche for innovation technology. The concept of a sociological niche and the concept of innovation share similarities to accept the existence of technology if there is an appropriate balance of mix between technical development and socio-organizational arrangements (Leeuwis & van den Ban, 2004).

Socio-ecological niche is a concept like ecological niche of an organism in classical ecological theory. The concept of socio-ecological niche is defined and used in research and development as a framework to integrate technologies in farming systems, to analyze the factors operating at various levels and delineate a procedure to identify compatible technology development. A niche is defined as the region n-dimensional and multidimensional space and hyper-volume of environmental factors affecting the welfare of species (Hutchinson, 1957). The concept of the socio-ecological niche in research and technology development can be tailored according to the specific needs of farming systems.

The concepts of socio-ecological niche and innovation or complete technology share similarities. The niche defined by a legume innovation technology combine several factors to fulfill the needs of farmers to find a suitable socio-ecological niches (Leeuwis & van den Ban 2004). The definition of socio-ecological niche is dependent on the notions of farmers and cannot receive blanket recommendations for technology. Institutional support services are cross-cutting and relevant to technological innovation integrated in the socio-ecological niche and not shown as separate layer.

POLICYMAKING AND STRATEGIES

The societal ecosystems and environmental services recognition delivered by high nature value systems lead to public policies interventions aimed to empower rural communities, foster technological innovation, and promote multifunctional landscapes. Innovations production for the sustainability in the nation state leading to circulation of knowledge aimed to promote sustainability through innovations obtained from research in science and technology. Innovation in policymaking is tight to research in geography innovation leading to place based innovation approach (Tödtling & Trippl 2005; Barca et al. 2012). The geography of innovation literature is limited to the conditions, and skewed to kinds of innovations such as technology driven and market-based (Shearmur et al. 2016)

Policymaking must enact more diverse well-designed environmental policies regulations to stimulate firms to develop green technological innovation, considering the use of the latest technological innovations without implementing huge research and development funds, green subsidies, and green innovation protection policies (Percival et al. 1992). Policymaking can formulate and implement market-based environmental regulations and policies to leverage market forces and encourage the development of green innovation technologies (Zameer and Yasmeen 2022).

There are diverse types of designed innovation institutions including formal organizations and informal networks to facilitate information flow, resources, and knowledge between firms aimed to add economic value to the scientific knowledge and technological innovation generated in the research institutions and universities and transferred in terms of research results to market institutions. Institutional set-ups of successful regional economies support by programs of regional innovation policies such as innovation support agencies, technological funding schemes, science parks, cluster policies, smart specialization industries, etc. (McCann & Ortega-Argilés, 2013).

Regional innovation policy consists of horizontal measures in financial schemes, technology transfer, etc. Smart strategic specialization is a vertical non-neutral measure. Regional innovation policies are aimed to support regional endogenous innovation potential by the diffusion of innovative technologies supported by strategic alliances and partnerships between academic institutions, public research centers, small, medium-sized, and large enterprises in vertical co-operation and between enterprises themselves in horizontal co-operation.

Science and technology models of innovation focusing on entrepreneurial activities may be supported by the implementation of smart strategic specialization to achieve sustainable transition considering that policymaking pursuing economic competitiveness may contradicts with social well-being, ecological integrity, and environmental cleanness. Smart strategies specialization policies and cluster initiatives are part of the regional innovation policy. The place-based approach of regional development policy argue in favor of endogenous potential regionalization (Barca et al., 2012; OECD, 2011). Smart strategy specialization is a policy instrument to foster regional innovative capability predicated on conventional science and technology innovation models of policies 1.0 and 2.0 and the socio-ecological model of innovation has been less relevant and mentioned (Foray et al., 2009).

Innovation policymaking is framed by innovation policies 1.0, 2.0 and 3.0. Innovation policy 1.0 becomes part of science and innovation technology policy based on research and development (R&D) based innovation with an emphasis on commercialization of scientific and technological knowledge. Innovation policy 2.0 is the system with the objectives of economic growth, competitiveness, and job creation linked between discovery and commercial use of applied knowledge.

Innovation policy 3.0 is the explicit mobilization of science, technology and innovation knowledge for meeting sustainable societal needs addressing the Sustainable Development Goals, sustainable and inclusive societies (Schot & Steinmueller, 2018). Innovation policy 3.0 is under debate including various scalar and spatial variations (Fagerberg 2018). Innovation policy 3.0 mobilizes science, technology, and innovation for meeting the societal needs and the sustainable development goals addressed by the United Nations. Previous generations, innovation 1.0 and innovation 2.0 only motivated to resolve structural failures Coenen et al., 2015; Schot & Steinmueller, 2018).

Green technology development distinguishes the shared framework conditions from technology specific ones leading to support the green innovation policies. Diffusion of findings and innovations in scientific and technological research support policymaking processes facilitated by communication (Klabbers et al. 1996).

Sustainable Development Goals are required to implement along with major transformations in policies and technologies, modes of transformative innovations involving stakeholders in inclusive and multi-scale.

The historical context had shaped science and technology policies to incorporate innovation as the strategic component for scientific and technology development. Business incubators, technological parks, and other institutions of support of technological innovation environments aimed to increase the creation and development of innovative technologies for innovation in products and services and the promotion of innovative organizational performance.

An environment of innovation systems has the objective to transform scientific knowledge and innovation technology into organizational business, products and services with social value, economic growth and environmental sustainability facilitated through business incubators, technology parks, technology transfer offices and marketing. Private or public incubators located in universities and research centers, house traditional, technology or mixed firms linked to technological sectors in need of modernization and with resources limitations.

Innovation environment systems have objectives such as to promote technological knowledge innovation; encourage science-based knowledge innovations, establish relationships of cooperation and trust, provide return on capital investments, generate an organizational culture, etc. Innovation environments apply a broad set of objectives, some of which may be related to universities linking with firms for resources exploitation, creation and support of new technological firms' regional development policies and economic revitalization of depressed regions.

There are various similar terms to name a technological innovation park such as science park, research park, innovation park, innovation center, business park, etc. (González Álvarez, 1996; Monck et al., 1988; Löfsten & Lindelöf, 2001). The environment of a technological innovation park can provide the transformation of pure research into commercial products and services under the assumption that scientific research is the origin of technological innovation (Löfsten & Lindelöf, 2001). Innovation parks need to significantly increase the specialized knowledge and technological assets centered in the universities to attract firms and institutions to provide research facilities and critical mass of technicians and researchers helping to develop industrial clusters in areas such as information technology, biotechnology, telecommunications, biomedical equipment, pharmaceuticals, chemicals, plastics, fibers, education, etc. (Porter, 2001).

Technological innovation parks serve private and public interest promoting a favorable innovation environment through making risky decisions, exchanging technical and market information, investing and cooperation in infrastructure, and making efficient gains meeting the needs and expectations of regional and local communities (OECD, 2001a). Technological innovation parks on the public interest are a strategic alternative to industrial policy aimed to accelerate the regional development taking advantage of local characteristic with low level intervention and public resources which can be leveraged by a private interests motivated by a tangible cost/benefit proportion (Pimienta-Bueno, 1999).

Firms involved in technological innovation have as objectives to access to the agendas, laboratories and equipment of research institutes and universities through contracts, involve students in industrial practices and projects, recruit graduates, promote consulting and growth of technology-based parks and incubators, encourage synergy in scientific research, development, and technological innovation, etc. Identifying and attracting firms with innovation projects for joint innovation action are process required to provide and support with resources and capabilities, scientific knowledge and technological innovation, infrastructure and facilities, financing, equipment, technology transfer and dissemination to expand the potential of firms (Lalkaka, 1999).

Technological-based firms need resources and capabilities which can made available by research centers and university settings in an innovation environment with technological base such as incubators and technology parks facilitates the attraction of other actors and agencies such as government institutions promoting innovation, consulting agencies, financing and investing, etc. University-based local innovation systems as a model are a source of inspiration acting as mechanisms for the technology generation, dissemination and transfer as their raison d'être innovation and entrepreneurial activity in an environment of technology-based companies. Firms in local innovation systems change their profile since a university based local innovation system is surrounded by emerging technology-based firms with the research centers and institutes as the sources of technology innovation and access to qualified resources and capabilities.

Generation of an innovation dynamic environment for firms can enable regional economic growth and social value based on a scientific and technological agenda for research institutions and universities based on research and development to attend and meet the demands of firms, communities, and society. These changes of the innovation dynamics contribute positively to the transformation of the research university as an institution capable of creating and disseminating scientific knowledge and technological innovation to benefit the society linking the specificities of the regional vocation.

Governments and organizations are focusing on establishing socio-ecological economy system through the intensification of economic and management institutional transformations supported by science and technology innovation policies aimed to improve the quality of the local community's development. Government policies support local innovation systems and innovation removing barriers and promoting the transformation of scientific and technological advances into improvement of economic performance and enhancement of living conditions (Wiig & Wood, 1995). The different government instances are expected to contribute to enable the policymaking processes to support innovation, the creation and formation of local innovation systems, the development of technological innovation capabilities.

The local innovation system has as a central element the research and higher education center as the technological resources provider and systems innovation. Research institutes and universities are active actors in creation, dissemination and application of scientific knowledge and technological innovation that may have the objectives to encourage and facilitate links with firms and industries, access to the state-of the art R&D activities, transfer of technological innovation commercialization of academic research and technological innovation, encourage the emergence of spin-off and start-up industries, create jobs and consulting opportunities, generate return on investments, etc.

The concept of local innovation models and systems demonstrate the advantages in relation to the conceptualization of university-based innovation systems in local contexts. Local innovation systems are community-driven in line with local demands of the community and better understanding of local environments, requiring more planning, technological sophistication, and financial support (Schlapfer & Marinova, 2001; de Melo, 2001). Local Innovation System builds on local economic and social institutional structures aimed to the scientific and technological knowledge and skills in the local communities. The local innovations systems are a form in which the university is the larger provider of technological resources for innovation.

Local university based innovation systems are characterized by their research and development institutes and centers in a research and development university with an infrastructure, laboratories, and the interactions of students with scientists and technologists with converged objectives, access to resources and communication which become more efficient under an institutional framework. Local innovation systems research university based have a prominent function in research and development of creation of scientific knowledge and innovation technology acting as elements of an innovation structure activities of Research and Development institutions, firms, anchor and collaborating companies aimed to generate new ventures, university start-up and emerging research university.

Under the local innovation systems approach, local and community development strategies seek the creation and formation of innovation skills for economic growth, social development and environmental sustainability supported by research centers, university and training institutions linked to the relationships with the productive sector. This innovation linkages are aimed to stimulate the creation, construction and formation of organizational capabilities and competencies leading to the formation and development of innovation networks and cooperation in the production, dissemination, transfer, and use of scientific and technological knowledge (D'Avila Garcez, 2000; de Melo, 2001). Local innovation systems may organize technological innovation events aimed to attract potential users and financial sources to entrepreneurs in technological innovation

Research centers and universities are engines of socio-technological innovation, knowledge creation and transfer to the regional economic growth (Saxenian, 1994; Kenney, 2000; Druilhe and Garnsey, 2000). Research institutes and universities can unite research projects around local community common goals, creating scientific and technological innovation knowledge, disseminating, transferring, and using it, training qualified work, and becoming a decisive component of socio-economic performance in some regions effecting societal changes in an economy of knowledge.

Local industrial systems are the result of an evolution process centered around the university as the genesis. Stanford University was instrumental in the establishment of the Silicon Valley becoming as the model for regional development from the agricultural sector to the leader in emerging companies and start-ups in microelectronic and information technology with leading roles in dictating the technological innovation trends markets for performance. This region is home of highly specialized firms of technological innovation organized to recombine the different subsystems formed by specialized suppliers.

Local innovation system by The Massachusetts Institute of Technology – (MIT) as the spatial concentrations of companies, including specialized suppliers of equipment and services and customers, and associated non-commercial institutions, universities, research institutes, training institutions, standard setting agencies, local trade associations, regulatory agencies, technology transfer agencies, trade associations, relevant government agencies, and departments that combine to create new products and/or services in lines of business (Lester, 2005).

In England, University of Cambridge has become the sources of technological innovation developed, abilities and capabilities to create technology intensive firms in their environment and interacting with the market. The policy of the university free professors to exploit their inventions protecting their intellectual property encouraging to create new firms in business incubators and technological parks aimed to launch start-ups.

Local innovation systems give support to multiple goals of agents and actors giving evidence that there is a localized physical proximity for relationships of trust transactions to share scientific and technological knowledge perspectives in local environments that are attractive to produced results in socio-technical innovation and entrepreneurial activities. The local innovation system must be able to converge multiple institutional objectives, manage resources to preserve integrity and provide feedback to raise the identification, identity, coordination, and governance aimed to service the creation, diffusion, spread and use the scientific and technological knowledge among the market, non-market organizations and anchor companies with the greatest influence.

The feedback loop of the local innovation system produced is explicit on the environmental factors produced through the analysis of performance. The proximity of its physical and interest agents facilitates to install processes of creation, dissemination, transfer, and use of scientific and technological innovation knowledge in the innovative activities of organizations. Local innovation Systems provide their communities with scientific and technological knowledge that suit their needs.

Local innovation systems have similar characteristics the national systems of innovation aimed to transform scientific and technological advances into goods and services with potential economic use and social value. Local innovation systems focus more rationally on regional and communities' interests (Schlapfer & Marinova 2001). National and local governments, authorities and development agencies have as objects technological innovation to stimulate the creation and formation of technology-based companies, generate employment, improve performance, reduce decline and imbalances in research and development activities in depressed regions, reproduce international experiences and technological innovation transfers to local reality.

The concept of National Innovation Systems builds on national institutions is constituted by a set of agencies and organizations of a society that promote the transformation of scientific and technological advances of goods and services, facilitate learning and transfer processes. These national institutions develop to become the lobby of powerful groups of interest to meet the emerging economic, social, cultural, and ecological needs and to promote the development and implementation of innovative technologies with long-term benefits.

Innovative technology transfer is an imperative to invest considering the local context in less developed countries (Schlapfer & Marinova, 2001) to be compatible with local indigenous sources of knowledge.

Transition management is a process of variation and selection leading to the gradual transformation of societies from existing innovative technologies and practices by coevolutionary steering (Lachman, 2013; Kemp, Loorbach, & Rotmans, 2007). The transition management of a societal system is a cyclical process that includes strategic levels and networks to initiate visions and paths upon transition agenda implemented at tactical level focusing on structural institutional regulatory barriers, economic conditions, specific innovative technologies, physical infrastructures, consumer routines, etc. leading to long-term goals.

The cooperation perspective (Melo, 1997) provided through collaboration agreements is a strategic skill to develop technologies and complementary skills as it has been considered in technological research (Price, 1996) and technological training of the technological innovation processes through the innovation networks (González Álvarez, 1997; OECD, 2001c; Porter, 2001; de Melo, 2002). Organizational, processes and product arrangements conducted in cooperatives and modernization poles of traditional technology sectors incorporate international advances in development efforts in innovation technologies through joint and shared efforts among institutions, government agents and other actors to generate technological knowledge. These efforts of cooperation are essential for small enterprises (González Alvarez, 1997).

CONCLUSION

The aims of this study is to analyze the relationships between policymaking and strategies for socio-ecological, environmental, and green socio-technological innovation assuming that the environment of a socio-ecological and socio-technological innovation are fostered by policymaking and strategies to provide the transformation of pure research on scientific knowledge into innovative commercial products and services, leading to the conclusions that policymaking and strategies accelerates the creation and development processes of socio-ecological, environmental, and green socio-technological innovation. Social and technological innovation models in high natural value systems, environmental and ecosystems services must be stimulated linked by rewarding the stakeholders by delivery to society. Technological innovation must adapt the capabilities and develop where there are information resources, institutions, organizations and individuals with knowledge and competences to make available to society.

The historical context involving the legal frameworks, government and financing agencies and business and academic communities shaped by science and technology incorporates innovation as the strategic component for science and technological development. Creation and development of technological knowledge and innovation policy in resource management is a context dependent variable and institutional process. In intensive competitive environments, technological innovation in organizational innovation is an imperative for the stagnant survival. Industrial ecology combines technological innovation changes in processes and products and organizational innovation. Technological innovation as a source of competitive advantage for organizations has advantages on the components of social capital leading to facilitate socio structural change in organizations and society. Applying the socio-ecological niche concept, technological innovation products can be rationalized based on biophysical performance, socio-cultural and economic issues fitting in the socio-ecological systems.

The socio-ecological transition needs further exploration to contribute to the different sustainability dimensions to increase the capabilities for flourishing and reconciling with economic growth, employment, productivity, and technological innovations. Transformation is a long-term perspective that integrates social and technological innovations with forms of cooperation, innovations in socio-ecological sustainability and the interests of the action's groups with knowledge, social and human resources. Depending on the societal paradigm with impacts on productive organizations requires the cooperation of necessary skills to overcome the arising difficulties from innovative technologies to obtain better opportunities and advantages offered by technological innovation. Institutions supporting socio-ecological innovation play a relevant role in raising cooperation for sustainable development and green growth in organizations facilitating the flow and transfer of technological innovation knowledge and ideas, information and human resources and nature to enhance the environment and maintain their vitality.

From the conception to the large-scale adoption, environmental innovation using green technologies remains a challenge. Upscaling of green technology innovations is a complicated process. Green financing process of small-, medium- and large-sized enterprises need to be supported by environmental government policies and regulations to increase their capabilities of green technological innovation. The public policies and objectives of technological innovation should be known, clearly analyzed, and interpreted to be adapted organizational activities subject to coordination structures of the established environment in specific locations in the context of local innovation systems. Further research needs to be conducted in technological innovation and change with institutional entrepreneurship. Future research must analyze the contingency of environmental regulations as the moderators of green finance on green technological innovation.

REFERENCES

Asheim, B., Grillitsch, M., & Trippl, M. (2017). Introduction: Combinatorial knowledge bases, regional innovation, and development dynamics. *Economic Geography*, *93*(5), 429–435. doi:10.1080/00130095.2017.1380775

Ayvazyan, S. A., Afanase, M. Y., & Rudenko, V. A. (2012). Some questions of specification of three-factor models of the company's production potential, taking into account intellectual capital. *Applied Economics*, *3*(27), 56–66.

Babenko, V.; Perevozova, I.; Mandych, O.; Kvyatko, T.; Maliy, O. & Mykolenko, I. (2019). World informatization in conditions of international globalization: factors of influence. *Global. J. Environ. Sci. Manage.*, *5*(SI), 172-179. doi:10.22034/gjesm.2019.SI.19

Barca, F., McCann, P., & Rodríguez-Pose, A. (2012). The case for regional development intervention: Place-based versus place-neutral approaches. *Journal of Regional Science*, *52*(1), 134–152. doi:10.1111/j.1467-9787.2011.00756.x

Bauman, Z. (2000). *Liquid Modernity*. Polity.

Beck, U., & Beck-Gernsheim, E. (2002). *Individualization: Institutionalized Individualism and Its Social and Political Consequences* (1st ed.). SAGE Publications Ltd.

Carayannis, E. G., & Campbell, D. F. J. (2010). Triple Helix, Quadruple Helix and Quintuple Helix and how do knowledge, innovation and the environment relate to each other? A proposed framework for a trans-disciplinary analysis of sustainable development and social ecology. *International Journal of Social Ecology and Sustainable Development*, *1*(1), 41–69. https://www.igi-global.com/bookstore/article.aspx?titleid=41959. doi:10.4018/jsesd.2010010105

Carayannis, E. G., & Campbell, D. F. J. (2011). Open Innovation Diplomacy and a 21st Century Fractal Research, Education, and Innovation (FREIE) Ecosystem: Building on the Quadruple and Quintuple Helix Innovation Concepts and the "Mode 3" Knowledge Production System. *Journal of the Knowledge Economy*, *2*(3), 327–372. doi:10.1007/s13132-011-0058-3

Coenen, L., Hansen, T., & Rekers, J. V. (2015). Innovation policy for grand challenges: An economic geography perspective. *Geography Compass*, *9*(9), 483–496. doi:10.1111/gec3.12231

Coenen, L., & Morgan, K. (2019). Evolving geographies of innovation: Existing paradigms, critiques, and possible alternatives. *Norsk Geografisk Tidsskrift.*

D'avila Garcez, C. M. (2000). Sistemas Locais de Inovação na Economia do Aprendizado: Uma abordagem conceitual. *Revista do BNDES*, *14*(7), 351–366.

Daly, H. E., & Farley, J. (2004). *Ecological Economics: Principles and Applications*. Island Press.

de Melo, A. (2001). *The Innovation Systems of Latin America and the Caribbean*. Inter-American Development Bank. Agosto.

de Melo, M.A.C. (1977b). *Articulated incrementalism: a strategy for planning (with special reference to the design of an information system as an articulative task).* [Tese de Doutorado, University of Pennsylvania, Filadélfia].

de Melo, M. A. C. (1997). *Processo de Planejamento e as Inovações Tecnológicas e Sociais: uma perspectiva sócio-ecológica*. Anais do 5o. Seminário de Modernização Tecnológica.

de Melo, M. A. C. (2002a). *Enriquecendo a Atuação de Incubadora de Emrpesas. Tecnologia e Inovação: experiências de gestão nas micro e pequenas empresas*. PGT/USP.

de Melo, M.A.C. (2002b). Inovação e Modernização Tecnológica e Organizacional nas MPMEs: o domínio interorganizacional. *Seminário Internacional: Políticas para Sistemas Produtivos Locais de MPME*: 2002

Dezembro.

Dias, A. B., Rosentha, L. D., & de Melo, M. A. C. (2000) Enriquecendo a Atuação de Incubadoras de Empresas. XXI Simpósio de Gestão da Inovação Tecnológica. Co-autores: Adriano Batista Dias e David Rosenthal. São Paulo: 2000.

Dimitrova, A. (2013). *Literature review on fundamental concepts and definitions, objectives, and policy goals as well as instruments relevant for socio-ecological transition*, (WWWforEurope Working Paper, 40.).

Druilhe, C., & Garnsey, E. (2000). The incubation of Academic Spin-off Companies. *Proceedings of the High Technology Small Firm Conference 2000*, University of Twente, Enschede, The Netherlands.

European Commission. (2012). *Guide to Research and Innovation Strategies for Smart Specialisations (RIS 3)*. Publications Office of the European Union.

Fagerberg, J. (2018). Mobilizing innovation for sustainability transitions: A comment on transformative innovation policy. *Research Policy*, *47*(9), 1568–1576. doi:10.1016/j.respol.2018.08.012

Geels, F. (2002). Technological transitions as evolutionary reconfiguration processes: A multilevel perspective and a case-study. *Research Policy*, *31*(8-9), 1257–1274. doi:10.1016/S0048-7333(02)00062-8

Gibson, C., Head, L., & Carr, C. (2015). From incremental change to radical disjuncture: Rethinking everyday household sustainability practices as survival skills. *Annals of the Association of American Geographers*, *105*(2), 416–424. doi:10.1080/00045608.2014.973008

González Álvarez, M. D. (1997). *Processos de Planejamento nos Pólos Tecnológicos: um enfoque adaptativo. Tese de Doutorado, Departamento de Engenharia Industrial*. PUC-Rio.

Gouverneur, J., & Netzer, N. (2014). Take the Wheel and Steer! Trade Unions and the Just Transition. In *State of the World 2014. State of the World*. Island Press., doi:10.5822/978-1-61091-542-7_21

Haberl, H., Winiwarter, V., Andersson, K., Ayres, R. U., Boone, C., Castillo, A., Cunfer, G., Fischer-Kowalski, M., Freudenburg, W. R., Furman, E., Kaufmann, R., Krausmann, F., Langthaler, E., Lotze-Campen, H., Mirtl, M., Redman, C. L., Reenberg, A., Wardell, A., Warr, B., & Zechmeister, H. (2006). From LTER to LTSER: Conceptualizing the socio-economic dimension of long-term socio-ecological research. *Ecology and Society*, *11*(2), 13. www.ecologyandsociety.org/vol11/iss2/art13/. doi:10.5751/ES-01786-110213

Heets, V. M., & Seminozhenko, V. P. (2006). Innovative Prospects of Ukraine, Kh.: Constant.

Henderson, R. (1994). Managing Innovation in the Information Age. *Harvard Business Review*, *72*(01), 100–107.

Hollands, R. G. (2008). Will the real smart city please stand up? Intelligent, progressive, or entrepreneurial? *City (London, England)*, *12*(3), 303–320. doi:10.1080/13604810802479126

Jensen, M. B., Johnson, B., Lorenz, E., & Lundvall, B. Å. (2007). Forms of knowledge and modes of innovation. *Research Policy*, *36*(5), 680–693. doi:10.1016/j.respol.2007.01.006

Juujärvi, S., & Pesso, K. (2013). Actor Roles in an Urban Living Lab: What Can We Learn from Suurpelto, Finland? *Technology Innovation Management Review*, *11*(3), 22–27. http://timreview.ca/article/742. doi:10.22215/timreview/742

Karvonen, A., Cugurullo, F., & Caprotti, F. (Eds.). (2019). *Inside Smart Cities: Place, Politics and Urban Innovation*. Routledge.

Kemp, R., Loorbach, D., & Rotmans, J. (2007). Transition management as a model for managing processes of co-evolution towards sustainable development. *International Journal of Sustainable Development and World Ecology*, *14*(1), 78–91. doi:10.1080/13504500709469709

Kenney, M. (Ed.). (2000). *Understanding Silicon Valley: the anatomy of an entrepreneurial region*. Stanford University Press. doi:10.1515/9781503618381

Kitchin, R. (2014). The real-time city? Big data and smart urbanism. *GeoJournal*, *79*(1), 1–14. doi:10.1007/s10708-013-9516-8

Klabbers, J. H. G., Swart, R. J., Janssen, R., Van Vellinga, P., & van Ulden, A. P. (1996). Climate science and climate policy: Improving the science/policy interface. *Mitigation and Adaptation Strategies for Global Change*, *1*(1), 73–93. doi:10.1007/BF00625616

Klebanova, T.S. & Kizima, N.A. (2010). *Models of estimation, analysis and forecasting of social and economic systems*. Monogr. Kh.: FOP Liborkina L.M., VD "INZHEK".

Kleiner, G. B. (2013). System Economics and System-Oriented Modeling. *Economics and Mathematical Methods*, *3*, 71–93.

Krausmann, F., Fischer-Kowalski, M., Grunbuhel, C., & Krausmann, F. (2008). Socio-ecological regime transitions in Austria and the United Kingdom. *Ecological Economics*, *65*(1), 187–201. doi:10.1016/j.ecolecon.2007.06.009

Lachman, D. (2013). A survey and review of approaches to study transitions. *Energy Policy*, *58*, 269–276. doi:10.1016/j.enpol.2013.03.013

Lalkaka, R., & Abetti, P. (1999). Business Incubation and Entreprise Support Systems in Restructuring Countries. *Creativity and Innovation Management*, *8*(3), 197–209. doi:10.1111/1467-8691.00137

Leeuwis, C., & van den Ban, A. (2004). Communication for Innovation in Agricultural and Rural Resource Management. Building on the Tradition of Agricultural Extension. Oxford: Blackwell Science (in cooperation with CTA).

Lester, R. K. (2005). *Universities, Innovation, and the Competitiveness of Local Economies A Summary Report from the Local Innovation Systems Project.* Industrial Performance Center Massachusetts Institute of Technology.

Löfsten, H., & Lindelof, P. (2001). Science parks in Sweden — Industrial renewal and development? *R & D Management*, *5*(3), 309–322. doi:10.1111/1467-9310.00219

Luo, Y., Salman, M., & Lu, Z. (2021). Heterogeneous impacts of environmental regulations and foreign direct investment on green innovation across different regions in China. *The Science of the Total Environment*, *759*, 143744. doi:10.1016/j.scitotenv.2020.143744 PMID:33341514

Marxt, C., & Brunner, C. (2012). *Analyzing and improving the national innovation system of highly developed countries: the case of Switzerland Technol.* Forecast. Soc. Chang., doi:10.1016/j.techfore.2012.07.008

McCann, P., & Ortega-Argilés, R. (2013). Modern regional innovation policy. *Cambridge Journal of Regions, Economy and Society*, *6*(2), 187–216. doi:10.1093/cjres/rst007

Monck, C. S. P., Porter, R. B., Quintas, P. R., & Storey, D. J. (1988). *Science Parks and the Growth of High Technology Firms*. Routledge.

OECD. (1999a). *Managing National Innovation Systems*. OECD.

OECD. (2001a). Innovative Clusters: drivers of national innovation systems. Paris: OECD, 2001a

OECD. (2001b). *Drivers of Growth: information technology, innovation, and entrepreneurship. Science and Industry Outlook*. OECD.

OECD. (2001c). *Innovative Networks: co-operation in national innovation systems*. OECD.

OECD. (2011). *Regions and innovation policy, OECD Reviews of regional innovation*. Author.

Oladapo, B. I., Ismail, S. O., Afolalu, T. D., Olawade, D. B., & Zahedi, M. (2021). Review on 3D printing: Fight against COVID-19. *Materials Chemistry and Physics*, *2021*(258), 123943. doi:10.1016/j.matchemphys.2020.123943 PMID:33106717

Papachristos, G., Sofianos, A., & Adamides, E. (2013). System Interactions in Socio-technical Transitions: Extending the Multi-level Perspective. *Environmental Innovation and Societal Transitions*, *7*, 53–69. doi:10.1016/j.eist.2013.03.002

Pelto, P. J. (1973) The Snowmobile revolution: technology and social change in the arctic. Cummings Publishing Company, Menlo Park, CA (revised Waveland Press, Prospects Heights, IL 1987.

Percival, R. V., Miller, A. S., Schroeder, C. H., & Leape, J. P. (1992). *Environ-mental regulation: law, science, and policy*. Little, Brown, and Company.

Pimenta-Bueno, J.A. (1999). Parque de Inovação Tecnológica e Cultural da Gávea: A Visão da PUC-Rio. *Documento Interno, 23*.

Pizzol, M., & Andersen, M. S. (2022). Green tech for green growth? Insights from Nordic environmental innovation. In *Circular Economy Oriented Business Model Innovations: A European Perspective*. Springer. doi:10.1007/978-3-031-08313-6_8

Porter, M. (1980). *Competitive Strategy: techniques for analyzing industries and competitors*. Free Press.

Porter, M. (2001). *Clusters of Innovation: regional foundations of U.S. competitiveness. Council on Competitiveness*. Monitor Group.

Price, R. M. (1996). Technology and Strategic Advantage. *California Management Review*, *38*(3), 38–56. doi:10.2307/41165842

Ramazanov, S., Antoshkina, L., Babenko, V., & Akhmedov, R. (2019). Integrated model of stochastic dynamics for control of a socio-ecological-oriented innovation economy. *Periodicals of Engineering and Natural Sciences*, *7*(2). http://pen.ius.edu.ba

Ramazanov, S. K. (2018). *Prediction and management of innovative economics based on the integral static model in phase space*.

Rockström, J., Schellnhuber, H. J., Hoskins, B., Ramanathan, V., Schlosser, P., Brasseur, G. P., Gaffney, O., Nobre, C., Meinshausen, M., Rogelj, J., & Lucht, W. (2016). The world's biggest gamble. *Earth's Future*, *10*(10), 465–470. doi:10.1002/2016EF000392

Saxenian, A. (1994). *Regional Advantage: culture and competition in Silicon Valley and Route 128*. Harvard Business Press.

Schlapfer, A., & Marinova, D. (2001). *Local Innovation Systems: nature, importance and role*. Conference Proceedings: International Summer Academy on Technological Studies: user involvement in technological innovation, Deutschlandsberg. Austria.

Schot, J., & Rip, A. (1997). The past and future of constructive technology assessment. *Technological Forecasting and Social Change*, *54*(2–3), 251–268. doi:10.1016/S0040-1625(96)00180-1

Schot, J., & Steinmueller, W. E. (2018). Three frames for innovation policy: R&D, systems of innovation and transformative change. *Research Policy*, *47*(9), 1554–1567. doi:10.1016/j.respol.2018.08.011

Shearmur, R., Carrincazeaux, C., & Doloreux, D. (Eds.). (2016). *Handbook on the Geographies of Innovation*. Edward Elgar. doi:10.4337/9781784710774

Solou, R. (2002). Theory of growth. Panorama of the economic thought of the end of the twentieth century. SPb.: Econ. Shk.

The European Sustainable Development Strategy (2001). *The potential for sustainable development of 2001*. A sustainable Europe for a better World // enterprise of chemical industry. M.: Publishing house Europe – Sustainable Development. SDS 2005-2010.

Tidd, J. (1997). *Managing Innovation: integrating technological, market and organizational change*. John Wiley & Sons.

Tödtling, F., & Trippl, M. (2005). One size fits all? Towards a differentiated regional innovation policy. *Research Policy*, *34*(8), 203–1219. doi:10.1016/j.respol.2005.01.018

Trist, E. L. (1976). *A Concept of Organizational Ecology: an invited address to the three Melbourne universities*.

von Schomberg, R. (2012). Prospects for technology assessment in a framework of responsible research and innovation. Dusseldorp, M. & Beecroft, R. (eds.) Technikfolgen abschätzen lehren, 39–61. Wiesbaden: VS Verlag für Sozialwissenschaften. doi:10.1007/978-3-531-93468-6_2

Wiig, H., & Wood, M. (1995). *What Comprises a Regional Innovation System? An empirical study*. Studies in Technology, Innovation and Economic Policy Group.

Zameer H, Yasmeen H (2022) *Green innovation and environmental awareness driven green purchase intentions*. Mark Intell Plan.

Chapter 13
Solutions and Exploratory Ideas for Environmental Sustainability

Jansirani D.
P.S.R. Engineering College, India

Ramalakshmi K.
P.S.R. Engineering College, India

Karunya Devi K.
P.S.R. Engineering College, India

Mohamed Rizwana A.
P.S.R. Engineering College, India

ABSTRACT

In light of the problems posed by fast global development and its effects on the environment, environmental sustainability has become a crucial concern. A possible route for sustainable development lies in the integration of environmental genomics, as standard methodologies might not be sufficient to solve these complex difficulties. Advanced genomic technologies are used in the interdisciplinary field of environmental genomics to research the biodiversity, relationships, and functions of species in their natural settings. The purpose of this study is to provide novel approaches and exploratory concepts for sustainable development using environmental genomics.

INTRODUCTION

In order to evaluate the health and resilience of ecosystems, we first talk about how environmental genomics can be used. By monitoring genetic diversity and species dynamics, researchers can gain valuable insights into how climate change and human activity affect ecosystems. Furthermore, identifying keystone species and their genetic adaptations can help in developing focused conservation efforts to safeguard endangered ecosystems. In the subsequent section, we investigate how environmental genomics might improve bioremediation efforts. In order to create more efficient and envi-

DOI: 10.4018/979-8-3693-3330-3.ch013

ronmentally friendly bioremediation methods for damaged areas, it is now possible to recognize and design microbial communities with particular genomic features. The possible benefits of this strategy include the restoration of contaminated areas and the reduction of the negative effects of industrial and agricultural operations. Furthermore, we discuss the increasing worries about sustainable agriculture and food security. By strengthening crop selection and boosting crop resistance to environmental challenges, environmental genomics can support precision farming. Additionally, by fostering the development of biofertilizer and bio pesticides that are environmentally friendly, the study of soil's microbial communities can help reduce the need for chemical inputs. In addition, we emphasize the importance of environmental genomics in monitoring and forecasting newly developing infectious illnesses. The impact of disease outbreaks on the health of individuals and the environment can be reduced by developing early detection and containment measures by tracking the genetic modifications of pathogens and how they interact with hosts and vectors. Ultimately, we discuss the political and ethical consequences of the use of environmental genomics. Concerns like data privacy, access to genetic resources, and the equitable distribution of benefits across many populations must be addressed as this subject develops.

ENVIRONMENTAL GENOMICS IN SUSTAINABILITY DEVELOPMENT

By providing creative answers and insights into resource management, ecosystem resilience, and environmental protection, environmental genomics plays a critical role in advancing sustainability development. To investigate the genetic makeup of species in their natural habitats and comprehend how they interact with the environment, this interdisciplinary field blends genomics, ecology, and bioinformatics. Environmental genomics greatly aids sustainability initiatives by concentrating on the following major areas.

RESTORATION AND ECOLOGICAL MANAGEMENT

Management of ecosystems is based on "our best understanding of the ecological interactions and processes necessary to sustain ecosystem composition, structure, and function"(Christensen et al., 1996). The idea of ecosystem management came into existence in the latter half of the 20th century as people started to wonder whether biodiversity and crucial ecosystem functions, such as the provision of water and commodities, soil productivity, carbon sequestration, and recreational and cultural resources, could be maintained in light of current societal trends toward population growth and expanding resource use. The following concepts are essential to the idea of ecosystem management:

- Sustainability of resources and species population viability;
- Importance of spatiotemporal connectivity, for instance between the levels of the ecological hierarchy (species to ecosystems); across large landscapes and administrative boundaries; and through time;
- Ecological integrity, or the degree to which an ecosystem is healthy.
- The preservation of ecological dynamics, including disturbance regimes;

- Adaptive and Cooperative Management; and
- Recognition of Humans and Their Values as Integral Ecosystem Components (Grumbine, 1994)

The primary focus of ecosystem management is on expansive, largely undeveloped, and independent landscapes (Aplet, 1999). Under increasing human pressure, severely fragmented and changed landscapes make up an increasing amount of the Earth's surface. However, human usage of ecosystems typically leads in some kind of soil and ecosystem degradation. According to the SER (2004), ecological restoration is "the process of aiding the recovery of an ecosystem that has been degraded, damaged, or destroyed." High ecological integrity and resilience, ecological connectedness between the restoration site and the greater landscape, preservation of native biodiversity and management of non-native species, and a trend toward ecosystem self-sustainability are characteristics of successful restoration (SER, 2004).

Restoration of the composition, structure, and/or function of important forested ecosystems has been undertaken by human cultures for decades, if not millennia. This is because the urge to repair and restore degraded landscapes is an innate human feature (Vallejo, 2009). There has been a lengthy history of large-scale forest restoration programs, but the vast majority of these initiatives have been private, unreported, and small-scale. Early brute-force examples include the English kings' reservation of royal forests in the Middle Ages (Young, 1979), the late 19th-century imperial-ordered restoration of the Brazilian forests surrounding Rio de Janeiro, and the Spanish general afforestation program, which replanted nearly 3 million ha of forests, primarily between 1940 and 1975 during the Franco dictatorship (Valbuena-Carrabaa et al., 2010). Even though restoration programs are more targeted, voluntary, and scientifically based now, they are still risky and frequently expensive endeavours. The difficulty is increased by the fact that many, if not most, restoration initiatives are now seeing goal-post shifting due to global change. Geologists have declared the Anthropocene, a new geological epoch, as a result of the unprecedented level of human-mediated disturbance on Earth and the addition of 230,000 people per day to the planet's population of 7.6 billion people (World Population Clock, https://www.worldometers.info/world-population/). No ecosystems on Earth are still pristine in the Anthropocene, if they ever were, since even the most isolated havens of nature are affected by a dirty atmosphere and warming oceans.

Designing regimens for forest management and restoration for constantly changing conditions is a challenging task. Traditional ecosystem management and restoration ecology prioritizes determining the proper state of things (ecological integrity, proper functioning condition, etc.), frequently by drawing on historical ecology (Landres et al., 1999;). Reference targets are typically defined as "the way things were before humans messed them up." Even if it can be challenging to reconstruct more than bits and pieces of the hypothetical reference state, it is possible to create a concrete target condition, create project plans, and identify a successful path (Egan and Howell, 2001). Since the supposedly static environmental baselines that connected the past to the present have changed into directional trends driven by human influences (e.g., atmospheric C content, air temperature, ocean acidity, changed disturbance regimes, etc. (Safford et al., 2012), the usefulness of such concrete reference conditions is becoming increasingly questioned. In these circumstances, a more nuanced strategy is required, where the degree of ecosystem alteration, socio-economic concerns, and the availability of workable intervention choices all interact to drive management and restoration response (Hobbs et al., 2014).

EFFECTIVENESS IN ENGINEERING OUR IMMEDIATE ENVIRONMENT OF HETEROGENEOUS MICROBIAL ACTIVITY

When working with natural microbial communities, one should consider the hypothesis that bacteria also operate within a certain kind of market economy, in addition to humans (Werner et al., 2014), it has been demonstrated that microorganisms engage in collaborative behaviour with the host as well as with other (micro)organisms. The gut ecology of humans and nearly all invertebrates can be viewed as a prominent example of this (Hooper et al., 2002;). The fact that bacteria may trade essential nutrients with one another to expedite the microbial community's general expansion highlights the striking similarities between economics and microbiological behaviour (Tasoff et al., 2015). One instance of how multiple species work together to complete a task more accurately and efficiently than they would individually in an open system, where the introduction of new microorganisms can affect microbial assembly and function in the receiving engineered or natural ecosystem, is the assembly of anode biofilms in microbial fuel cells (Mei and Liu, 2019;).

Using mixed microbial communities typically allows for the accomplishment of more effective and efficient microbial processes since the optimal species/community combinations will be selected and implemented. Microbial biotechnologists must prioritize understanding how to manage and enhance these cooperative communities, or microbiomes, as they exist in soils, aquatic environments, built reactor systems, and even within the body if they are to meet the Sustainable Development Goals of the United Nations. In order to achieve this goal, it is our responsibility as scientists to create a collection of useful tools and solutions that will enable practitioners and decision-makers to choose the best microbial biotechnology. Rewilding to restore ecosystems using invertebrates and microbes is an illustration of such a strategy (Contos et al., 2021).

FUNDAMENTAL GUIDELINES FOR SOIL MICROBIAL RESOURCE MANAGEMENT

Soil ecosystems: Problems and Opportunities Although there is a wealth of information available on soil microbiology, there are currently no active methods for shaping the microbial population or directing it to perform a specific function. The exchange of greenhouse gases CO2, CH4, and NOx is one of the well-documented soil ecosystem services (Lal, 2004). This service is strongly dependent on microbial activity and their interactions with the surrounding soil environment and land use (Galbally et al., 2010;). By retaining precipitation, removing impurities from groundwater, and promoting plant development for energy and crops through appropriate nutrient delivery, soil ecosystems, for example, significantly contribute to ecosystem services. These ecosystem services are predicted to have an economic value in 2021 that is at least equal to 50% of the global GNP (Costanza et al., 1997, 2017). Ecosystem services have a monetary value of around €5,000 per person each year (Statista, 2021) when compared to the 7.9 billion people who make up the world's population today (www.worldometers.info/worldpopulation/). In order to maintain and even improve these services, we must consider the benchmarks and the resources required. We need to research how much improving the microbial community in soil ecosystems can be done without running the danger of having unfavourable, adverse effects.

In order to achieve the Sustainable Development Goals, microbial communities can be designed and stimulated to provide the ecosystem services found in soil. (1) The close interaction between plants and microbes, known as the rhizosphere, can promote plant development for energy crops, food and feed, and groundwater generation. This can be accomplished, for instance, by managing the microbes' top predators. (2) The inorganic and organic matter produced by microorganisms, including those in grasslands, and plants (shown by 1*), which aid in the formation of necromass and the storage of carbon (represented by 2*). By getting rid of CO2, CH4, N2O, and other air pollutants, effective management of soil microorganisms can reduce atmospheric concentrations of greenhouse gases.

BIOREMEDIATION IN ENVIRONMENTAL SUSTAINABILITY

The process of decomposing and detoxifying dangerous compounds in the environment into carbon dioxide, water, microbial biomass, and metabolites by products that are less toxic than the parent substance is known as bioremediation. This procedure makes use of living things including plants, algae, fungus, and bacteria. These bacteria may be native to the polluted area or they may be discovered and brought in for bio-remediation from outside the site. Microorganisms transform and break down these pollutants through their metabolic activities, using them as building blocks for further

Figure 1. Soil-air-water ecosystem management scheme

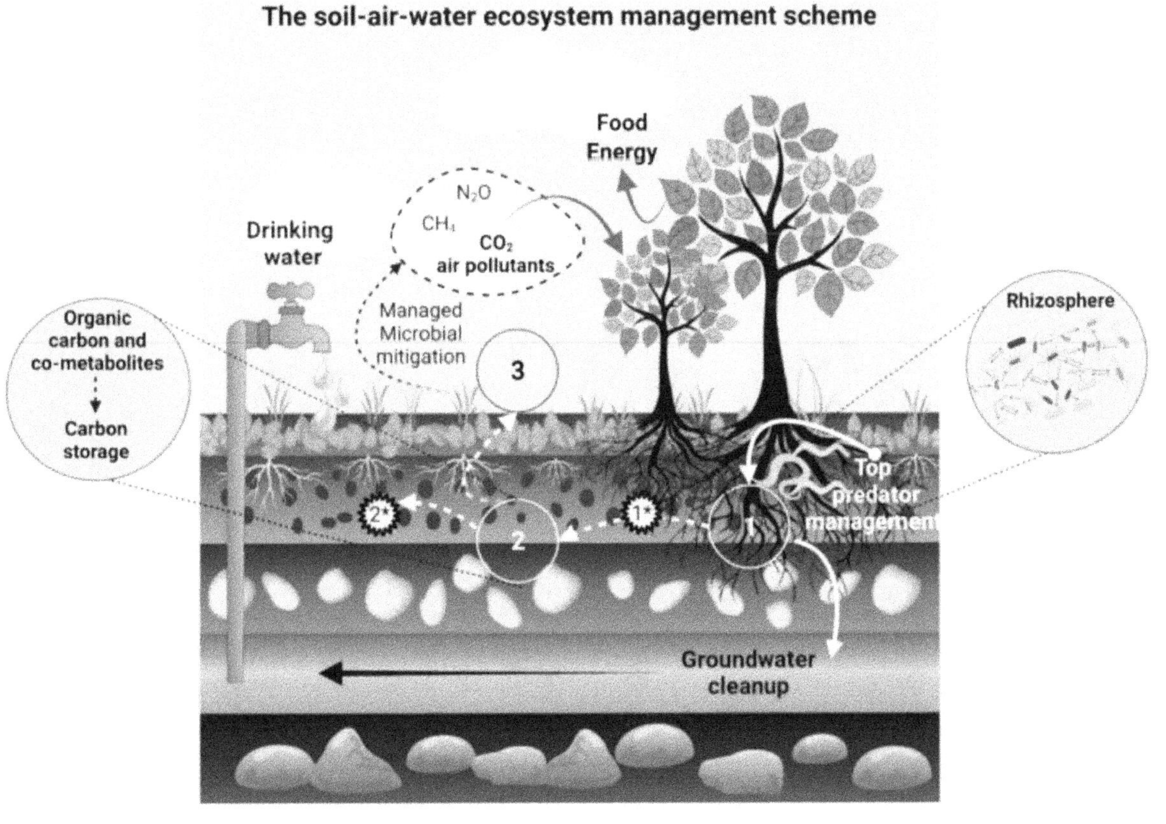

development. It is sometimes feasible to introduce more potential bacteria into the polluted region since full degradation of a pollutant requires the cooperation of many bacteria. We call this method "bio augmentation." The biodegradation process depends on favourable environmental conditions, the type and solubility of the pollutant, and the pollutant's bioavailability to microbes. As a result, the environmental conditions are controlled or altered to allow for sufficient microbial growth and, as a result, quick and efficient biodegradation.

MICROBES USED IN BIOREMEDIATION

Microbes may be found in many different environments, such as salt lakes, deserts, thermal springs, glaciers, and seas. Isolating microbes capable of degrading materials from contaminated environments, such as landfills, heavy metal-polluted sites, petroleum-contaminated sites, pesticide-contaminated sites from farming operations, and wastewater treatment plants, can aid in the degradation of various pollutants. This means that harmful pollutants can be changed into less harmful or innocuous substances by the metabolic activity of bacteria (Tiwari and Singh, 2014). Microorganisms use the dangerous pollution as a source of carbon and energy in both aerobic and anaerobic conditions. Due to the uneven distribution of pollutants and soil microorganisms, surfactants can be used to make the pollutant available to or in touch with the bacteria for effective pollutant breakdown.

The aerobic bacteria *Mycobacterium, Alcaligens, Sphingomonas*, and Pseudomonas are well-known for their ability to destroy illustrations that demonstrate the concept of bioremediation. Alkanes and

Figure 2. Illustration of bioremediation principle

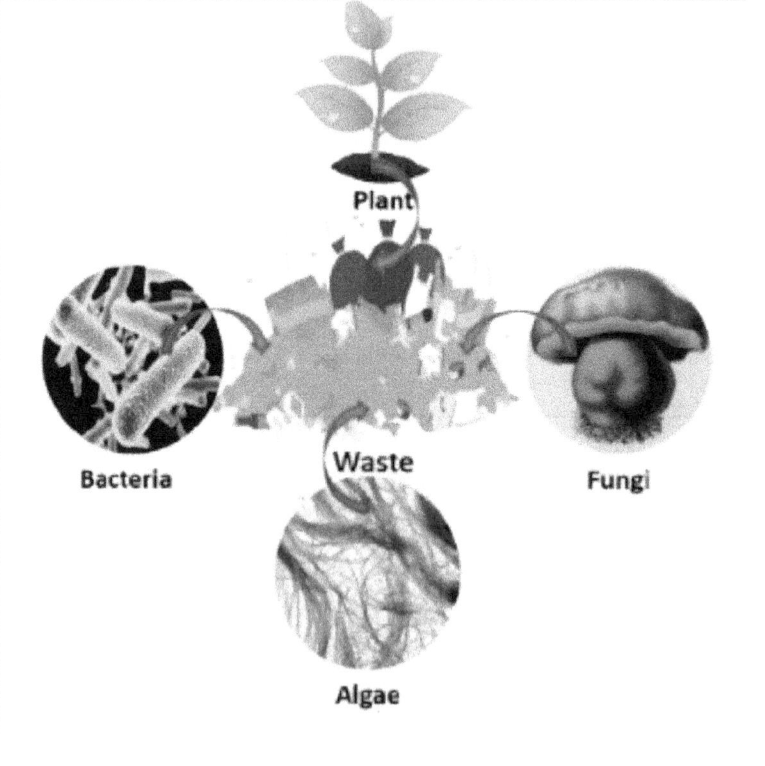

polycyclic aromatic hydrocarbons are hydrocarbons, thus the concepts of pesticide bioremediation apply. A few aerobic methylotrophs are also known to break down chlorinated aliphatic substances, including dichloroethane and trichloroethylene. It is known that anaerobic bacteria may break down PCBs, chloroform, and trichloroethylene, a chlorinated solvent, among other things. Certain toxic and persistent pollutants have also been documented to be successfully remedied by a few fungal species, including *Phanerochaete chrysosporium*. (Harekrushna and Kumar, 2012).

CLIMATE CHANGE MITIGATION AND ADAPTATION

Climate change mitigation and adaptation are two interconnected strategies aimed at addressing the challenges posed by global warming and its impacts on the environment. Environmental genomics is a field that plays a significant role in understanding and addressing climate change. Let's explore each of these topics:

- Climate Change Mitigation: Climate change mitigation refers to actions taken to reduce or prevent the emission of greenhouse gases (GHGs) into the atmosphere, thereby limiting the extent of global warming. The primary goal of mitigation is to stabilize the climate by reducing the concentration of GHGs, such as carbon dioxide (CO2), methane (CH4), and nitrous oxide (N2O), in the atmosphere. Some key strategies for climate change mitigation include:
    - Transitioning to Renewable Energy: Shifting from fossil fuels (coal, oil, and natural gas) to renewable energy sources (solar, wind, hydro, geothermal) reduces carbon emissions from energy production.
    - Energy Efficiency: Improving energy efficiency in buildings, transportation, and industries reduces energy consumption and associated emissions.
    - Afforestation and Reforestation: Planting trees and restoring forests help sequester CO2 from the atmosphere.
    - Sustainable Agriculture: Implementing practices that reduce methane emissions from livestock and nitrous oxide emissions from fertilizers.
    - Carbon Capture and Storage (CCS): Capturing CO2 emissions from industrial processes and power plants and storing them underground to prevent their release into the atmosphere.
- Climate Change Adaptation: Climate change adaptation involves adjusting to the changing climate and its impacts to minimize negative consequences and exploit opportunities that may arise. As climate change has already started to manifest, adaptation measures become necessary to protect vulnerable communities and ecosystems. Some adaptation strategies include:
    - Building Resilient Infrastructure: Constructing buildings and infrastructure that can withstand extreme weather events like floods, storms, and heatwaves.
    - Water Management: Implementing strategies for water conservation and managing water resources to deal with changing precipitation patterns.
    - Biodiversity Conservation: Preserving and restoring ecosystems to support wildlife and maintain ecosystem services that are critical for human well-being.
    - Disaster Preparedness and Early Warning Systems: Developing systems to predict and respond to climate-related disasters.

- Public Health Measures: Addressing the health risks associated with climate change, such as heat-related illnesses and the spread of vector-borne diseases.

Environmental Genomics and its Role: Environmental genomics, also known as metagenomics, is a branch of genomics that focuses on studying the genetic material (DNA, RNA) recovered directly from environmental samples, such as soil, water, and air. It provides insights into the biodiversity and functional potential of microbial communities present in various ecosystems.

Environmental genomics plays a crucial role in climate change mitigation and adaptation in several ways:

Assessing Ecosystem Health: By analysing the genetic diversity of microbial communities, researchers can monitor the health and resilience of ecosystems and identify changes that might be influenced by climate change.

Bioremediation and Carbon Sequestration: Understanding the genetic potential of microbial communities allows scientists to identify and manipulate specific organisms that can assist in bioremediation (breaking down pollutants) and carbon sequestration in soils.

Adaptation Strategies: Environmental genomics helps identify resilient species and genetic traits that can aid in developing crops and other organisms better adapted to changing climate conditions.

Monitoring Environmental Changes: Genomic tools enable continuous monitoring of environmental changes, helping researchers understand the effects of climate change on ecosystems in real-time.

Bioenergy Production: Environmental genomics contributes to the identification and engineering of microorganisms involved in bioenergy production processes, which can serve as a renewable energy source and reduce GHG emissions.

Overall, the integration of environmental genomics with climate change mitigation and adaptation efforts holds great promise in understanding and managing the impacts of global warming on our planet.

SUSTAINABLE AGRICULTURE

Agriculture is a comprehensive phrase that includes all facets of raising crops (for food and fiber), raising livestock, fishing, forestry, etc.

By utilizing modern technology, mechanization, increased use of fertilizers and pesticides, and expansion of irrigation facilities, food and fiber yield has grown. Due to these modifications, less labor was needed to produce the majority of the food and fiber.

Despite the good effects of these improvements, they have also led to some major environmental and social issues, such as topsoil degradation, the depletion and contamination of groundwater and other water resources, and the loss of farm laborers as a result of farm machinery's replacement. The need to promote "sustainable agriculture" is developing as a result of the growing negative effects of contemporary agriculture. The production of food, fiber, plant, or animal products using farming practices that preserve the environment, public health, and welfare of people and animals is known as sustainable agriculture.

Many environmentally sound agricultural practices are included into sustainable agriculture, which also provides creative and financially viable opportunities for farmers, laborers, consumers, decision-makers, and many other people involved in the overall food chain.

Figure 3. Concept of sustainable agriculture

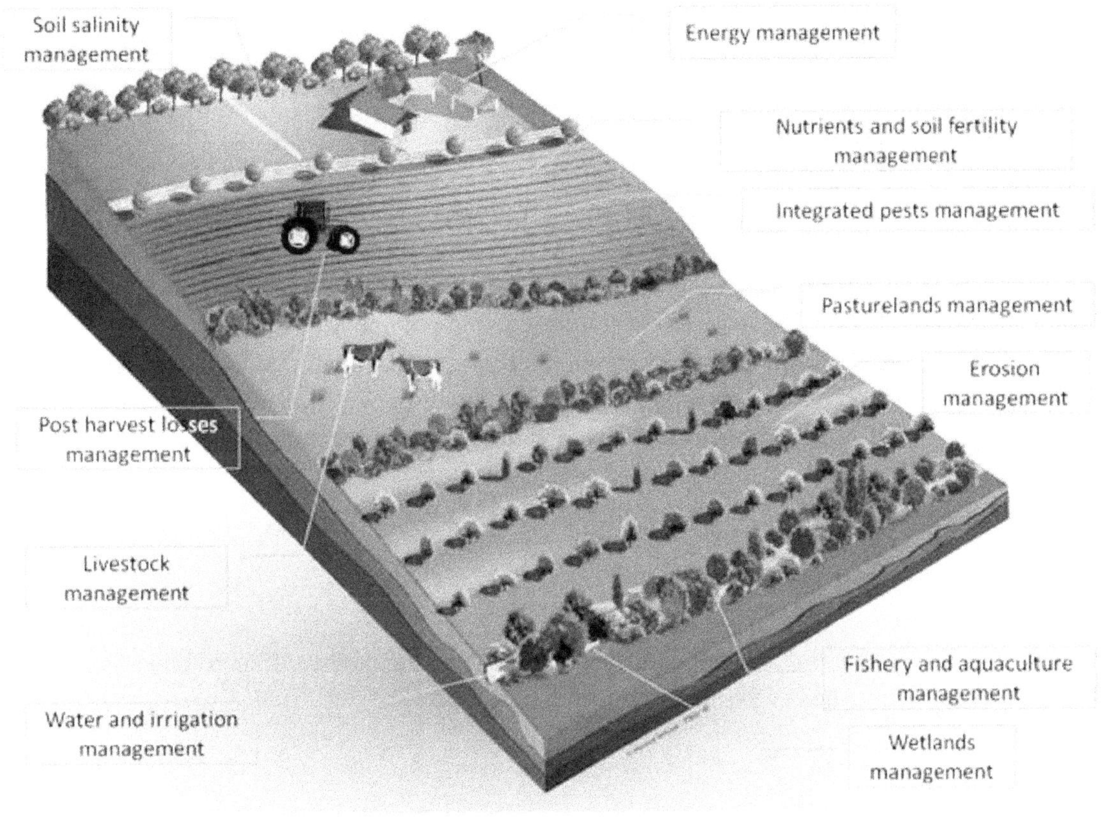

Sustainable agricultural practices preserve productivity and profitability while using the least amount of energy and hazardous substances. Think of organic farming.

Thus, sustainable agriculture is one that,

- supports profitable production
- protects environmental quality
- uses natural resources efficiently
- provides consumers with affordable, high-quality products
- decreases dependency on non-renewable resources
- enhances the quality of life for farmers and rural communities and
- will last for generations to come.

Methods of Sustainable Agriculture

The local geography (topography), soil type and condition, local climate, pests, local inputs, and the farmer's objectives must all be taken into account during planning.

Next, the grower (farmer) must decide on the best procedures. Sustainable agriculture uses a number of techniques, including:

cultivating methods to improve economic and biological stability.
to meet the need, a selection of enhanced varieties.
using the right tillage technique to manage the soil.

SUSTAINABLE AQUACULTURE

Our expanding population might be fed in large part through sustainable aquaculture.

Currently, 42% of the seafood we eat comes from farms, yet there are no laws defining what "good" aquaculture looks like just yet. Our food supply is significantly influenced by aquaculture, so it must be practiced sustainably. The OF is particularly interested in a number of closed-system technologies, such as recirculating tanks, raceways, flow-through systems, and inland ponds. Numerous species of fish, shellfish, and aquatic plants are supported by these systems. We encourage efforts to prevent the environmental and food safety problems associated with open pen aquaculture, despite the obvious advantages (health and other benefits) of closed-system aquaculture systems that have been acknowledged. We intend to make both domestic and international efforts.

Methods of Aquaculture

Currently, aquaculture is mostly practiced using four different methods: Offshore Open Pens, Closed Land-Based Pens, Open Ancient Systems.

OFFSHORE OPEN PENS

With the exception of shellfish mariculture, the near-shore aquaculture systems have traditionally been used to grow shellfish, salmon, and other carnivorous finfish. These systems are generally regarded as the least sustainable and most environmentally harmful type of aquaculture. The intrinsic "open to the ecosystem" aspect of these systems makes it incredibly challenging to address issues with faecal waste, interactions with predators, introduction of non-native/exotic species, excess inputs (food, antibiotics), habitat damage, and disease transmission. Additionally, the current technique of continuing down the shoreline after disease outbreaks that render the pens inoperable cannot be supported by coastal waters.

CLOSED LAND-BASED SYSTEMS

Recirculating aquaculture systems (RAS), which are land-based "closed" systems, are gaining popularity as a long-term sustainable aquaculture option in both developed and developing countries. In developing nations, small, affordable closed systems are being designed, while more expensive, larger, and commercially viable alternatives are being developed in affluent nations. These self-sufficient systems frequently permit efficient polyculture approaches to co-growing plants and animals. They are especially regarded as sustainable when they employ renewable energy, ensure that almost all of their water is recovered, and concentrate on growing both omnivores and herbivores.

OPEN ANCIENT SYSTEMS

Fish aquaculture is nothing new; many cultures have been engaged in it for many years. Ancient Egyptians cultivated tilapia as part of their sophisticated irrigation system, and Hawaiians were able to farm a variety of species such as milkfish, mullet, prawns, and crab. Ancient Chinese cultures fed silkworm dung and nymphs to carp raised in pools on silkworm farms. Aquaculture was practiced by the Mayans and some Native American societies in North America, according to archaeologists (www.enaca.org).

CIRCULAR ECONOMY AND WASTE MANAGEMENT

The linear "take, make, dispose" approach of conventional economic systems has issues, which the circular economy and waste management both seek to overcome. Both ideas put a heavy emphasis on lowering waste output, encouraging resource efficiency, and reducing how much consumption and production affect the environment. Let's examine each of these ideas in turn:

Circular Economy

By maximizing the value and utility of goods, resources, and materials by preserving them for as long as feasible, the circular economy is an alternative economic model. The following are the key tenets of a circular economy:

Figure 4a. Offshore open pens

Figure 4b. Closed land-based aquaculture

Figure 4c. Open ancient aquaculture

- Design for Longevity and Reuse: To encourage reuse, products are made to be strong, simple to fix, and upgradeable.
- Reusing things and recycling materials at the end of their useful lives are two ways to reduce waste. This strategy is known as "Reduce, Reuse, Recycle."

- Resource Efficiency: This is the process of maximizing the use of resources while reducing waste, energy use, and emissions during the production and consumption processes.
- Extended Producer Responsibility (EPR): Encouraging producers to be accountable for their goods throughout their full lifecycle, including collection, recycling, and correct disposal.
- Promoting the sharing of goods and services to increase their use and lessen the necessity for individual ownership.
- In order to create a more sustainable and regenerative economic structure, the circular economy seeks to uncouple economic growth from resource use and environmental damage.

Waste Management

The collection, transportation, processing, recycling, and disposal of waste products produced by human activity are referred to as waste management. Minimizing environmental pollution, reducing the amount of waste sent to landfills and incineration plants, and recovering valuable resources from trash are all goals of effective waste management systems. Significant elements of waste management include:

Waste reduction: Promoting waste prevention and reduction through consumer behavior changes, product design, and awareness campaigns.

Recycling, on the other hand, is the process of gathering and processing waste items like paper, plastic, glass, and metals to turn them into fresh goods or raw resources.

Composting: Using composting to manage organic waste (such food scraps and yard waste) and produce nutrient-rich soil amendments.

Garbage-to-Energy: Using techniques like anaerobic digestion or incineration, non-recyclable garbage is transformed into energy.

Proper landfill management is essential for reducing their negative effects on the environment and promoting the use of landfill gas for energy production.

INTEGRATING CIRCULAR ECONOMY AND WASTE MANAGEMENT

Waste management and the circular economy are two ideas that go hand in hand. The effective reuse, recycling, or repurposing of resources and products depends on waste management in a circular economy. The formation of trash can be considerably decreased by implementing the concepts of the circular economy, and precious resources can be kept in use for extended periods of time, obviating the need for raw material extraction.

Furthermore, waste management procedures are essential in facilitating the shift to a circular economy because they offer the processes and infrastructure required for efficient material recovery and recycling. More resources can be recovered from abandoned products by improving waste management systems, closing the loop and promoting a more resource-wise and sustainable economy.

Both the circular economy and waste management are essential strategies to address the environmental challenges of our time, including resource depletion, pollution, and climate change. They offer opportunities for businesses, governments, and individuals to work together toward a more sustainable and prosperous future.

Figure 5. Circular economy

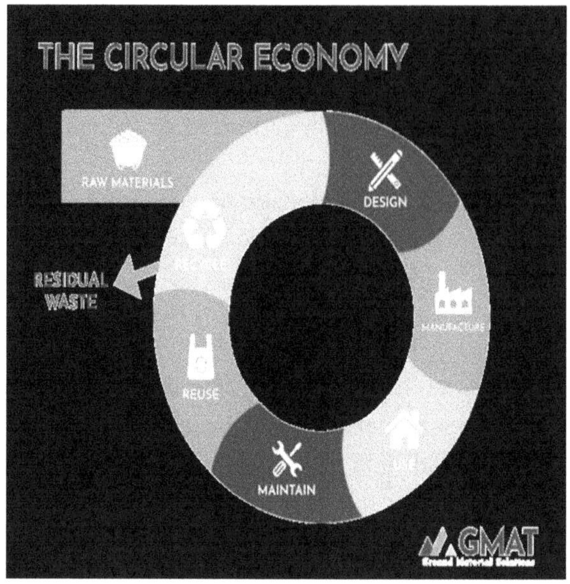

PUBLIC AWARENESS AND EDUCATION IN SUSTAINABILITY

In order to create a society that is more environmentally conscious and sustainable, public awareness and education on sustainable development are essential. We can empower people and communities to take proactive steps to safeguard the environment and advance social and economic well-being by educating and involving the public. The following are some essential components of public education and awareness for sustainable development:

Informative Campaigns: To raise public understanding of the value of sustainability and its effects on various facets of life, governments, non-governmental organizations (NGOs), and other stakeholders should conduct public awareness campaigns. To reach a large audience, these campaigns might make use of a variety of communication platforms, including radio, television, social media, and public events.

School curriculum: From primary schools to colleges, incorporate sustainability-related issues into the curriculum. This includes educating pupils on issues like resource management, climate change, environmental preservation, and sustainable development concepts. Students' comprehension of these ideas can be strengthened even further by promoting project-based learning and outdoor activities.

Workshops and Training: Planning seminars, workshops, and training sessions to involve various societal groups, such as professionals, decision-makers, and community leaders. These occasions can offer useful information and resources for putting sustainable principles into reality across a range of industries, including agriculture, energy, and urban planning.

Public Events and Exhibitions: Organizing fairs, events, and displaying highlight green products, eco-friendly technologies, and effective sustainability programs. These gatherings offer chances for direct interaction with the public and motivate them to develop more environmentally friendly habits.

Working together with the media to expose sustainability-related problems and success stories. Working with journalists and content producers can promote sustainability issues through news stories, documentaries, and special features.

Citizen Science Initiatives: Encouraging citizen science projects that involve the public in data collection and monitoring of environmental parameters. Involving citizens in scientific research fosters a sense of ownership and responsibility towards their local environment.

Encouraging community involvement in decision-making processes and sustainable development activities that have an immediate influence on their surroundings. Projects that are led by the community frequently have greater odds of success and long-term sustainability.

Encouraging companies and organizations to issue sustainability reports that openly demonstrate their effects on the environment and society. This procedure pushes organizations to use more sustainable practices and fosters responsibility.

Rewards & Incentive Programs: Offering rewards and recognition to people, organizations, and communities who make great efforts to embrace sustainable practices. Recognition programs can act as constructive criticism and encourage others to imitate them.

Combining these strategies will help us create a society that values sustainable growth and understands how important it is for both human well-being and the health of the environment. A more affluent and sustainable future is made possible by the broad adoption of sustainable behaviours, which are sparked by public awareness and education.

INTERNATIONAL COLLABORATION AND DATA SHARING FOR ENVIRONMENTAL SUSTAINABILITY

International collaboration and data sharing are essential pillars for advancing environmental sustainability on a global scale. Addressing complex environmental challenges requires collective efforts, shared knowledge, and coordinated actions among nations and stakeholders. Here's how international collaboration and data sharing contribute to environmental sustainability:

- Knowledge Exchange and Learning: Collaborative efforts enable countries and organizations to share research findings, best practices, and successful initiatives. By learning from each other's experiences, countries can accelerate progress and avoid reinventing the wheel, making the most of limited resources.
- Global Problem Solving: Environmental issues often transcend national borders, such as climate change and biodiversity loss. International collaboration allows countries to work together on these global challenges, pooling expertise and resources to find effective solutions that benefit all parties involved.
- Data Standardization and Harmonization: Sharing environmental data across borders requires standardization and harmonization of data collection methods, formats, and quality assurance protocols. This ensures that data from different sources can be integrated and compared, leading to more robust and reliable assessments and decision-making.
- Early Warning and Rapid Response: In the face of environmental disasters or emerging threats, international data sharing enables early warning systems and rapid response mechanisms. Timely

information exchange helps affected regions prepare for and mitigate the impacts of natural disasters and environmental crises.

- Assessing Progress Towards Sustainable Goals: International collaboration and data sharing facilitate the tracking and monitoring of progress towards global sustainability goals, such as the United Nations' Sustainable Development Goals (SDGs). Sharing data on environmental indicators allows countries to assess their performance and adjust policies and actions accordingly.
- Identifying Global Trends and Patterns: By pooling and analysing data from different regions, international collaboration helps identify global trends and patterns in environmental change. This insight is crucial for understanding the interconnectedness of ecosystems and guiding global conservation efforts.
- Resource Management and Allocation: Many environmental challenges require international cooperation for effective resource management. Shared data on resource availability, such as water, fisheries, or forests, helps countries make informed decisions about allocation and sustainable use.

International collaboration and data sharing are indispensable components of environmental sustainability. By promoting mutual learning, coordinated actions, and data-driven decision-making, countries can work together to safeguard the environment and create a more sustainable and resilient future for all.

CONCLUSION

In conclusion, "Solutions and Exploratory Ideas for Environmental Sustainability through Environmental Genomics" emphasizes the transformational potential of environmental genomics in addressing the pressing issues of environmental sustainability. This multidisciplinary discipline provides cutting-edge and data-driven solutions that go beyond conventional techniques, paving the path for a more resilient and environmentally balanced future. We acquire previously unheard-of insights into the genetic variety, interactions, and adaptations of species within their natural settings through the integration of cutting-edge genomic technology. Nevertheless, embracing environmental genomics for sustainability necessitates addressing ethical concerns and promoting equitable access to genetic resources. Responsible data sharing and collaboration among nations and communities are essential to ensure that the benefits of environmental genomics reach all corners of the world.

REFERENCES

Aplet, G. H. (1999). On the nature of wildness: Exploring what wilderness really protects. *Denver University Law Review*, 347–367. https://digitalcommons.du.edu/dlr/vol76/iss2/4/

Christensen, N. L., Bartuska, A. M., Brown, J. H., Carpenter, S., D'Antonio, C., Francis, R., Franklin, J. F., MacMahon, J. A., Noss, R. F., Parsons, D. J., Peterson, C. H., Turner, M. G., & Woodmansee, R. G. (1996). The report of the ecological society of America committee on the scientific basis for ecosystem management. *Ecological Applications*, *6*(3), 665–691. doi:10.2307/2269460

Contos, P., Wood, J. L., Murphy, N. P., & Gibb, H. (2021). Rewilding with invertebrates and microbes to restore ecosystems: Present trends and future directions. *Ecology and Evolution*, *11*(12), 7187–7200. doi:10.1002/ece3.7597 PMID:34188805

Costanza, R., d'Arge, R., de Groot, R., Farber, S., Grasso, M., Hannon, B., Limburg, K., Naeem, S., O'Neill, R. V., Paruelo, J., Raskin, R. G., Sutton, P., & van den Belt, M. (1997). The value of the world's ecosystem services and natural capital. *Nature*, *387*(6630), 253–260. doi:10.1038/387253a0

Costanza, R., de Groot, R., Braat, L., Kubiszewski, I., Fioramonti, L., Sutton, P., Farber, S., & Grasso, M. (2017). Twenty years of ecosystem services: How far have we come and how far do we still need to go? *Ecosystem Services*, *28*, 1–16. doi:10.1016/j.ecoser.2017.09.008

Egan, D., & Howell, E. A. (Eds.). (2001). The Historical Ecology Handbook: A Restorationist's Guide to Reference Ecosystems. Island Press. doi:10.5751/ES-00468-070111

Galbally, I., Meyer, C. P., Wang, Y.-P., & Kirstine, W. (2010). Soil–atmosphere exchange of CH4, CO, N2O and NOx and the effects of land-use change in the semiarid Mallee system in Southeastern Australia. *Global Change Biology*, *16*(9), 2407–2419. doi:10.1111/j.1365-2486.2010.02161.x

Grumbine, R. E. (1994). What is ecosystem management? *Conservation Biology*, *8*(1), 27–38. doi:10.1046/j.1523-1739.1994.08010027.x

Harekrushna, S., & Kumar, D.C. (2012). A review on bioremediation. *International Journal of chemistry and environment, 2*(1), 13-21.

Hobbs, R. J., Higgs, E., Hall, C. M., & Carol, M. (2014). Managing the whole landscape: Historical, hybrid, and novel ecosystems. *Frontiers in Ecology and the Environment*, *12*(10), 557–564. doi:10.1890/130300

Hooper, L. V., Midtvedt, T., & Gordon, J. I. (2002). How host-microbial interactions shape the nutrient environment of the mammalian intestine. *Annual Review of Nutrition*, *22*(1), 283–307. doi:10.1146/annurev.nutr.22.011602.092259 PMID:12055347

Lal, R. (2004). Soil carbon sequestration impacts on global climate change and food security. *Science*, *304*(5677), 1623–1627. doi:10.1126/science.1097396 PMID:15192216

Landres, P. B., Morgan, P., & Swanson, F. J. (1999). Overview of the use of natural variability concepts in managing ecological systems. *Ecological Applications*, *9*, 1179–1188.

Mei, R., & Liu, W. T. (2019). Quantifying the contribution of microbial immigration in engineered water systems. *Microbiome*, *7*(1), 144–156. doi:10.1186/s40168-019-0760-0 PMID:31694700

Safford, H. D., Hayward, G., Heller, N., & Wiens, J. A. (2012). Climate change and historical ecology: can the past still inform the future? J.A. Wiens, G. Hayward, H.D. Safford, C.M. Giffen (Eds.), Historical Environmental Variation in Conservation and Natural Resource Management. John Wiley and Sons, New York, NY.

Tasoff, J., Mee, M. T., & Wang, H. H. (2015). An economic framework of microbial trade. *PLoS One*, *10*(7), e0132907. doi:10.1371/journal.pone.0132907 PMID:26222307

Tiwari, G., & Singh, S. P. (2014). Application of bioremediation on solid waste management: A review. *Journal of Bioremediation & Biodegradation*, *5*(6), 248–1172.

Valbuena-Carabaña, M., de Heredia, U. L., Fuentes-Utrilla, P., González-Doncel, I., & Gil, L. (2010). Historical and recent changes in the Spanish forests: A socio-economic process. *Review of Palaeobotany and Palynology*, *162*(3), 492–506. doi:10.1016/j.revpalbo.2009.11.003

Vallejo, V. R. (2009). In S. Bautista, J. Aronson, & V. R. Vallejo (Eds.), *Problems and perspectives of dryland restoration* (pp. 13–22). Land Restoration to Combat Desertification, Fundacion Centro de Estudios Ambientales del Mediterráneo.

Werner, G. D. A., Strassmann, J. E., Ivens, A. B. F., Engelmoer, D. J. P., Verbruggen, E., Queller, D. C., Noë, R., Johnson, N. C., Hammerstein, P., & Kiers, E. T. (2014). Evolution of microbial markets. *Proceedings of the National Academy of Sciences of the United States of America*, *111*(4), 1237–1244. doi:10.1073/pnas.1315980111 PMID:24474743

Young, C. R. (2015). *The Royal Forests of Medieval England*. University of Pennsylvania Press.

Chapter 14
Ecological Footprint and Its Enhancing Factors in SAARC Countries

Ajay Kumar Singh
https://orcid.org/0000-0003-0429-0925
Department of Humanities and Social Sciences, Graphic Era University (Deemed), Dehradun, India

Ravindra Singh
https://orcid.org/0000-0003-3062-1812
Department of Humanities and Social Sciences, Graphic Era University (Deemed), Dehradun, India

Santosh Singh
https://orcid.org/0000-0003-2448-4794
School of Liberal Arts and Management, DIT University, Dehradun, India

ABSTRACT

High economic development is caused to increase ecological footprint (EF). Increasing EF creates an impediment for sustainable development. Previous studies applied several methods and indicators to examine EF in different countries. However, very few studies could examine the originating factors of EF using empirical models. Hence, this chapter measures EF for SAARC countries during 2000 – 2022. It explains the comparative performance of SAARC countries in EF using a time trend analysis. Thereupon, it also examines the impact of certain indicators on EF in SAARC countries using robust empirical analysis. There seems to be significant diversity in EF due to disparity in economic development and its enhancing factors in SAARC countries. EF increases as industrialization, greenhouse gas, inflation, GDP per person employed and age dependency ratio increase. Digitalization, renewable energy consumption, technological change and life expectancy rate showed a negative impact on EF.

DOI: 10.4018/979-8-3693-3330-3.ch014

INTRODUCTION

The global countries are increasing the level of investment, consumption, productivity of resources, industrialization and other activities to attain their desired tasks. Moreover, the prime goal of above-mentioned activities is to increase the living standard of people (Saleem et al., 2019). High economic growth is essential to increase energy, food, shelter and living standard, and it is positive to fulfil diverse needs of individuals (Singh et al., 2023a). Therefore, most countries have achieved significant growth in economic development and social welfare (Kızılgöl & Öndes, 2022). While high economic growth and rising population are responsible for increasing water pollution, environment degradation, air pollution, acid rain, soil quality degradation, loss of biodiversity, and sound pollution, etc. (Dietz et al., 2007; Saboori et al., 2012; Singh & Singh, 2024). These are also responsible for environmental calamities and have an extensive burden on ecological services (Ray & Ray, 2011). Further, these have an adverse impact on environmental and natural habits (Bilsborrow, 1992).

Depletion of natural resources and EF are increasing due to population growth (Sharma et al., 2020; Dogan et al., 2022). Thus, it is solely accountable to increase EF (Kazemzadeh et al., 2023). Globalization is supportive to enhance competition, trade benefits, and transfer of technologies across countries (Figge et al., 2016; Rahman et al., 2019). Globalization is increasing pressure on the environment, and it is also causing increased GHGs emissions and global warming that are harmful for the global ecosystem (Kumar et al., 2015; Saleem et al., 2019). Hence, above-mentioned issues are of great concern for national and global policy makers (Rahman et al., 2019).

The current population of South Asia is around 1.9 billion and it includes around 24.89% of world population. Therefore, South Asia has more difficulty to maintain the stability in biodiversity due to over population that is harmful for biocapacity and ecological footprint (EF) (Kazemzadeh et al., 2023). Biocapacity means that quantity of ecological assets that help to produce food, clothing, shelter, and absorb carbon emissions. For instance, Bhutan has the highest value of EF (i.e., 4.4 million gha per person) and Afghanistan has lowest values of EF (i.e., 0.7 million gha per person) in South Asia. While the value of EF for India is estimated 1.16 million gha per person. Figure 1 shows an increasing EF due to rising activity worldwide. Figure 2 explores the factors affecting EF. Acid rain, forest deforestation, environment degradation and several other activities are causes to increase EF (Singh et al., 2019). Hence, environmental sustainability may be a proper solution to maintain biodiversity and reduce EF (Singh et al., 2019).

Footprint means how much of ecosystem is essential to produce desired goods of the present population, and adjust waste materials. Hence, EF and biodiversity are adversely affected due to increase in food consumption, energy, transportation, and goods and services (Mancini et al., 2016). Hence, demand of ecosystem services by the rising population will increase extensive problems for environmental development. The concept of EF has gained prominence as a comprehensive measure in calculating human impact on ecology (Wackernagel & Rees, 1996).

The measurement of EF incorporates complex methodologies that integrate data on resource consumption, land use, and waste generation by humans during a period (Galli et al., 2020). Such results for these different aspects of human impact on the environment provides a more hostile, comprehensive assessment of sustainability as compared to other indicators like gross domestic product (GDP) or environmental impact assessments (Dooren et al., 2018; Zambrano-Monserrate et al., 2020). EF measures the decline in the common property of nature (Wiedmann & Barrett, 2010). The analytical findings of EF have diverse applications in various fields, including policy-

Figure 1. The trends in all typed of EF world-wide

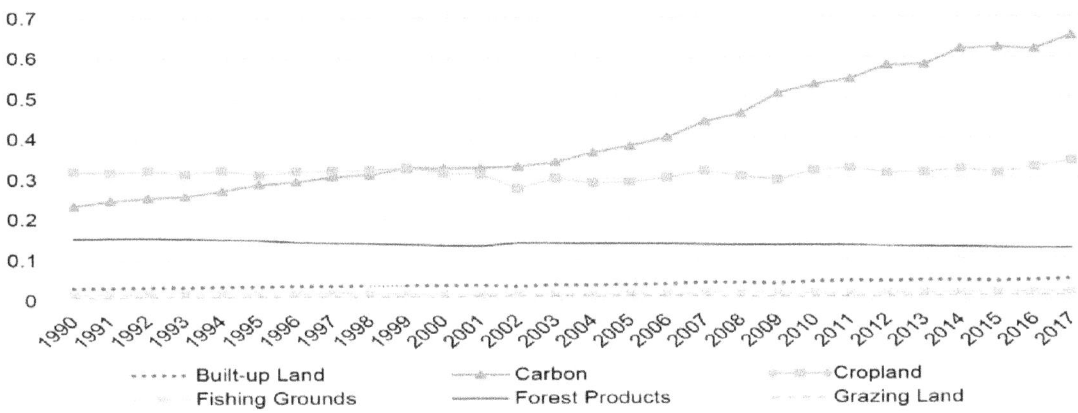

Figure 2. Factors contributing EF

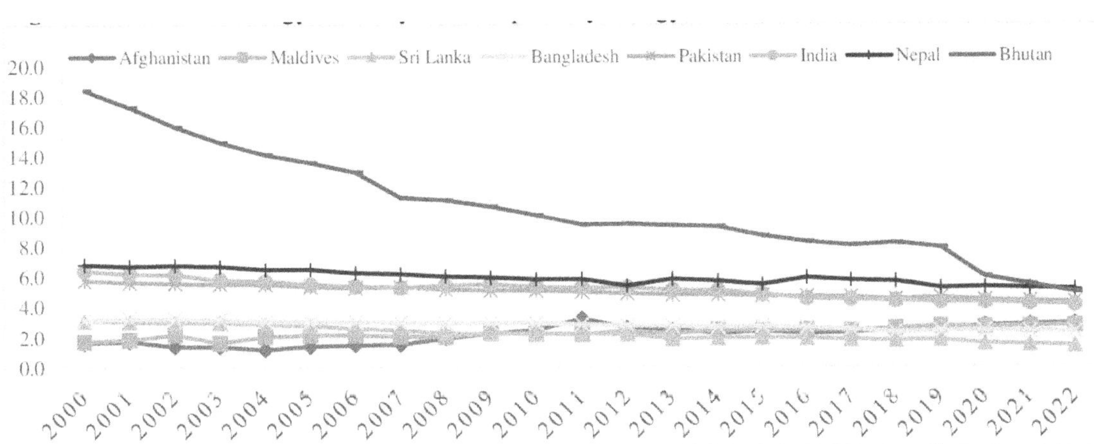

making, urban planning, corporate sustainability, and education. However, there is still uncertainty about the implication of EF on economic activities and vice-versa (Fiala, 2008; Stoeglehner & Narodoslawsky, 2008).

Many economists estimated the EF using different methods in academic research. Most studies provided a clear understanding of EF and its measurement (Kazemzadeh et al., 2023). Few studies estimated Atkinson Index and Theil Index to explain the performance of EF (Duro & Teixidó-Figueras, 2013). Previous studies applied several methods to examine the environmental degradation as an EF across different countries (Rahman et al., 2019). Despite that, EF and its measurement are not defined appropriately in policy documents and existing literature due to its multifaceted interconnection with human development, globalization, technological development, financial development and demographic indicators (Zhang & Lin, 2012; Figge et al., 2016; Sabir & Gorus, 2019; Mishra & Dash, 2022; Mehmood et al., 2023). However, very few studies could examine the originating factors of EF using empirical models. Therefore, the present study measures the answer of below questions:

- What is an effective and reliable measurement of EF in SAARC countries?
- What is the relative performance of EF in SAARC countries?
- What policy initiative may be favorable to reduce EF in SAARC countries?
- Which are the key activities that increase EF in SAARC countries?
- Which social-economic indicators are more responsible for increasing EF?

The chapter has achieved following objectives:

- To estimate EF for SAARC countries.
- To provide comparative performance of SAARC countries in EF
- To observe originating factors of EF in SAARC countries.

REVIEW OF LITERATURE

The theoretical perspective of EF is clear in existing studies (Baabou et al., 2017; Kızılgöl & Öndes, 2022; Han & Dalaibaatar, 2023; Magazzino, 2024). Mancini et al. (2016) explained the clarity and transparency in EF and its affecting factors as per the review of literature. Ulucak and Lin (2017) suggested that EF can be understood as a cumulative predictor of ecological damage caused by human development. EF signifies the available ecological assets required for the current population including dairy, seafood, and infrastructure (Baabou et al., 2017). It is also a measurement of human pressure on ecological services. Therefore, EF can be linked with to increase sustainable development (Han & Dalaibaatar, 2023). EF is a best measure of ecological deficit (Kızılgöl & Öndes, 2022). Rapid human activity in agriculture, dairy and fishing creates more pressure on resources and ecological sustainability (Wei et al., 2023). Such exploitation is creating a huge burden on the global ecology and hindering the well-being of the masses (Danish, 2019). Over time, human concern about the magnitude of problems in the biosphere has increased the attention of policy makers to achieve global sustainability (Kitzes et al., 2008). The main strength of the EF is to summarize the human influence on different types of bio-productivity as a whole and to provide some estimation of over-consumption with communicable results to the audience (Wiedmann & Barrett, 2010). Moreover, such an overshoot condition clearly demands strategies and intervention to bring the unpleasant part of the human economy under control. Scientific research community is looking for a substantial solution to reduce CO_2 emissions (contraction and convergence) (Magazzino, 2024). These dense demographic countries are controlling human consumption which has critical implications on environmental sustainability (Mishra & Dash, 2022).

The Environmental Kuznets Curve (EKC) theory is extensively applied to observe the relationship of environmental degradation and economic development (Dietz et al., 2007; Basarir & Arman, 2014; Aşıcı & Acar, 2016; Ayeche et al., 2016; Aşıcı & Acar, 2016). However, environmental degradation and per capita GDP do not show a uniform association with each other across countries. Most scholars reported a U-shaped and inverted U-shaped association between the mentioned indicators (Kızılgöl & Öndes, 2022). Kasperowicz (2015) noticed a negative association of GDP with environmental degradation. The contribution of human capital is also responsible for EF in South Asia and it shows that these countries are not adopting environmental education (Mehmood et al., 2023).

Most scholars also adopted environmental performance index (EPI) to assess environmental relationship with social, economic and human development (Samimi et al., 2011; Figge et al., 2016). It is noted a causality between EF and economic complexity (Wan et al., 2022). In recent decades most scholars are preferring the EF to measure the overall environmental degradation. EF, therefore, has a positive relationship with economic growth, development of human capital, and urbanization (Mehmood et al., 2023). Human capital development, urbanization, and biocapacity also promote EF. It is also essential to increase economic and social success (Han & Dalaibaatar, 2023). While EF is increasing due to multiple reasons (Mehmood et al., 2023). Trade, globalization, technology transfer and foreign direct investment (FDI) also increases EF (Rahman et al., 2019; Kazemzadeh et al., 2023). Poverty is also responsible for increasing EF and environmental degradation (Ray & Ray, 2011).

Population growth is cause to increase EF as it enhances extensive pressure on natural resources, and increases soil erosion and air and water pollution, and reduces biodiversity (Bilsborrow, 1992; Singh & Singh, 2024). Technological change and foreign trade also show a positive impact on EF (Sabir & Gorus, 2019). In China, Magazzino (2024) also found that trade openness and urbanization are responsible for increasing EF in China. While Kızılgöl & Öndes (2022) examined the role of economic growth, trade, urbanization and renewable consumption on EF in OECD countries. It reported that FDI has also increased EF. Technological change has an insignificant impact on environmental quality and EF. Demand for energy consumption is increasing due to a change in consumption pattern (Ray and Ray, 2011). It stated that economic growth is an important cause for EF. Financial development also determines EF (Majeed & Mazhar, 2019).

EF increases as energy consumption increases (Saleem et al., 2019). For instance, the United States contributes 14%, China 28%, and India 7% to the world's CO2 emissions in the atmosphere (GCB, 2020). The global contribution to fossil fuel CO_2 emissions is reported to be 51% from these economies. Energy is the primary driver of CO_2 emissions, and India has quickly grown economically by using extensive quality of energy. India's primary consumption of energy increased by 8% in 2018. Fossil fuels contribute to nearly 80% of the total electricity generated worldwide. In addition, India has 2nd rank in renewable energy after China (IEA, 2018). Population density, urbanization, industry value added and energy use are also causes to increase EF (Aşıcı & Acar, 2016; Magazzino, 2024). Yilanci and Pata, 2020 concluded that the EF in ASEAN countries and found a nonlinear trend in EF. Life expectancy and education are useful to reduce EF (Dietz et al., 2007). Kızılgöl and Öndes (2022) investigated the role of foreign trade on EF in 31 OECD countries. Kazemzadeh et al. (2023) claimed that energy consumption is caused to increase EF. Magazzino (2024) explained the association of EF with electricity consumption, and GDP in China.

RESEARCH METHODS

Study Area and Source of Data

This chapter estimates the ecological footprint (EF) for SAARC countries. The statistics of most suitable and pertinent indicators for these countries are available during 2000 – 2022. Hence, this chapter considers this time period for proposed statistical and empirical analysis. The required data is derived from world development indicators (the World Bank), The Food and Agriculture Organization of the

United Nations (FAO), The Organisation for Economic Co-operation and Development (OECD) and The World Intellectual Property Organization (WIPO).

Theoretical Foundation for Estimating EF

EF refers to the measurement of environmental degradation and loss of biodiversity. The diverse demand can be fulfilled using the environment and natural resources like air, water, land, etc. in production activities. Therefore, indicators and measurement of EF are clear in the academic literature (Mancini et al., 2016; Rahman et al., 2019). Most studies used multiple indicators related to biodiversity for measuring EF for a specific country (Dietz et al., 2007). Saleem et al. (2019) considered CO_2, N_2O and CH_4 as proxy variables in BRICS countries. Zhang and Lin (2012); Basarir and Arman (2014); Kasperowicz (2015) also used similar variables in empirical investigation. Supply and demand of natural resources are also used to estimate EF.

EF includes all human activities that are responsible for environmental degradation. Thus, exact measurement of EF is not possible. Despite that, it is essential to measure the environmental and ecological degradation to prevent these from human and natural activities (Saleem et al., 2019). It is imperative to increase natural resource sustainability and sustainable development (Singh et al., 2021). Most scholars used diverse techniques to deduct the value of EF. Thereupon, different indicators are adopted by the existing researchers for measuring the EF. Hence, existing scientists and researchers followed different indicators and methods to calculate EF. The authors of this study could not consider the underlined set of indicators in the estimation of EF. Global Footprint Network published the data of EF and its associated drivers across countries. However, this organization does not have data of EF for all SAARC countries (YUEFI & GFN, 2023).

This study, therefore, includes per capita CO_2 emissions and per capita GDP to estimate EF for SAARC countries. The ratio of CO_2 emission with per capita GDP infers how much the quantity of environmental degradation is increasing due to increase in production. Consequently, production in all sectors is responsible for increasing CO_2 emissions. Per capita CO_2 emission is a significant cause for economic development (Mehmood et al., 2023). Economic development increases as investment in all sectors increases. While investment promotes employment and demand for products. Accordingly, it helps to increase growth of manufacturing and industrial sectors that emit CO_2 emissions in the atmosphere. Hence increasing CO_2 emissions have an adverse implication on environmental development (Ray & Ray, 2011; Zhang & Lin, 2012). Thus, it is consistent to use the ratio of per capita GDP with CO_2 emission as a proxy indicator for EF. In this chapter, EF is calculated as:

$$(EcFo)_{ct} = \{(PCCO_2E)_{ct}\}/\{(PCGDP)_{ct}\} \quad (1)$$

Here *EcFo* is EF; $PCCO_2E$ is per capita CO_2 emissions; *PCGDP* is per capita GDP; c is specific country in SAARC; and t is time (i.e., 2000 – 2022) in equation (1).

Explanation of Variables

This study assessed the EF enhancing factors. Thus, EF is considered as a dependent variable (*DV*) in this research. Industrialization is a cause to increase total GHGs emissions (Singh et al., 2019).

Industry value added, therefore, is used to capture the impact of industrialization on EF (Aşıcı & Acar, 2016). Forest rents and total natural resources rents also help to reduce EF. Energy intensity and renewable energy consumption helps to reduce EF (Majeed & Mazhar, 2019; Dogan et al., 2022; Kazemzadeh et al., 2023). FDINI is causing to increase EF . Inflation is the cause of increasing EF. Technological change is useful to reduce EF. Thus, scientific technical research articles are used to evaluate the implication of technological change on EF (Sabir & Gorus, 2019). It also works as a significant driver to increase transfer of scientific knowledge among the researchers. Digitalization and digital technologies are also beneficial to reduce EF and environmental degradation (Singh & Jyoti, 2023b). Hence, individuals using the internet are used to examine the role of digitalization in EF. Employment opportunities may also increase environmental degradation (Dogan et al., 2022). Wage and salaried workers, vulnerable employment, and GDP per person employed are included as *IV*s in the regression equations. Life expectancy rate and age dependency ratio also have a substantial effect on EF (Dietz et al., 2007). Magazzino (2024) noted that EF is projected to increase as life expectancy increases. Average precipitation helps to maintain the water availability on the earth (Ashraf & Singh, 2022). Hence, above-mentioned indicators are considered as *IV*s in the regression equations (Table 1).

Table 1. Summary of the DV and IVs

Description of *DV/IVs*	Code	Unit
CO_2 emissions (metric tons per capita)	*CO2EMPC*	metric tons
GDP per capita (constant 2015 US$)	*GDPPC*	US$
Ecological Footprint [CO2 emission per capita Kg/Per capita GDP constant US$] = *DV*	*ECFO*	Ratio
Industry (including construction), value added (% of GDP)	*INVAAD*	%
Total greenhouse gas emissions (kt of CO_2 equivalent)	*TOGHGSEM*	kt of CO2 equivalent
Forest rents (% of GDP)	*FOREGDP*	%
Total natural resources rents (% of GDP)	*TONARERE*	%
Energy intensity level of primary energy (MJ/$2017 PPP GDP)	*ENINLEPREN*	MJ/$2017 PPP GDP
Renewable energy consumption (% of total final energy consumption)	*REENCO*	%
Foreign direct investment, net inflows (% of GDP)	*FODIINNEIN*	%
Inflation GDP deflator (annual %)	*INGDPDEAN*	%
Scientific and technical journal articles	*SCTEJOAR*	Number
Individuals using the Internet (% of population)	*INUSIN*	%
Wage and salaried workers total (% of total employment)	*WASAWOTO*	%
Vulnerable employment total (% of total employment)	*VUEMTO*	%
GDP per person employed (constant 2017 PPP $)	*GDPPPE*	PPP $
Life expectancy at birth, total (years)	*LIEXBITO*	Years
Age dependency ratio (% of working-age population)	*ADRPWAP*	%
Average precipitation in depth (mm per year)	*AVPRDEPY*	mm

Source: Indicators are s\elected as per previous studies.

Empirical Analysis

This study developed various forms of regression equations to measure the implications of different indicators related to social, economic, environmental, financial, and technological change on EF using panel data of SAARC countries. Country-wise panel data include the EF as *DV* and other variables as *IV*s in the regression equations. Linear regression equation is used to observe the linear dependency of EF with *IV*s. Non-linear regression equation is considered to examine the non-linear relationship of EF with *IV*s. The equation is also adopted to recognize the U-shaped or hilly-shaped association of EF with *IV*s (Saleem et al., 2019). Finally, the elasticities of *IV*s with EF are calculated using a log-linear regression equation. Neagu (2020) also adopted the above-mentioned models. The chapter assumes that *EF* is function of following variables:

$$(EOFO) = f\{INVAAD, TOGHGSEM, FOREGDP, TONARERE, ENINLEPREN, REENCO, FODIINNEIN, INGDPDEAN, SCTEJOAR, INUSIN, WASAWOTO, VUEMTO, GDPPPE, LIEXBITO, ADRPWAP, AVPRDEPY\} \quad (2)$$

The explanation of indicators is with Table 1. The linear regression equation is adopted as:

$$(EOFO)_{ct}=\alpha_0+\alpha_1(INVAAD)_{ct}+\alpha_2(TOGHGSEM)_{ct}+\alpha_3(FOREGDP)_{ct}+\alpha_4(TONARERE)_{ct}+\alpha_5(ENINLEPREN)_{ct}+\alpha_6(REENCO)_{ct}+\alpha_7(FODIINNEIN)_{ct}+\alpha_8(INGDPDEAN)_{ct}+\alpha_9(SCTEJOAR)_{ct}+\alpha_{10}(INUSIN)_{ct}+\alpha_{11}(WASAWOTO)_{ct}+\alpha_{12}(VUEMTO)_{ct}+\alpha_{13}(GDPPPE)_{ct}+\alpha_{14}(LIEXBITO)_{ct}+\alpha_{15}(ADRPWAP)_{ct}+\alpha_{16}(AVPRDEPY)_{ct}+\mu_{ct} \quad (3)$$

Here, α_0 is constant coefficient; $\alpha_1, \alpha_1..., \alpha_{16}$ are regression coefficient of *IV*s; *c* is cross-sectional country; *t* is time; and is error term in equation (3). This equation shows the linear relationship of ecological footprint (*EOFO*) with the variables. The log-linear regression equation is adopted in following form:

$$(EOFO)_{ct}=\beta_0+\beta_1 ln(INVAAD)_{ct}+\beta_2 ln(TOGHGSEM)_{ct}+\beta_3 ln(FOREGDP)_{ct}+\beta_4 ln(TONARERE)_{ct}+\beta_5 ln(ENINLEPREN)_{ct}+\beta_6 ln(REENCO)_{ct}+\beta_7 ln(FODIINNEIN)_{ct}+\beta_8 ln(INGDPDEAN)_{ct}+\beta_9 ln(SCTEJOAR)_{ct}+\beta_{10} ln(INUSIN)_{ct}+\beta_{11} ln(WASAWOTO)_{ct}+\beta_{12} ln(VUEMTO)_{ct}+\beta_{13} ln(GDPPPE)_{ct}+\beta_{14} ln(LIEXBITO)_{ct}+\beta_{15} ln(ADRPWAP)_{ct}+\beta_{16} ln(AVPRDEPY)_{ct}+\varphi_{ct} \quad (4)$$

Here, β_0 is constant coefficient; $\beta_1, \beta_2, ..., \beta_{16}$ are elasticities of respective *IV*s; *ln* is natural logarithm; and φ_{ct} is error-term in equation (4).

The original and square term of all *IV*s are comprised in the non-linear regression equation to observe the association of *EF* with undertaken variables. The model is executed as:

$$(EOFO)_{ct}=\xi_0+\xi_1(INVAAD)_{ct}+\xi_2(INVAAD^2)_{ct}+\xi_3(TOGHGSEM)_{ct}+\xi_4(TOGHGSEM^2)_{ct}+\xi_5(FOREGDP)_{ct}+\xi_6(FOREGDP^2)_{ct}+\xi_7(TONARERE)_{ct}+\xi_8(TONARERE^2)_{ct}+\xi_9(ENINLEPREN)_{ct}+\xi_{10}(ENINLEPREN^2)_{ct}+\xi_{11}(REENCO)_{ct}+\xi_{12}(REENCO^2)_{ct}+\xi_{13}(FODIINNEIN)_{ct}+\xi_{14}(FODIINNEIN^2)_{ct}+\xi_{15}(INGDPDEAN)_{ct}+\xi_{16}(INGDPDEAN^2)_{ct}+\xi_{17}(SCTEJOAR)_{ct}+\xi_{18}(SCTEJOAR^2)_{ct}+\xi_{19}(INUSIN)_{ct}+\xi_{20}(INUSIN^2)_{ct}+\xi_{21}(WASAWOTO)_{ct}+\xi_{22}(WASAWOTO^2)_{ct}+\xi_{23}(VUEMTO)_{ct}+\xi_{24}(VUEMTO^2)_{ct}+\xi_{25}(GDPPPE)_{ct}+\xi_{26}(GDPPPE^2)_{ct}+\xi_{27}(LIEXBITO)_{ct}+\xi_{28}(LIEXBITO^2)_{ct}+\xi_{29}(ADRPWAP)_{ct}+\xi_{30}(ADRPWAP^2)_{ct}+\xi_{31}(AVPRDEPY)_{ct}+\xi_{32}(AVPRDEPY^2)_{ct}+\eta_{ct} \quad (5)$$

Here, ξ_0 is constant coefficient; ξ_1, ξ_2, ..., ξ_{16} are coefficients of respective *IV*s; and η_{ct} is error-term in equation (5).

Selection of Appropriate Model and Use of Statistical Software

This study compiles selected dependent and independent variables in a panel of SAARC countries during 2000-2022. The chapter follows different statistical tests to observe the strength of empirical results and scientific processes to select a consistent model. The statistical values of variance inflation factors (*VIF*) are estimated to recognize the presence of multicollinearity in *IV*s (Ashraf & Singh, 2022). Ramsey RESET test is used to check the validity of the functional form of the model. *AIC* and *BIC* are applied to recognize the feasibility of regression equations (Kumar et al., 2017). Hausman specification and Breusch and Pagan Lagrangian multiplier tests are applied to check the suitability of fixed effect and random effect model, respectively (Aşıcı & Acar, 2016). Cross-sectional dependence is detected using Breusch-Pagan LM test and Pesaran's test (Neagu, 2020; Singh et al., 2022b; Dogan et al., 2022). Modified Wald test and Wooldridge test are preferred to examine the presence of groupwise heteroskedasticity and autocorrelation, respectively in the panel data (Singh & Jyoti, 2023a). Feasible generalized square (*FGLS*) estimation model is used to estimate coefficients *IV*s (Singh et al., 2022a). The above-mentioned analysis is done using SPSS and STATA statistical software.

DISCUSSION ON DESCRIPTIVE RESULTS

Trend in the Variables

Figure 3 and Figure 4 show the trend in GDP per capita and CO_2 emissions, respectively for SAARC regions. The progress of EF for these countries is given in Figure 5. While Figure 13 shows the trend in inflation GDP deflator (annual %) in SAARC countries during 2000 – 2022.

Figure 3. Trend in GDP per capita in SAARC countries

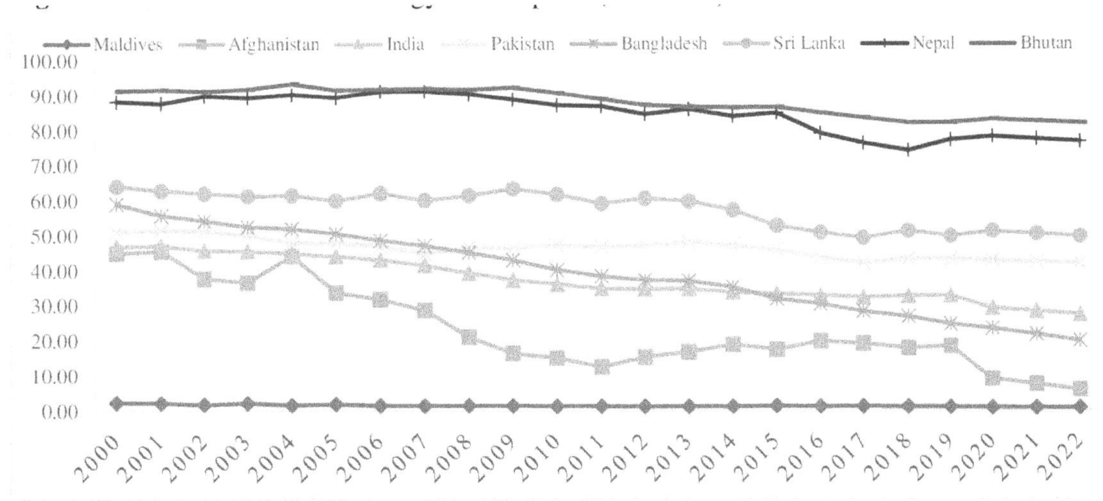

Figure 4. CO2 emissions (metric tons per capita) in SAARC countries

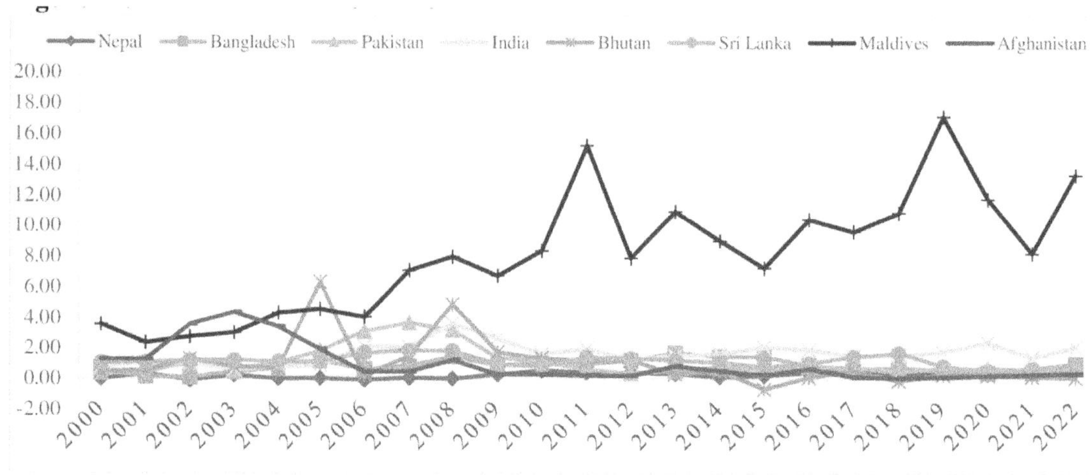

Figure 5. Relative performance of EF in SAARC countries

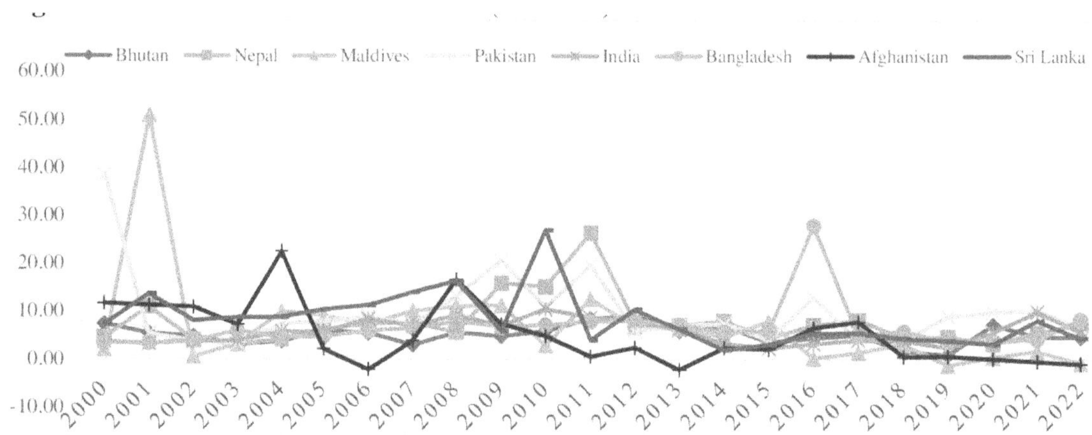

India is facing several challenges due to over population stress, high economic growth, intensification of the agricultural sector, urbanization and industrialization (Ray & Ray, 2011; Jyoti et al., 2023). India has the largest share of land area of 3287000 km^2, largest population, extensive urbanization, heavy industrialization and large dependency of their population on the agricultural sector. India, therefore, is a largest contributor of GHGs in SAARC countries (Figure 7). The EF has consistently increased due to above-mentioned reasons in India. India has a large number of research organizations in SAARC countries. Thus, it has the largest number of scientific and technical journal articles in SAARC (Figure 14). Here, this variable is used to show the progress of SAARC countries in technological change. Thereupon, it has the largest number of internet users (Figure 14). Moreover, the latest technological advancements enrich overall productivity, resulting in resource-intensive practices, further intensifying the ecological footprint. The EF analysis has significant implications for sustainability in India.

Pakistan seems to have the most contribution of EF in SAARC. However, it is also observed that the EF began to decline in Pakistan after 2000. Pakistan also has a 2nd largest supply of natural resources among other economies in SAARC. Contribution of Bangladesh in EF in SAARC is provided in Figure 5. EF is still stagnant in Bangladesh during 2000 – 2022. Bangladesh has the 4th largest land area (i.e., (147570 km^2) in SAARC. It shows that the capacity of Bangladesh to supply natural resources is stagnant during 2000 – 2022. The population growth of Bangladesh has declined from 1.8% in 2002 to 1.1% in 2014. Decreasing population growth is a reason to reduce EF in Bangladesh.

EF has consistently declined in Bhutan during 2000 – 2022. Bhutan has 2nd smallest land area (i.e., 38394km^2) in SAARC countries. Therefore, the capacity of supply of natural resources is not enough in Bhutan. Accordingly, EF is declining in Bhutan. Bhutan has the largest industry value added in the SAARC countries (Figure 6). Bhutan has the largest value of forest rents as a percentage of GDP (Figure 8). This country also has the largest value of total natural resources rents (Figure 9). Moreover, Bhutan has a better position in energy intensity (Figure 10), and renewable energy consumption (Figure 11) that help to reduce its environmental degradation.

The contribution of EF in Maldives is minimal in SAARC countries during 2000 - 2022. Also, it is similar throughout the timeline. The land area (i.e. 297.8 km^2) of Maldives is the lowest in SAARC countries. Maldives has the largest GDP per capita (Figure 3) and it also has the largest value of per capita CO_2 emission (Figure 4). Maldives has the largest number of wage and salaried workers as a % of total employment (Figure 16). Hence, it is reported that the capacity of supply of natural resources is very low in Maldives. Maldives seems to be a more successful country to attract FDINI (Figure 12). Maldives has the largest share of vulnerable employment in SAARC countries (Figure 17). It has a largest value of GDP per person employed (Figure 18), and it has a better position in life expectancy at birth (Figure 19) in the SAARC region.

Figure 5 indicates the contribution of Afghanistan in EF and it seems largest in SAARC economies. The land area (i.e., 652237 km^2) of Afghanistan is 3rd largest in SAARC (World Bank). It infers that Afghanistan has a great capacity to supply natural resources to fulfil the economy's needs as compared to other countries in SAARC. This country has the largest value of age dependency ratio in SAARC (Figure 20). There is a small contribution of Sri Lanka in EF in SAARC. Also, EF is declined as compared to the initial stages of the time period of 2000 – 2022. The land area of Sri Lanka is 65610 km^2 (World Bank), which is 3rd smallest in SAARC region.

The contribution of Nepal in EF is relatively larger than Maldives. Nepal has a lower contribution of EF among SAARC countries (except, Maldives). The land area of Nepal is 147181 km2 (World Bank) which is 5th largest in SAARC countries. The country does not have much urbanization and industrialization. Hence the capacity of supply of natural resources of this country is in a better position among the SAARC countries.

Statistical Summary of the Variables

The basic statistician property of variables is given in Table 2. The minimum, maximum and mean values infer that SAARC countries have substantial variation in all variables. Thereupon, all variables (except, *INVAAD*, *LIEXBITO*, and *AVPRDEPY*) are not found abnormal in nature as per the statistical values of skewness. The results of Cronbach's alpha test is also included in this table.

Figure 6. Trend in Industry (Including Construction), Value added (% of GDP) in SAARC

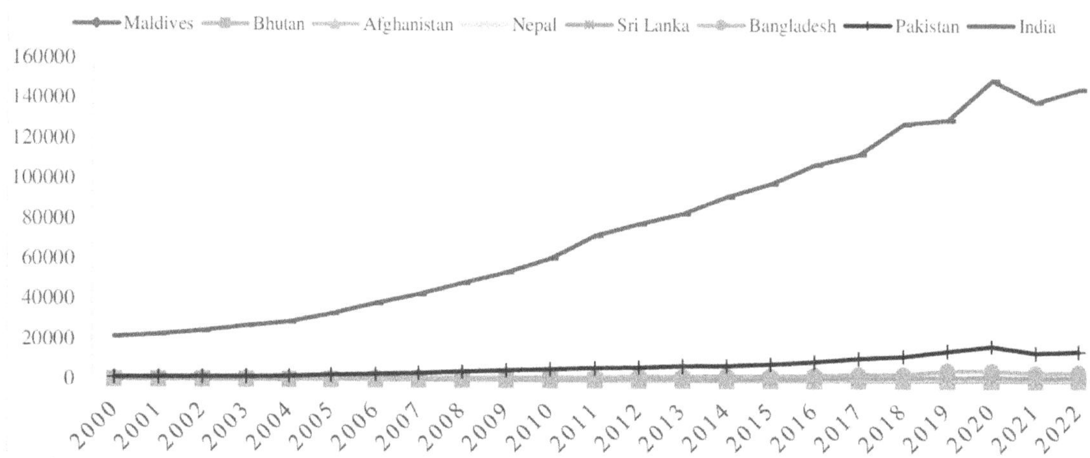

Figure 7. Trend in Total Greenhouse Gas Emissions (kt of CO2 equivalent) in SAARC

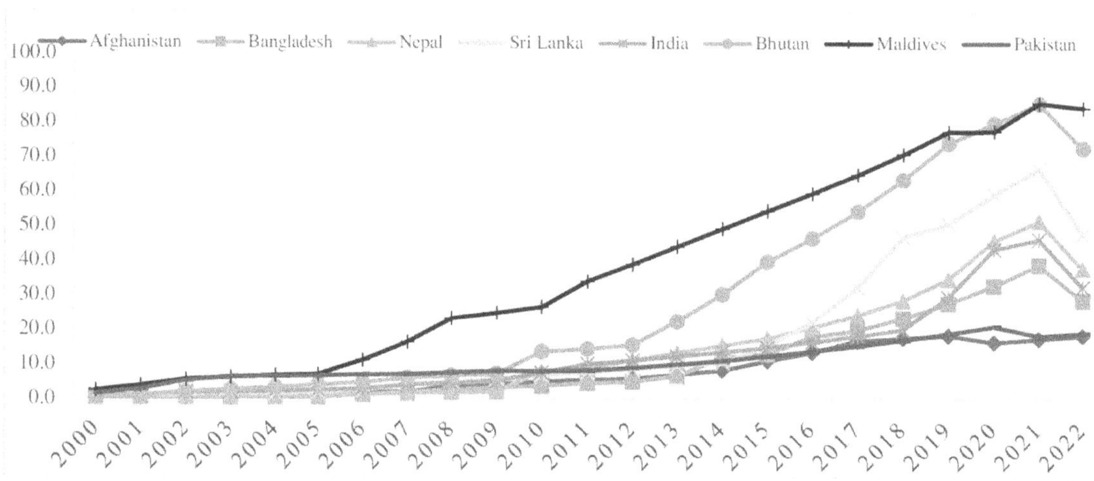

Correlation Coefficients Between Variables

The association of *EF* with *IVs* is estimated through Karl-Pearson correlation coefficient analysis (Table 3). The estimates demonstrate that EF is negatively associated with *INVAAD*, *FOREGDP*, *TONARERE*, *REENCO*, *FODIINNEIN*, *INGDPDEAN*, *VUEMTO*, *LIEXBITO*, *ADRPWAP*, and *AVPRDEPY*. Hence, *EF* is likely to decrease as forest rent, total natural resource rent, renewable energy consumption, foreign direct investment, inflation, digitalization, vulnerable employment, life expectancy rate, age dependency ratio and average precipitation increases.

Figure 8. Trend in forest rents (% of GDP) in SAARC

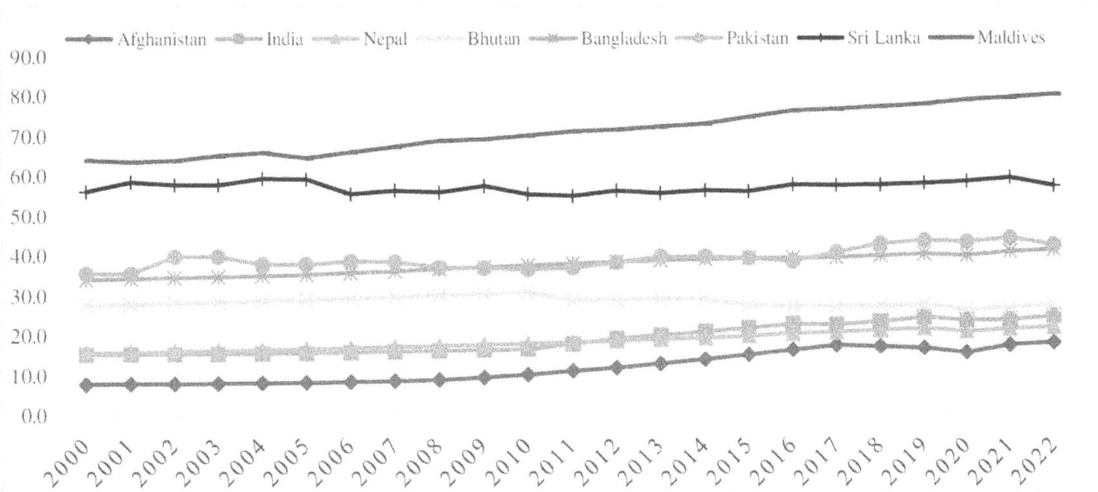

Figure 9. Trend in total natural resources rents (% of GDP) in SAARC

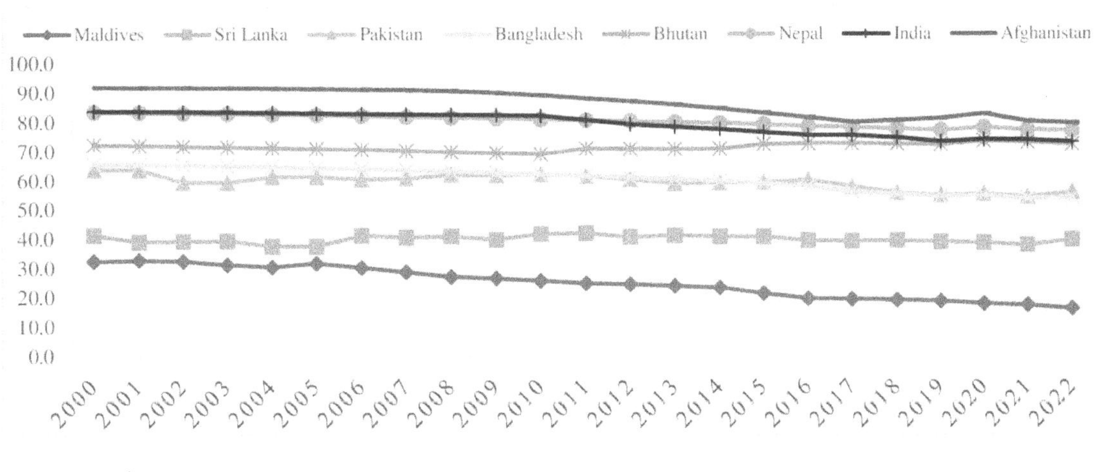

DISCUSSION ON EMPIRICAL RESULTS

The statistical issues in panel data are estimated through various processes (Table 4). Log-linear regression model seems more suitable to produce the coefficient of *IV*s with EF. The *F*-values under *Ramsay RESET* test claimed that the functional form of all models is observed well-defined. Moreover, the results of Hausman specification and Breusch and Pagan Lagrangian multipliers indicate that random effect model (REM) and fixed effect model (FEM) can be considered to estimate the coefficient of *IV*s. The REM and FEM are found inappropriate due to the presence of groupwise heteroskedasticity and autocorrelation (Kızılgöl & Öndes, 2022). Therefore, the coefficient of *IV*s with *EF* is calculated through *FGLS* estimation model.

The coefficient of *IV*s with EF is provided in Table 5. It is stated previously that log-linear regression equation produces more consistent results. Industrialization is a prime source of CO_2

Figure 10. Trend in energy intensity level of primary energy in SAARC

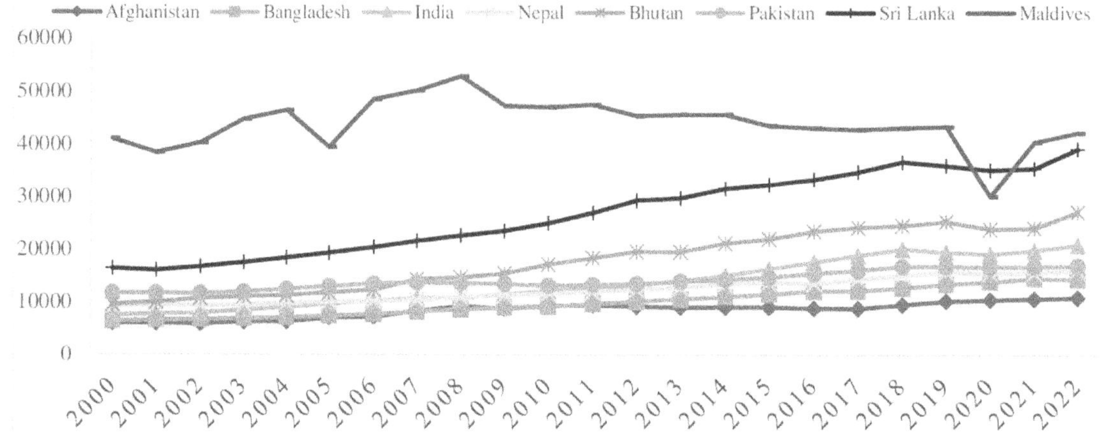

Figure 11. Trend in renewable energy consumption (% of total) in SAARC

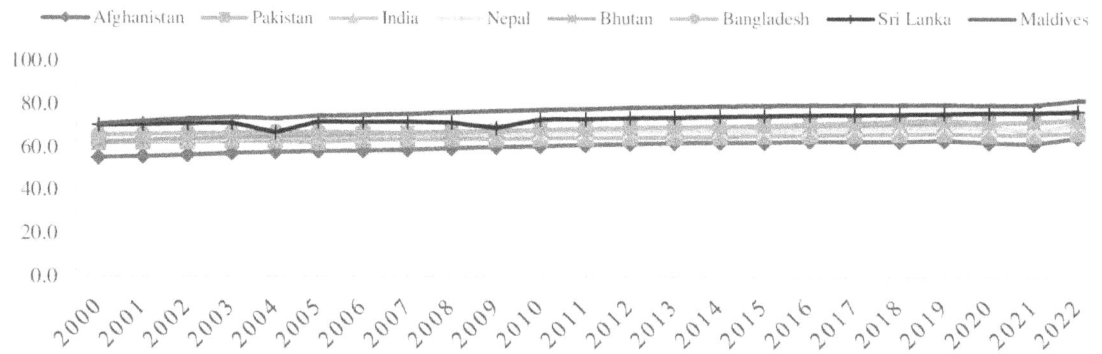

Figure 12. Trend in FDINI in SAARC

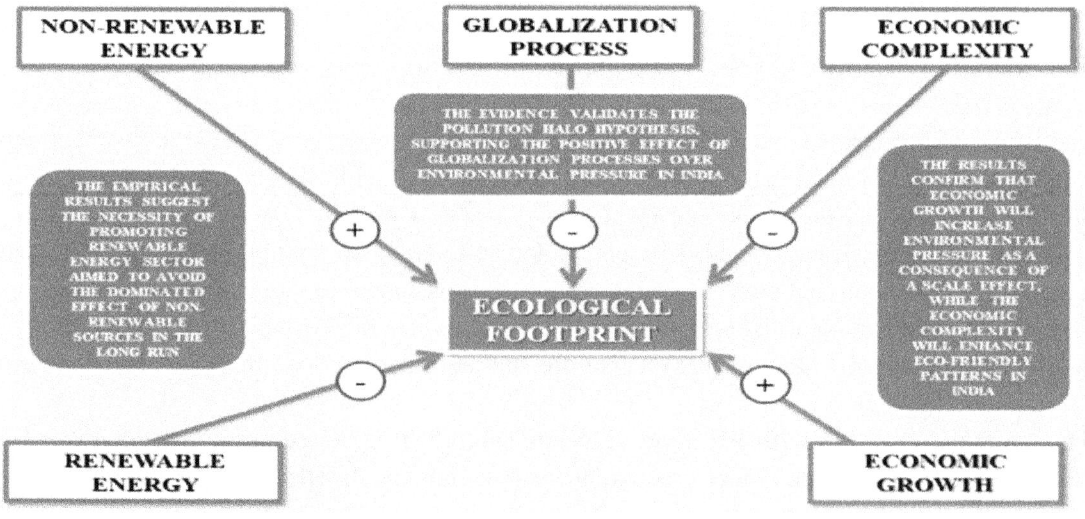

Figure 13. Trend in inflation GDP deflator (annual %) in SAARC

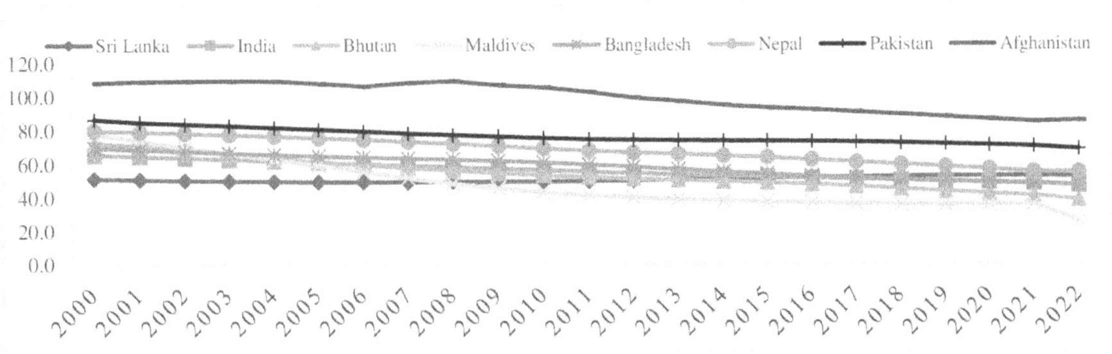

Figure 14. Trend in technological progress in SAARC

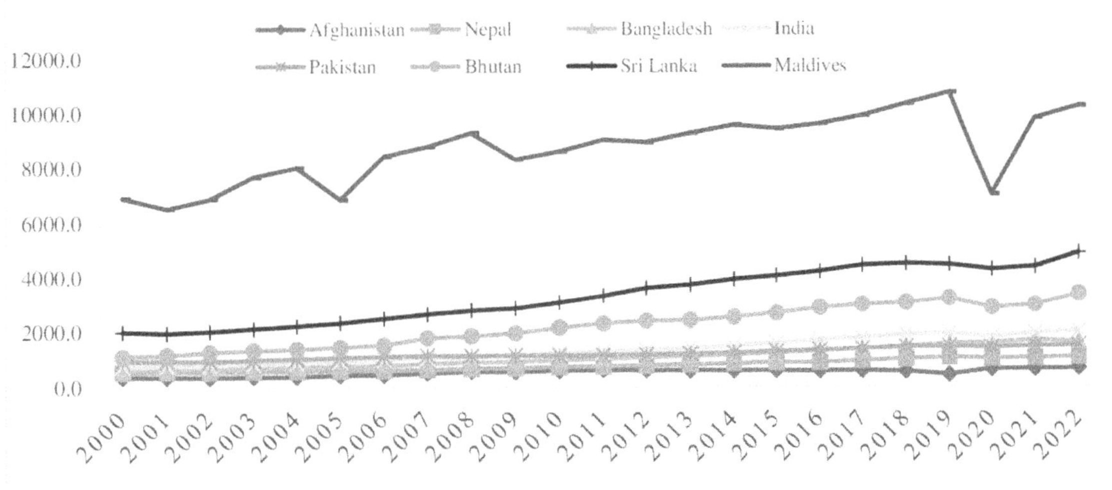

Figure 15. Progress of digitalization in SAARC

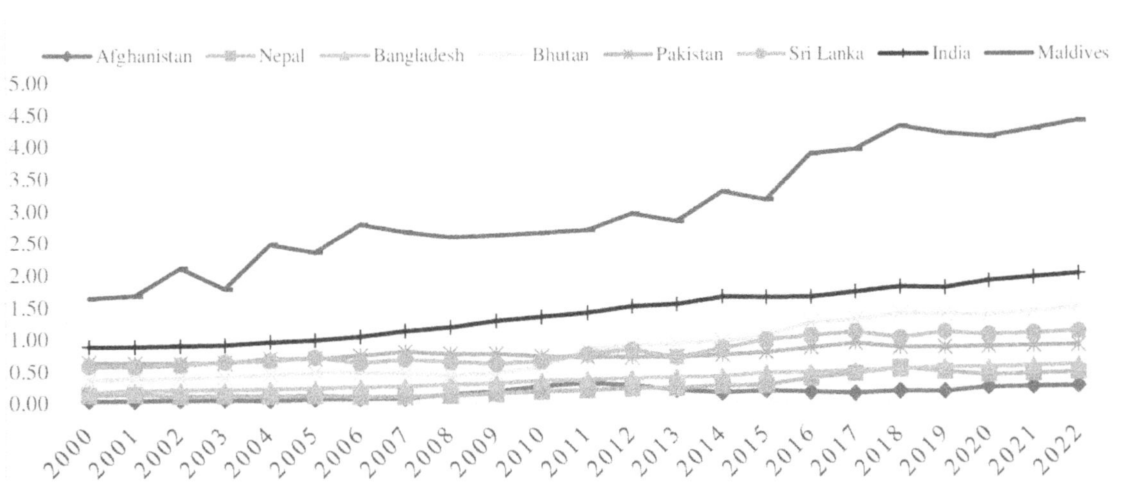

Figure 16. Progress of wage and salaried workers in SAARC

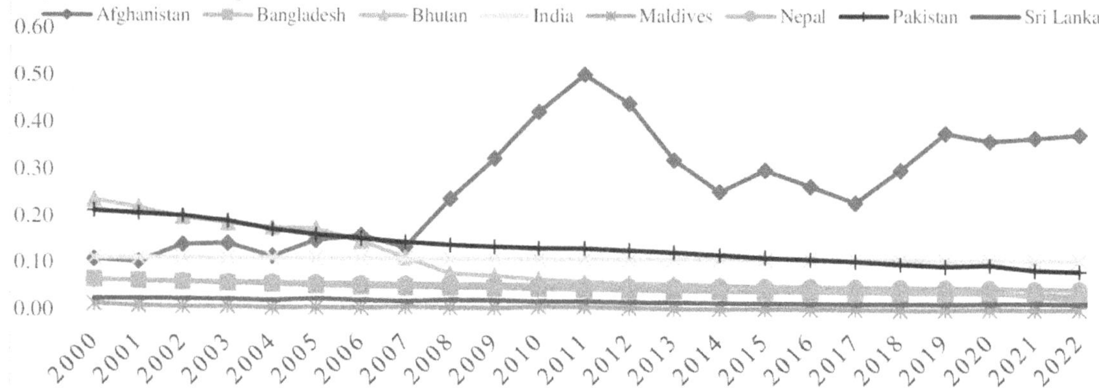

Figure 17. Trend in vulnerable employment total (% of total employment) in SAARC

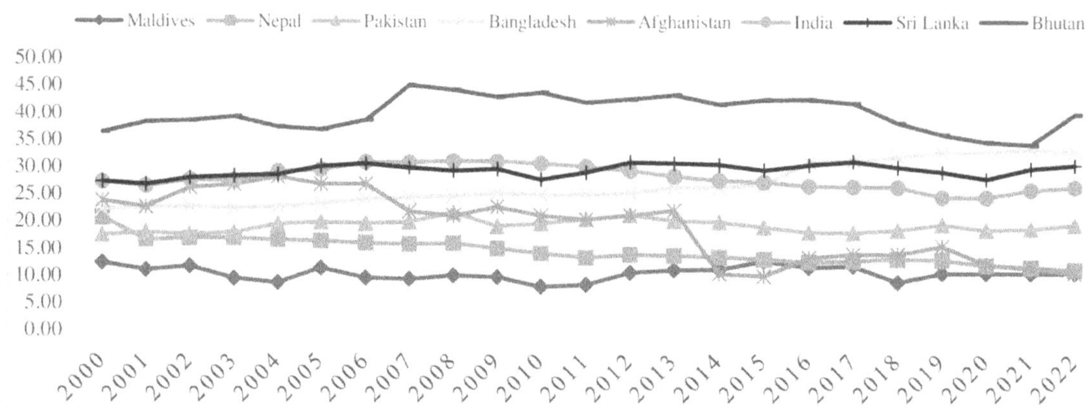

Figure 18. Trend in GDP per person employed (constant 2017 PPP $) in SAARC

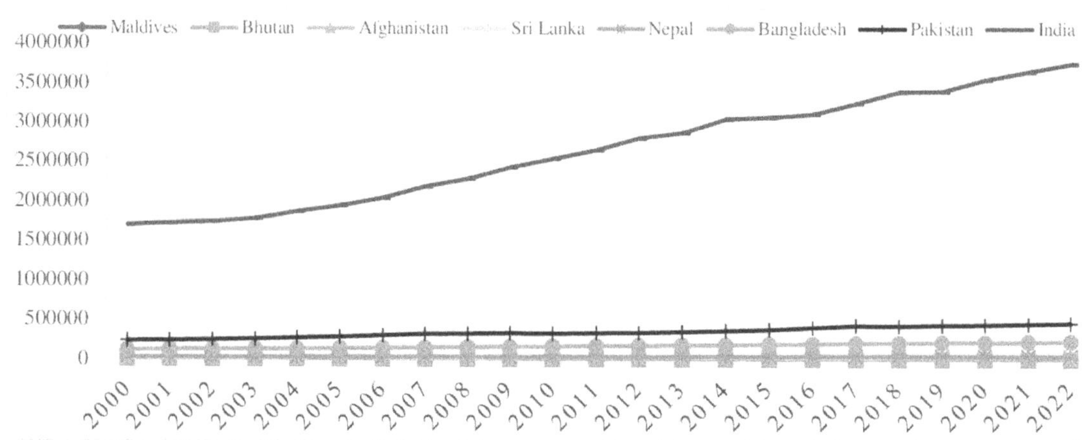

Figure 19. Trend in life expectancy rate in SAARC

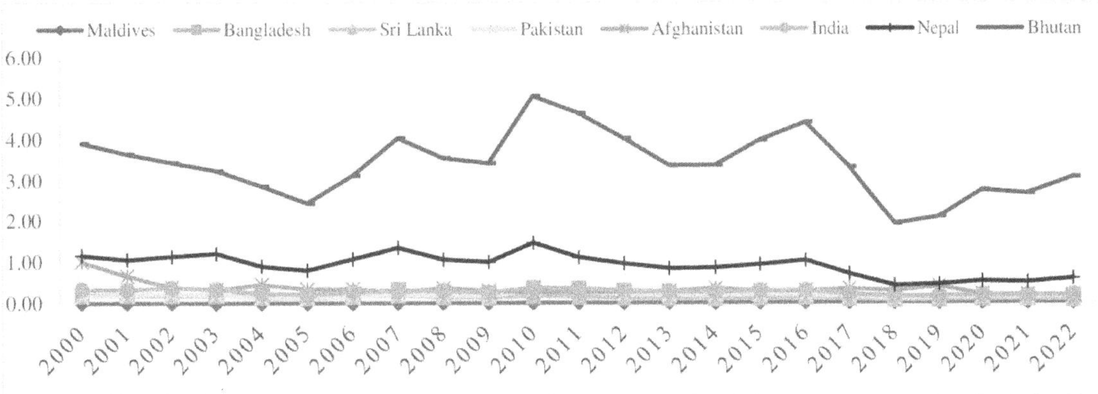

Figure 20. Trend in age dependency ratio in SAARC

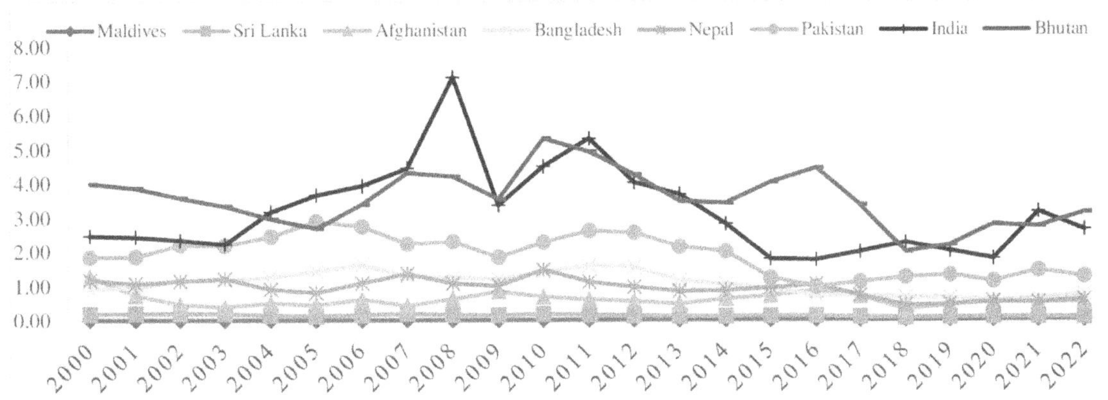

emissions and GHGs emissions and environmental degradation. The positive coefficient of industry value added with EF indicates that industrial development is responsible for increasing EF. Furthermore, the scientific research community has accepted that GHGs emissions are also accountable for global warming (Rahman et al., 2019). Thus, it showed a negative implication on environmental development and ecosystem services. Accordingly, there is reported a positive coefficient of total GHGs with EF. The estimate reveals that EF increases due to increase in GHGs emissions. There is also found a negative coefficient of forest and natural resources rent with EF. Hence, forest and natural resources rent help to reduce EF. Therefore, global countries should impose more rent on forestry and natural resources to increase protection of environmental indicators and ecological services.

Energy intensity measures its usages for multiple purposes in the production sectors. Therefore, energy intensity may be useful to minimize CO_2 emissions. The results showed that EF declines as energy intensity increases. Therefore, renewable energy consumption is useful to reduce EF. It means that EF is expected to decline as renewable energy consumption increases. Therefore, it is suggested that consumers must be shifted towards renewable energy to reduce EF. FDINI is a crucial to improve

Table 2. Basic statistical property of the variables

Statistics	*Min*	*Max*	*Mean*	*SD*	*Skewness*	*α-Value*
ECFO	0.001	0.502	0.091	0.093	1.918	0.799
INVAAD	10.052	106.795	45.298	17.866	0.402	0.782
TOGHGSEM	6521.18	3743147.00	412514.20	882976.70	2.513	0.788
FOREGDP	0.042	148.249	8.383	27.678	3.873	0.813
TONARERE	0.223	4.379	1.371	0.863	1.453	0.818
ENINLEPREN	0.186	106.007	8.155	20.228	3.293	0.790
REENCO	5.966	101.550	61.420	17.292	-1.278	0.788
FODIINNEIN	0.064	50.893	4.772	5.141	4.559	0.806
INGDPDEAN	0.104	163.130	13.272	30.751	3.331	0.807
SCTEJOAR	1.000	133.290	15.481	18.189	4.067	0.814
INUSIN	0.003	120.948	9.328	23.491	3.030	0.816
WASAWOTO	7.802	67.900	22.989	12.586	1.370	0.783
VUEMTO	51.701	5012.040	474.240	798.092	3.621	0.808
GDPPPE	1972.00	149212.60	20861.88	26686.65	3.12	0.78
LIEXBITO	55.298	114.370	76.971	11.574	0.389	0.777
ADRPWAP	66.000	2019.526	475.858	503.232	1.389	0.777
AVPRDEPY	87.474	1761.198	807.566	506.760	0.499	0.793

Source: Estimated by authors.

Table 3a. Karl-Pearson correlation coefficient among the variables

Variables	*ECFO*	*INVAAD*	*TOGHGSEM*	*FOREGDP*	*TONARERE*	*ENINLEPREN*	*REENCO*	*FODIINNEIN*
ECFO	1	-0.359 [b]	0.080	-0.129 [a]	-0.236 [b]	0.067	-0.655 [b]	-0.325 [b]
INVAAD	-0.359 [b]	1	0.169 [a]	0.647 [b]	-0.318 [b]	0.07	0.415 [b]	0.04
TOGHGSEM	0.080	0.169 [a]	1	-0.107	-0.033	0.908 [b]	0.456 [b]	0.05
FOREGDP	-0.129 [a]	0.647 [b]	-0.107	1	-0.186 [b]	-0.087	0.125 [a]	-0.132 [a]
TONARERE	-0.236 [b]	-0.318 [b]	-0.033	-0.186 [b]	1	-0.068	0.376 [b]	0.111
ENINLEPREN	0.067	0.07	0.908 [b]	-0.087	-0.068	1	0.463 [b]	0.055
REENCO	-0.655 [b]	0.415 [b]	0.456 [b]	0.125 [a]	0.376 [b]	0.463 [b]	1	0.162 [a]
FODIINNEIN	-0.325 [b]	0.04	0.050	-0.132*	0.111	0.055	0.162 [a]	1
INGDPDEAN	-0.244 [b]	0.079	-0.138 [a]	-0.074	-0.09	-0.104	-0.128 [a]	0.520 [b]
SCTEJOAR	0.559 [b]	-0.338 [b]	0.083	-0.153*	-0.254 [b]	0.095	-0.584 [b]	-0.095
INUSIN	-0.062	-0.111	-0.151 [a]	-0.108	-0.255 [b]	-0.076	0.03	-0.157 [a]
WASAWOTO	-0.013	0.368 [b]	0.716 [b]	-0.077	-0.072	0.651 [b]	0.417 [b]	0.064
VUEMTO	-0.228 [b]	0.281 [b]	0.013	-0.098	-0.249 [b]	-0.032	-0.042	0.011
GDPPPE	0.006	0.189 [b]	0.891 [b]	-0.061	-0.177 [b]	0.935 [b]	0.454 [b]	0.136 [a]
LIEXBITO	-0.415 [b]	0.491 [b]	0.305 [b]	0.280 [b]	0.125 [a]	0.270 [b]	0.419 [b]	0.438 [b]
ADRPWAP	-0.130 [a]	0.538 [b]	0.723 [b]	0.163 [a]	-0.299 [b]	0.659 [b]	0.393 [b]	-0.047
AVPRDEPY	-0.292 [b]	0.436 [b]	0.177 [b]	0.064	-0.367 [b]	0.176 [b]	0.232 [b]	0.303 [b]

Table 3b. Karl-Pearson correlation coefficient among the variables

VARIABLES	INGDPDEAN	SCTEJOAR	INUSIN	WASAWOTO	VUEMTO	GDPPPE	LIEXBITO	ADRPWAP	AVPRDEPY
ECFO	-0.244 [b]	0.559 [b]	-0.062	-0.013	-0.228 [b]	0.006	-0.415 [b]	-0.130 [a]	-0.292 [b]
INVAAD	0.079	-0.338 [b]	-0.111	0.368 [b]	0.281 [b]	0.189 [b]	0.491 [b]	0.538 [b]	0.436 [b]
TOGHGSEM	-0.138 [a]	0.083	-0.151 [a]	0.716 [b]	0.013	0.891 [b]	0.305 [b]	0.723 [b]	0.177 [b]
FOREGDP	-0.074	-0.153 [a]	-0.108	-0.077	-0.098	-0.061	0.280 [b]	0.163 [a]	0.064
TONARERE	-0.09	-0.254 [b]	-0.255 [b]	-0.072	-0.249 [b]	-0.177 [b]	0.125*	-0.299 [b]	-0.367 [b]
ENINLEPREN	-0.104	0.095	-0.076	0.651 [b]	-0.032	0.935 [b]	0.270**	0.659 [b]	0.176 [b]
REENCO	-0.128 [a]	-0.584 [b]	0.030	0.417 [b]	-0.042	0.454 [b]	0.419 [b]	0.393 [b]	0.232 [b]
FODIINNEIN	0.520 [b]	-0.095	-0.157 [a]	0.064	0.011	0.136 [a]	0.438 [b]	-0.047	0.303 [b]
INGDPDEAN	1	0.077	-0.138 [a]	0.081	0.202 [b]	0.116	0.591**	-0.095	0.524 [b]
SCTEJOAR	0.077	1	-0.046	0.115	0.09	0.109	-0.099	0.042	-0.098
INUSIN	-0.138 [a]	-0.046	1	-0.209 [b]	0.052	-0.02	-0.450 [b]	-0.150 [a]	0.331 [b]
WASAWOTO	0.081	0.115	-0.209 [b]	1	0.297 [b]	0.741 [b]	0.402 [b]	0.693 [b]	0.196 [b]
VUEMTO	0.202 [b]	0.090	0.052	0.297 [b]	1	0.052	0.156 [a]	0.553 [b]	0.113
GDPPPE	0.116	0.109	-0.02	0.741 [b]	0.052	1	0.409 [b]	0.681 [b]	0.405 [b]
LIEXBITO	0.591 [b]	-0.099	-0.450 [b]	0.402 [b]	0.156 [a]	0.409 [b]	1	0.368 [b]	0.408 [b]
ADRPWAP	-0.095	0.042	-0.150 [a]	0.693 [b]	0.553 [b]	0.681 [b]	0.368 [b]	1	0.204 [b]
AVPRDEPY	0.524 [b]	-0.098	0.331 [b]	0.196 [b]	0.113	0.405 [b]	0.408 [b]	0.204 [b]	1

Source: Estimated by authors. **Note:** [b] infers that correlation coefficient is statistically significant at 1% significance level, and [a] infer that correlation coefficient is statistically significant at 5% significance level.

Table 4. Results of statistical issues in the panel data

Statistical test/Regression equation	Log-linear	Linear	Non-linear
Mean *VIF*	9.85	11.57	212.93
AIC	-303.5632	-646.5698	-811.1164
AIC	-121.8639	-591.9159	-711.4534
Ramsey RESET test using powers of the independent variables [*F* – value]	82.54*	21.67*	17.94*
Ramsey RESET test using powers of the fitted values of DV [*F* – value]	80.06*	5.37*	27.99*
Hausman fixed random [Chi^2 - *value*]	0.00*	0.00*	0.00*
Breusch and Pagan Lagrangian multiplier test for random effects [$Chibar^2$- *value*]	0.00*	0.00*	0.00*
Breusch-Pagan LM test of independence [Chi^2]	93.981*	56.275*	31.559*
Pesaran's test of cross-sectional independence	1.965**	0.413*	1.511*
Modified Wald test for groupwise heteroskedasticity in fixed effect regression model [Chi^2- value]	2925.71*	3443.21*	2272.00*
Wooldridge test for autocorrelation in panel data [*F* – value]	93.428*	273.809*	762.372*

Source: Estimated by authors.

economic development, while it is cause to increase EF. Ayeche et al. (2016); Sabir and Gorus (2019) described a positive impact of FDINI on EF. Though, this research could not observe a similar impact of FDINI on EF in the SAARC countries.

Table 5. Impact of IVs on EF

Model	Log-linear regression			Linear regression			Non-linear regression		
Number of obs.	184			184			184		
Number of groups	8			8			8		
Wald Chi^2	1142.01			904.81			2878.47		
Prob > Chi^2	0.000			0.000			0.000		
Log likelihood	-121.8639			340.2849			436.5582		
ECFO [DV]	***Reg. Coef.***	***Std. Err.***	***P>|z|***	***Reg. Coef.***	***Std. Err.***	***P>|z|***	***Reg. Coef.***	***Std. Err.***	***P>|z|***
INVAAD	0.942	0.269	0.000	0.003	0.001	0.000	0.005	0.001	0.000
INVAAD^2							0.000	0.000	0.001
TOGHGSEM	0.458	0.082	0.000	0.000	0.000	0.143	0.000	0.000	0.004
TOGHGSEM^2							0.000	0.000	0.003
FOREGDP	-0.093	0.048	0.053	-0.001	0.000	0.002	-0.001	0.000	0.004
FOREGDP^2							0.000	0.000	0.000
TONARERE	-0.140	0.119	0.236	0.039	0.007	0.000	0.041	0.014	0.003
TONARERE^2							-0.004	0.003	0.229
ENINLEPREN	-0.064	0.057	0.265	0.001	0.001	0.233	0.003	0.001	0.023
ENINLEPREN^2							0.000	0.000	0.028
REENCO	-2.416	0.181	0.000	-0.007	0.001	0.000	-0.018	0.001	0.000
REENCO^2							0.000	0.000	0.000
FODIINNEIN	-0.041	0.065	0.533	-0.001	0.001	0.181	0.001	0.001	0.405
FODIINNEIN^2							0.000	0.000	0.461
INGDPDEAN	0.032	0.064	0.617	-0.001	0.000	0.000	-0.001	0.000	0.187
INGDPDEAN^2							0.000	0.000	0.231
SCTEJOAR	-0.545	0.096	0.000	0.000	0.000	0.346	-0.002	0.001	0.016
SCTEJOAR^2							0.000	0.000	0.016
INUSIN	-0.024	0.031	0.440	0.001	0.000	0.012	-0.003	0.001	0.000
INUSIN^2							0.000	0.000	0.000
WASAWOTO	0.314	0.205	0.127	0.001	0.001	0.076	0.011	0.002	0.000
WASAWOTO^2							0.000	0.000	0.000
VUEMTO	-0.466	0.125	0.000	0.000	0.000	0.001	0.000	0.000	0.696
VUEMTO^2							0.000	0.000	0.790
GDPPPE	0.282	0.172	0.100	0.000	0.000	0.002	0.000	0.000	0.742
GDPPPE^2							0.000	0.000	0.002
LIEXBITO	-1.362	0.560	0.015	0.001	0.001	0.237	-0.011	0.004	0.001
LIEXBITO^2							0.000	0.000	0.001
ADRPWAP	0.057	0.151	0.707	0.000	0.000	0.539	0.000	0.000	0.000
ADRPWAP^2							0.000	0.000	0.000
AVPRDEPY	-0.017	0.175	0.925	0.000	0.000	0.076	0.000	0.000	0.000
AVPRDEPY^2							0.000	0.000	0.000
Con. Coef.	4.121	2.256	0.068	0.318	0.036	0.000	0.815	0.132	0.000

Source: Estimated by authors.

Development of technology helps to upsurge production and growth of the manufacturing sector. Therefore, its impact on EF may be positive. If technological change is helping to increase green growth, then it may be effective to reduce environmental degradation and EF. Therefore, global countries should promote green development to reduce EF (Wei et al., 2023). Otherwise, technological change is caused to increase EF (Dogan et al., 2022). The coefficient of scientific and technical journal articles with EF is detected negative. Hence, it claimed that technological change and technology transfer are beneficial to reduce EF (Jyoti & Singh, 2023). The significance of digitalization in EF is also appeared negative. Hence digitalization seems useful to reduce EF and environmental degradation (Singh & Jyoti, 2023b). The impact of wage and salaried workers on EF is observed positive. Subsequently, EF declines and vulnerable employment increases in SAARC countries (Aly, 2020; Jyoti & Singh, 2023).

The regression coefficients of GDP per person, employed and age dependency ratio with EF appeared positive. The results infer that EF may increase due to increase in GDP per person employed and age dependency ratio. The role of life expectancy rate is useful to reduce EF. Precipitation is a vital natural resource that increases the water in the ground. Our result claimed that EF declines as average precipitation increases.

Furthermore, EF has a linear and non-linear association with *IV*s as per the results of non-linear regression equation. EF has a linear relationship with total greenhouse gas emissions, vulnerable employment, GDP per person employed, and average precipitation. Forest rent, renewable energy consumption, inflation GDP deflator, scientific and technical journal articles, individuals using the internet, life expectancy rate, and age dependency ratio show a hilly-shaped relationship with EF. While EF has a U-shaped relationship with industry value added, total natural resource rent, energy intensity, foreign direct investment net inflows, and wage and salaried workers.

CONCLUSION

Per capita CO_2 emission and GDP are used to estimate EF in SAARC countries in this study. EF is calculated as a ratio of per capita CO_2 emission and per capita GDP of individual countries in SAARC during 2000 – 2022. Per capita GDP helps to increase growth of economic activities. Whereas CO_2 emission indicates environmental degradation. Per capita CO_2 emissions also measure the exploitation in natural resources due to increase in production scale. The estimated values of EF indicate that Afghanistan has the largest value of EF among the SAARC during 2000 – 2022. Hence, the country seems unable to reduce EF and environmental degradation. India and Pakistan have 2nd and 3rd positions in EF. While Nepal is reported as a 4th largest contributor in environmental degradation. Moreover, there is also an increasing trend in EF in other SAARC countries like Bangladesh, Bhutan, Maldives, Bangladesh, and Sri Lanks after 2015. India has seen decline in EF with maximum contribution to combined EF in the SAARC countries. There is also a declining trend in EF in Bhutan. The EF in Nepal has also witnessed an incline after 2010. Nepal's contribution is least in EF. Bangladesh also faced a slowdown in EF. Sri Lanka faced a decline in EF throughout the time period. Maldives has the lowest contribution to EF in the SAARC region.

EF is positively associated with total greenhouse gas emissions, energy intensity and technological change. Whereas EF is negatively associated with forest rents, total natural resources rents, renewable energy consumption, FDINI, inflation, digitalization, vulnerable employment, life expectancy, age dependency ratio, and average precipitation. The regression results expose that industrialization, total greenhouse gas, wage and salaried workers, and age dependency ratio are causes to increase EF and environmental degradation. Forest rent, natural resource rent, energy intensity, renewable energy consumption, FDINI, inflation, digitalization, vulnerable employment, life expectancy rate, and average precipitation showed a negative impact on EF. Hence, these variables are found beneficial to reduce EF in SAARC countries. Also, there is reported a hilly-shaped association of EF with forest rent, renewable energy consumption, inflation, technological change, digitalization, life expectancy rate, and age dependency ratio. Hence, these variables would be responsible to increase EF up to a certain extent only. Furthermore, a U-shaped association of EF with industrialization, total natural resource rent, energy intensity, FDIINI, and wage and salaried workers, is noticed in SAARC countries. Hence, above-mentioned results are beneficial for policy makers to consider the above-mentioned variables in further policy implementation to reduce EF and environmental degradation in the SAARC region in future.

POLICY IMPLICATIONS

Economic development, urbanization, and industrialization are highly responsible for rising per capita CO_2 emissions that are harmful for environmental development and EF. While reducing EF is essential to increase the protection of ecosystem services and sustain the future of people (Kazemzadeh et al., 2023). Hence, SAARC countries should reduce per capita CO_2 and GHGs emissions to increase stability in biodiversity. While environmental technologies seem positive to increase biodiversity. Therefore, use of environmental technologies in the production should be promoted to reduce EF. The government should also encourage the business community to set up green businesses to reduce environmental degradation and EF. Researchers should develop eco-friendly technology, green technology and green innovation for industries to reduce CO_2 and GHGs emissions (Singh & Jyoti, 2023a). Green and appropriate technology would be useful to reduce EF. Clean technology and environmentally friendly production processes must be initiated to reduce EF (Majeed & Mazhar, 2019; Neagu, 2020). Higher academic and research institutions should create green entrepreneurial skills among the students (Singh et al., 2023b). The SAARC countries should also increase their production activities towards green growth. These countries should also work together to achieve SDGs in multiple dimensions and green growth. The global countries require multidisciplinary partnership and innovative methodologies to reduce EF. An integrated approach from various stakeholders of the society such as social, economic, and ecological perspectives can provide a comprehensive tool for policy makers with more informed decisions to foster global sustainability (Figge et al., 2016; Singh & Kumar, 2022). Use of renewable energy resources helps to reduce EF and greenhouse gas emissions (Sabir & Gorus, 2019; Dogan et al., 2022). R&D investment in the energy sector must be promoted to reduce EF (Majeed & Mazhar, 2019; Sharma et al., 2020). Ecological civilization, ecologically friendly and environmentally friendly ways help to reduce EF and achieve sustainable development (Wu et al., 2023; Han & Dalaibaatar, 2023). Introduction of environmental education in higher academic institutions would be effective to increase the awareness of people towards environmental issues and EF (Ray & Ray, 2011).

INNOVATION AND SIGNIFICANCE OF THE CHAPTER

This chapter estimates the EF as considering per capita GDP as a worthy indicator of economic development and per capita CO_2 emissions for SAARC countries during 2000 – 2022. Thereupon, it measures the implications of technological change, financial indicators, demographic change, digitalization, and average precipitation on EF using panel data analysis. The descriptive and empirical findings provide crucial policy implications to reduce EF and environmental degradation in SAARC countries. Finally, it suggests further scope of research.

LIMITATIONS AND FURTHER RESEARCH DIRECTIONS

It is noted that calculation of EF through CO_2 emission and GDP cannot give exact values of EF for SAARC countries. The estimated values of EF are helpful for policy makers to implement conducive policies to reduce EF in these countries. Next study, therefore, can consider other methods and more indicators to calculate EF for developing and developed countries. We also could not check the viability of estimated values of EF in this chapter. This is another limitation of this chapter. Furthermore, we could consider limited macro level indicators as *IVs* in the empirical investigations. Accordingly, it measures the impact of certain macro level indicators on EF. While EF also depends on multiple activities. Existing researchers can also measure the association of EF with sustainable development in further study.

REFERENCES

Aly, H. (2020). Digital transformation, development and productivity in developing countries: Is artificial intelligence a cure or a blessing? *Review of Economics and Political Science*, *7*(4), 238–256. doi:10.1108/REPS-11-2019-0145

Ashraf, S. N., & Singh, A. K. (2022). Implications of appropriate technology and farm inputs in the agricultural sector of Gujarat: Empirical analysis based on primary data. *Agricultural Economics Research Review*, *35*(2), 59–77. doi:10.5958/0974-0279.2022.00031.3

Aşıcı, A. A., & Acar, S. (2016). Does income growth relocate ecological footprint? *Ecological Indicators*, *61*(2), 707–7014. doi:10.1016/j.ecolind.2015.10.022

Ayeche, M. B., Barhoumi, M., & Hammas, M. A. (2016). Causal linkage between economic growth, financial development, trade openness and CO2 emissions in European countries. *American Journal of Environmental Engineering*, *6*(4), 110–122. http://article.sapub.org/10.5923.j.ajee.20160604.02.html

Baabou, W., Grunewald, N., Ouellet-Plamondon, C., Gressot, M., & Gallib, A. (2017). The ecological footprint of mediterranean cities: Awareness creation and policy implications. *Environmental Science & Policy*, *69*(1), 94–104. doi:10.1016/j.envsci.2016.12.013

Basarir, A., & Arman, H. (2014). The effects of economic growth on environment: An application of environmental Kuznets curve in United Arab Emirates. *The Online Journal of Science and Technology*, *4*(1), 53–59. https://tojsat.net/journals/tojsat/articles/v04i01/v04i01-07.pdf

Bilsborrow, R. E. (1992). Population growth, internal migration, and environmental degradation in rural areas of developing countries. *European Journal of Population*, *8*(1), 125–148. doi:10.1007/BF01797549 PMID:12158965

Danish, H., Hassan, S. T., Baloch, M. A., Mahmood, N., & Zhang, J. W. (2019). Linking economic growth and ecological footprint through human capital and biocapacity. *Sustainable Cities and Society*, *47*(1), 13–24. doi:10.1016/j.scs.2019.101516

Danish, Z. W., & wang, Z. (2019). Investigation of the ecological footprint's driving factors: What we learn from the experience of emerging economies. *Sustainable Cities and Society*, *49*(1), 1–12. doi:10.1016/j.scs.2019.101626

Dietz, T., Rosa, E. A., & York, R. (2007). Driving the human ecological footprint. *Frontiers in Ecology and the Environment*, *5*(1), 13–18. https://www.jstor.org/stable/20440554. doi:10.1890/1540-9295(2007)5[13:DTHEF]2.0.CO;2

Dogan, E., Majeed, M. T., & Luni, T. (2022). Revisiting the nexus of ecological footprint, unemployment, and renewable and non-renewable energy for South Asian economies: Evidence from novel research methods. *Renewable Energy*, *194*(1), 1060–1070. doi:10.1016/j.renene.2022.05.165

Dooren, V., Aiking, H., & Vellinga, P. (2018). In search of indicators to assess the environmental impact of diets. *The International Journal of Life Cycle Assessment*, *23*(1), 1297–1314. doi:10.1007/s11367-017-1371-2

Duro, J. A., & Teixidó-Figueras, J. (2013). Ecological footprint inequality across countries: The role of environment intensity, income and interaction effects. *Ecological Economics*, *93*(1), 34–41. doi:10.1016/j.ecolecon.2013.04.011

Fiala, N. (2008). Measuring sustainability: Why the ecological footprint is bad economics and bad environmental science. *Ecological Economics*, *67*(4), 519–525. doi:10.1016/j.ecolecon.2008.07.023

Figge, L., Oebels, K., & Offermans, A. (2016). The effects of globalization on ecological footprints: An empirical analysis. *Environment, Development and Sustainability*, *19*(1), 863–876. doi:10.1007/s10668-016-9769-8

Galli, A., Iha, K., Pires, S. M., Mancini, M. S., Alves, A. A., Zakai, G., Lin, D., Murthy, A., & Wackernagel, M. (2020). Assessing the ecological footprint and biocapacity of portuguese cities: Critical results for environmental awareness and local management. *Cities (London, England)*, *96*(1), 1–13. doi:10.1016/j.cities.2019.102442

GCB. (2020). *Global carbon budget report*. GCB. https://essd. copernicus.org/articles/12/3269/2020/essd-12-3269-2020.pdf

Han, J., & Dalaibaatar, E. (2023). A study on the influencing factors of China's ecological footprint based on EEMD-GaoDetector. *Sustainability (Basel)*, *15*(8), 1–15. doi:10.3390/su15086680

IEA. (2018). *Energy information administration. International Energy Outlook*. Paris, France: US Department of Energy. http://energyatlas.iea. org/#!/profile/WORLD/IND

Jyoti, B., & Singh, A. K. (2023). Does digitalization have a causal relationship with economic development? An experience from a country-wise panel data statistical analysis. In R. S. Nazlı & G. Sari (Eds.), *Perspectives on Society and Technology Addiction* (pp. 34–61). IGI Global., doi:10.4018/978-1-6684-8397-8.ch003

Jyoti, B., Singh, A. K., & Ashraf, S. N. (2023). Appropriate technologies and their implications in the agricultural sector. In P. H. Alex Khang (Ed.), *Advanced Technologies and AI-Equipped IoT Applications in High-Tech Agriculture* (pp. 65–87). IGI Global., doi:10.4018/978-1-6684-9231-4.ch004

Kasperowicz, R. (2015). Economic growth and CO2 emissions: The ECM analysis. *Journal of International Students*, *8*(3), 91–98. https://www.jois.eu/files/07_Kasperowicz.pdf

Kazemzadeh, E., Fuinhas, J. A., Salehnia, N., Koengkan, M., & Silva, N. (2023). Assessing influential factors for ecological footprints: A complex solution approach. *Journal of Cleaner Production*, *414*(15), 32–43. doi:10.1016/j.jclepro.2023.137574

Kirikkaleli, D., Adebayo, T. S., Khan, Z., & Ali, S. (2021). Does globalization matter for ecological footprint in Turkey? Evidence from dual adjustment approach. *Environmental Science and Pollution Research International*, *28*(1), 14009–14017. doi:10.1007/s11356-020-11654-7 PMID:33201501

Kitzes, J., Wackernagel, M., Loh, J., Peller, A., Goldfinger, S., Cheng, D., & Tea, K. (2008). Shrink and share: Humanity's present and future ecological footprint. *Philosophical Transactions of the Royal Society of London. Series B, Biological Sciences*, *363*(1491), 467–475. doi:10.1098/rstb.2007.2164 PMID:17652075

Kızılgöl, Ö., & Öndes, H. (2022). Factors affecting the ecological footprint: A study on the OECD countries. *The Science of the Total Environment*, *849*(25), 34–45. doi:10.1016/j.scitotenv.2022.157757 PMID:35926619

Kumar, A., Ahmad, M. M., & Sharma, P. (2017). Influence of climatic and non-climatic factors on sustainable food security in India: A statistical investigation. *International Journal of Sustainable Agricultural Management and Informatics*, *3*(1), 1–30. doi:10.1504/IJSAMI.2017.082917

Kumar, A., Sharma, P., & Ambrammal, S. K. (2015). Climatic effects on sugarcane productivity in India: A stochastic production function application. *International Journal of Economics and Business Research*, *10*(2), 179–203. doi:10.1504/IJEBR.2015.070984

Magazzino, C. (2024). Ecological footprint, electricity consumption, and economic growth in China: Geopolitical risk and natural resources governance. *Empirical Economics*, *66*(1), 1–25. doi:10.1007/s00181-023-02460-4

Majeed, M. T., & Mazhar, M. (2019). Financial development and ecological footprint: A global panel data analysis. *Pakistan Journal of Commerce and Social Sciences*, *13*(2), 487–514. http://hdl.handle.net/10419/201002

Mancini, M. S., Galli, A., Niccolucci, V., Lin, D., Bastianoni, S., Wackernagel, M., & Marchettini, N. (2016). Ecological footprint: Refining the carbon footprint calculation. *Ecological Indicators*, *61*(2), 390–403. doi:10.1016/j.ecolind.2015.09.040

Mehmood, U., Muhammad, U. A., & Muhammad, A. J. (2023). Associating economic growth and ecological footprints through human capital and biocapacity in south Asia. *WORLD (Oakland, Calif.)*, *4*(3), 598–611. doi:10.3390/world4030037

Mishra, A. K., & Dash, A. K. (2022). Connecting the carbon ecological footprint, economic globalization, population density, financial sector development, and economic growth of five South Asian countries. *Energy Research Letters*, *3*(2), 1–6. doi:10.46557/001c.32627

Neagu, O. (2020). Economic complexity and ecological footprint: Evidence from the most complex economies in the world. *Sustainability (Basel)*, *12*(21), 1–18. doi:10.3390/su12219031

Rahman, A., Radulescu, M., Ma, H., Dagar, V., Hussain, I., & Khan, M. K. (2019). The impact of globalization, energy use, and trade on ecological footprint in Pakistan: Does environmental sustainability exist? *Energies*, *14*(17), 1–16. doi:10.3390/en14175234

Ray, S., & Ray, I. A. (2011). Impact of population growth on environmental degradation: A case of India. *Journal of Economics and Sustainable Development*, *2*(8), 72–77. https://core.ac.uk/reader/234645493

Sabir, S., & Gorus, M. S. (2019). The impact of globalization on ecological footprint: Empirical evidence from the South Asian countries. *Environmental Science and Pollution Research International*, *26*(32), 33387–33398. doi:10.1007/s11356-019-06458-3 PMID:31522394

Saboori, B., Sulaiman, J. B., & Mohd, S. (2012). An empirical analysis of the environmental Kuznets curve for CO2 emission in Indonesia: The role of energy consumption and foreign trade. *International Journal of Economics and Finance*, *4*(2), 243–251. doi:10.5539/ijef.v4n2p243

Saleem, N., Rahman, S. R., & Jun, Z. (2019). The impact of human capital and biocapacity on environment, environmental quality measure through ecological footprint and greenhouse gases. *Journal of Pollution Effects & Control*, *7*(4), 1–13. doi:10.35248/2375-4397.19.7.237

Samimi, A. J., Kashefi, A., Salatin, P., & Lashkarizadeh, M. (2011). Environmental performance and HDI: Evidence from countries around the world. *Middle East Journal of Scientific Research*, *10*(3), 294–301. https://idosi.org/mejsr/mejsr10(3)11/3.pdf

Sharma, R., Sinha, A., & Kautish, P. (2020). Does renewable energy consumption reduce ecological footprint? *Evidence from eight developing countries of Asia. MPRA Paper No. 104277*. MPRA. https://mpra.ub.uni-muenchen.de/104277/

Singh, A. K., Issac, J., & Narayanan, K. G. S. (2019). Measurement of environmental sustainability index and its association with socio-economic indicators in selected Asian economies: An empirical investigation. *International Journal of Environment and Sustainable Development*, *18*(1), 57–100. doi:10.1504/IJESD.2019.098641

Singh, A. K., & Jyoti, B. (2023a). Appropriate technology and adaptation strategies mitigate the adverse impact of climate change on the agricultural sector: A case study in Gujarat, India. In P. H. Alex Khang (Ed.), *Advanced Technologies and AI-Equipped IoT Applications in High-Tech Agriculture* (pp. 389–416). IGI Global., doi:10.4018/978-1-6684-9231-4.ch023

Singh, A. K., & Jyoti, B. (2023b). Empirical evidence from different income group countries on the interdependency between digitalization and economic development. In Patricia Ordóñez de Pablos and Xi Zhang (Eds.), 5G, Artificial Intelligence, and Next Generation Internet of Things: Digital Innovation for Green and Sustainable Economies, (pp. 44-85). IGI Global. DOI: doi:10.4018/978-1-6684-8634-4.ch003

Singh, A. K., Jyoti, B., Kumar, S., & Lenka, S. K. (2021). Assessment of global sustainable development, environmental sustainability, economic development and social development index in selected economies. *International Journal of Sustainable Development and Planning*, *16*(1), 123–138. doi:10.18280/ijsdp.160113

Singh, A. K., Jyoti, B., & Mishra, V. K. (2023b). Policy Implications from selected countries for a sustainable future of the society through green entrepreneurship and economic development. In A. Mohammed El Amine, S. Ul-Durar, & H. Hattab (Eds.), *Entrepreneurship Ecosystems and Their Opportunities and Challenges* (pp. 129–162). IGI Global. doi:10.4018/978-1-6684-7140-1.ch006

Singh, A. K., & Kumar, S. (2022). Exploring the impact of sustainable development on social-economic, and science and technological development in selected countries: A panel data analysis. *Society & Sustainability*, *4*(1), 55–83. doi:10.38157/ss.v4i1.405

Singh, A. K., Kumar, S., & Jyoti, B. (2022a). Association of sustainable food security with climatic and non-climatic factors in Gujarat state of India: A district-wise panel data investigation. In: Leal Filho, W., Kovaleva, M., Popkova, E. (eds.), Sustainable Agriculture and Food Security. World Sustainability Series (pp. 259-282). Springer Cham. doi:10.1007/978-3-030-98617-9_16

Singh, A. K., Kumar, S., & Jyoti, B. (2022b). Does sustainable livelihood security have a relationship with climatic and geographical factors? evidence from a state-wise panel data investigation in India. *The IUP Journal of Applied Economics*, *21*(1), 7–35.

Singh, A. K., Sharma, A. K., & Jyoti, B. (2023a). Does economic development have a causal relationship with environmental degradation? Experience from different income group countries. In G. Sart (Ed.), *Considerations on Education for Economic, Social, and Environmental Sustainability* (pp. 300–333). IGI Global. doi:10.4018/978-1-6684-8356-5.ch015

Singh, A. K., & Singh, S. (2024). A case study from Kanpur Nagar, Uttar Pradesh for assessing the economic value of water and its determinants. In A. Karmaoui (Ed.), *Water-Soil-Plant-Animal Nexus in the Era of Climate Change* (pp. 212–242). IGI Global. doi:10.4018/978-1-6684-9838-5.ch010

Stoeglehner, G., & Narodoslawsky, M. (2008). Implementing ecological footprinting in decision-making processes. *Land Use Policy*, *25*(1), 421–431. doi:10.1016/j.landusepol.2007.10.002

Ulucak, R., & Lin, D. (2017). Persistence of policy shocks to ecological footprint of the USA. *Ecological Indicators*, *80*(1), 337–343. doi:10.1016/j.ecolind.2017.05.020

Wackernagel, M., & Rees, W. (1996). Our ecological footprint: reducing human impact on the earth. *New Society Publishers.* https://escholarship.org/content/qt7730w81q/qt7730w81q.pdf?t=q9ns62

Wan, X., Jahanger, A., Usman, M., Radulescu, M., Balsalobre-Lorente, D., & Yu, Y. (2022). Exploring the effects of economic complexity and the transition to a clean energy pattern on ecological footprint from the Indian perspective. *Frontiers in Environmental Science*, *9*(1), 1–17. doi:10.3389/fenvs.2021.816519

Wei, Z., Jian, Z., Sun, Y., Pan, F., Han, H., Liu, Q., & Mei, Y. (2023). Ecological sustainability and high-quality development of the yellow rive delta in China based on the improved ecological footprint model. *Scientific Reports*, *13*(1), 1–16. doi:10.1038/s41598-023-30896-2 PMID:36593249

Wiedmann, T., & Barrett, J. (2010). A review of the ecological footprint indicator—Perceptions and methods. *Sustainability (Basel)*, *2*(6), 1645–1693. doi:10.3390/su2061645

Wu, S., Wang, J., Jia, Y., & Wang, X. (2023). A comparative study on the ecological footprint of living consumption in northwest ethnic regions: 1980-2018. *Frontiers in Environmental Science*, *11*(1), 1–12. doi:10.3389/fenvs.2023.1046203

Yilanci, V., & Pata, U. K. (2020). Convergence of per capita ecological footprint among the ASEAN-5 countries: Evidence from a non-linear panel unit root test. *Ecological Indicators*, *113*(1), 13-26. 106178. doi:10.1016/j.ecolind.2020.106178

York University Ecological Footprint Initiative & Global Footprint Network (YUEFI & GFN). (2023). *Public data package of the national footprint and biocapacity accounts*. Produced for the Footprint Data Foundation and distributed by Global Footprint Network. https://data.footprintnetwork.org

Zambrano-Monserrate, M., Ruano, M. A., Ormeño-Candelario, V., & Sanchez-Loor, D. A. (2020). Global ecological footprint and spatial dependence between countries. *Journal of Environmental Management*, *272*(15), 33–42. doi:10.1016/j.jenvman.2020.111069 PMID:32854881

Zhang, C., & Lin, Y. (2012). Panel estimation for urbanization, energy consumption and CO2 emissions: A regional analysis in China. *Energy Policy*, *49*(1), 488–498. doi:10.1016/j.enpol.2012.06.048

A CALL TO ACTION FOR A BIODIVERSE FUTURE

The journey through "Biodiversity Loss Assessment for Ecosystem Protection" has illuminated the gravity of the challenge we face – a burgeoning biodiversity crisis with far-reaching consequences. We have explored a diverse arsenal of methodologies and approaches for assessing biodiversity loss, equipping ourselves with the knowledge needed to measure and understand this critical issue.

The information gleaned from these assessments paints a clear picture: the time for complacency is over. We must translate this knowledge into concrete action. This book serves as a call to arms, urging readers to become active participants in the fight to protect our planet's precious biodiversity.

From Assessment to Action

Biodiversity loss assessment provides the scientific foundation for effective conservation strategies. By identifying priority areas, pinpointing drivers of loss, and monitoring progress, we can ensure that our efforts are targeted and impactful. This book has highlighted the importance of cost-benefit analysis, ensuring that conservation solutions are not only ecologically sound but also economically viable.

Ultimately, the true test lies in the application of knowledge. We must translate the insights gleaned from assessments into tangible conservation actions. This includes:

- **Habitat Protection and Restoration:** Safeguarding existing habitats and restoring degraded ones is critical for providing sanctuaries for species and maintaining ecosystem functionality.
- **Sustainable Practices:** Promoting sustainable agriculture, forestry, and fishing practices will reduce human pressures on ecosystems and foster coexistence with biodiversity.
- **Policy and Regulation:** Effective environmental policies and regulations are essential to curb the drivers of biodiversity loss, such as pollution and unsustainable resource exploitation.
- **Education and Public Awareness:** Raising public awareness about the importance of biodiversity and fostering a sense of environmental stewardship is crucial for mobilizing support for conservation efforts.

A Shared Responsibility

The task of protecting biodiversity is not one that can be undertaken by scientists and conservationists alone. It is a collective responsibility shared by individuals, governments, corporations, and communities across the globe. We all have a role to play, from making informed consumer choices to advocating for policy change.

Looking Forward

"Biodiversity Loss Assessment for Ecosystem Protection" concludes on a note of hope. By embracing the tools and knowledge presented within these pages, we can move beyond mere assessment. We can become active stewards of our planet, fostering a future where vibrant biodiversity thrives and ecosystems continue to provide their irreplaceable services for generations to come. Let this book serve as a springboard for positive change, a catalyst for a future where humanity coexists in harmony with the natural world.

The future of biodiversity rests in our hands. Let us choose wisely.

Ashok K. Rathoure
Saraca Research Inc., India

Compilation of References

Abaza, H., Bisset, R., & Sadler, B. (2004). *Environmental Impact Assessment and Strategic Environmental Assessment: Towards an Integrated Approach. Economics and Trade Branch.* UNEP.

Abbasi, T., Sanjeevi, R., Anuradha, J., & Abbasi, S. A. (2013). Impact of Al 3+ on sludge granulation in UASB reactor. *Indian Journal of Biotechnology, 12*(2), 254–259.

Abbasi, T., Sanjeevi, R., Makhija, M., & Abbasi, S. A. (2012). Role of Vitamins B-3 and C in the Fashioning of Granules in UASB Reactor Sludge. *Applied Biochemistry and Biotechnology, 167*(2), 348–357. doi:10.1007/s12010-012-9691-y PMID:22549583

Adams, R. (2010). *Accounting Report.* Association of Chartered Accountants., www.accaglobal.com/

Aerts, R., Honnay, O., & Van Nieuwenhuyse, A. (2018). Biodiversity and human health: Mechanisms and evidence of the positive health effects of diversity in nature and green spaces. *British Medical Bulletin, 127*(5), 5–22. doi:10.1093/bmb/ldy021 PMID:30007287

Agarwal, N., Singh, G., & Singh, H. (2011). Present status of Icthyofaunal diversity of Garhwal Himalayan river Bhilangna and its tributaries with reference to changing environment. *Environment Conservation Journal, 12*(3), 101–108. doi:10.36953/ECJ.2011.120319

Ahmad, M., & Ali, A. (2023). Mapping the Future of Sustainable Development Through Cloud-Based Solutions: A Case Study of OpenStreetMap. In Promoting Sustainable Management Through Technological Innovation (pp. 153–176). IGI Global. doi:10.4018/978-1-6684-9979-5.ch011

Ahmad, M., & Khiyal, M. S. H. (2023). Assessment of land administration in Pakistan and the potential role of volunteered geographic information. In Handbook of Research on Driving Socioeconomic Development With Big Data. IGI Global. doi:10.4018/978-1-6684-5959-1.ch014

Ahmad, M. (2023a). Connecting People and Places: How Citizen Diplomacy and VGI Are Strengthening Disaster Response and Community Development. In S. Kavoğlu & E. Köksoy (Eds.), *Global Perspectives on the Emerging Trends in Public Diplomacy* (pp. 195–226). IGI Global. doi:10.4018/978-1-6684-9161-4.ch009

Ahmad, M. (2023b). Exploring the Role of OpenStreetMap in Mapping Religious Tourism in Pakistan for Sustainable Development. In V. João Pereira Domingues Martinho, J. Augusto Guerra da Rocha Nunes, M. Lúcia de Jesus Pato, & L. Andrade de Matos Castilho (Eds.), *Experiences, Advantages, and Economic Dimensions of Pilgrimage Routes* (pp. 23–40). IGI Global. doi:10.4018/978-1-6684-9923-8.ch002

Ahmad, M. (2023c). Leveraging Social Media Geographic Information for Smart Governance and Policy Making: Opportunities and Challenges. In C. Chavadi & D. Thangam (Eds.), *Global Perspectives on Social Media Usage Within Governments* (pp. 192–213). IGI Global. doi:10.4018/978-1-6684-7450-1.ch013

Ahmad, M. (2024a). Connecting the Dots: Harnessing OpenStreetMap for Big Data Analytics and Market Insights. In D. Darwish (Ed.), *Big Data Analytics Techniques for Market Intelligence* (pp. 329–347). IGI Global. doi:10.4018/979-8-3693-0413-6.ch013

Ahmad, M. (2024b). From Crisis to Opportunity: How VGI Is Transforming Disaster Response and Recovery Efforts. In S. Siyal (Ed.), *Using Crises and Disasters as Opportunities for Innovation and Improvement* (pp. 215–232). IGI Global., doi:10.4018/978-1-6684-9522-3.ch013

Ahmad, M. (2024c). The Role of Data Science and Volunteered Geographic Information in Enhancing Government Service Delivery: Opportunities and Challenges. In O. S. Ogunleye (Ed.), *Machine Learning and Data Science Techniques for Effective Government Service Delivery* (pp. 254–281). IGI Global.

Ahmad, M. (2024d). Unleashing Business Potential: Harnessing OpenStreetMap for Intelligent Growth and Sustainability. In S. Singh, S. S. Rajest, S. Hadoussa, A. J. Obaid, & R. Regin (Eds.), *Data-Driven Intelligent Business Sustainability* (pp. 177–198). IGI Global. doi:10.4018/979-8-3693-0049-7.ch013

Ahmad, M., Khayal, M. S. H., & Tahir, A. (2022). Analysis of Factors Affecting Adoption of Volunteered Geographic Information in the Context of National Spatial Data Infrastructure. *ISPRS International Journal of Geo-Information*, *11*(2), 120. doi:10.3390/ijgi11020120

Akçakaya, H. R. (2001). Linking population-level risk assessment with landscape and habitat models. *The Science of the Total Environment*, *274*(1–3), 283–291. doi:10.1016/S0048-9697(01)00750-1 PMID:11453303

Alcamo, J. (2002). *Scenarios as Tools for International Environmental Assessments. Experts Corner Report, Prospects and Scenarios No.5*. European Environment Agency.

Allen, E. L., Engeman, R., & Krupa, H. (1996). Evaluation of three relative abundance indices for assessing dingo population. *Wildlife Research*, *23*(2), 197–206. doi:10.1071/WR9960197

Alvarsson, J. J., Wiens, S., & Nilsson, M. E. (2010). Stress recovery during exposure to nature sound and environmental noise. *International Journal of Environmental Research and Public Health*, *7*(3), 1036–1046. doi:10.3390/ijerph7031036 PMID:20617017

Aly, H. (2020). Digital transformation, development and productivity in developing countries: Is artificial intelligence a cure or a blessing? *Review of Economics and Political Science*, *7*(4), 238–256. doi:10.1108/REPS-11-2019-0145

Amoyaw-Osei, Y. (1997). *Developing Strategic EA Capacity in a Developing Country: The Ghana Experience*. Paper to 17th Annual Meeting of the International Association for Impact Assessment, New Orleans, USA.

Anderson, N. (1959). Urbanism and Urbanization. *American Journal of Sociology*, *65*(1), 68–73. doi:10.1086/222628

Anderson, T. (1867). An enumeration of the Indian species of Acanthaceae. *Journal of the Linnean Society of London, Botany*, *9*(39), 425–454. doi:10.1111/j.1095-8339.1867.tb01308.x

Andersson, A. A., Tilley, H. B., Lau, W., Dudgeon, D., Bonebrake, T. C., & Dingle, C. (2021). CITES and beyond: Illuminating 20 years of global, legal wildlife trade. *Global Ecology and Conservation*, *26*, e01455. doi:10.1016/j.gecco.2021.e01455

Andersson, R., & Östlund, L. (2004). Spatial patterns, density changes and implications on biodiversity for old trees in the boreal landscape of northern Sweden. *Biological Conservation*, *118*(4), 443–453. doi:10.1016/j.biocon.2003.09.020

Ankitkumar, B. Rathod, Prashantkumar B Sathvara, Aviral Tripathi, J. Anuradha, Sandeep Tripathi, & R. Sanjeevi. (2022). Vegetational Cartography Analysis Utilizing Multi-temporal Ndvi Data Series: A Case Study from Rajkot District (Gujarat), India. *Tianjin Daxue Xuebao, 55*(04), 490–497. https://doi.org/ doi:10.17605/OSF.IO/UGJYM

Ankitkumar, B Rathod, Anuradha, J., Prashantkumar B Sathvara, Tripathi, S., & Sanjeevi, R. (2023). Vegetational Change Detection Using Machine Learning in GIS Technique: A Case Study from Jamnagar (Gujarat). *Journal of Data Acquisition and Processing*, *38*(1), 1046–1061. doi:10.5281/zenodo.7700655

Antrop, M. (2004). Landscape change and the urbanization process in Europe. *Landscape and Urban Planning*, *67*(1–4), 9–26. doi:10.1016/S0169-2046(03)00026-4

APHA. (1971). *Standard methods for the examination of water and waste water.* American Public Health, Association.

Aplet, G. H. (1999). On the nature of wildness: Exploring what wilderness really protects. *Denver University Law Review*, 347–367. https://digitalcommons.du.edu/dlr/vol76/iss2/4/

Aras, G., & Crowther, D. (2008). Developing sustainable reporting standards. *Journal of Applied Accounting Research*, *9*(1), 4–16. doi:10.1108/09675420810886097

Argyrou, V. (2005). *La lógica del ecologismo: Antropología, ecología y poscolonialidad.* Libros Berghahn.

Armitage, D., Mbatha, P., Muhl, E. K., Rice, W., & Sowman, M. (2020). Governance principles for community-centered conservation in the post-2020 global biodiversity framework. *Conservation Science and Practice*, *2*(2), e160. doi:10.1111/csp2.160

Aronson, R. B. (2001). The Limits of Detectability: Short-Term Events and Short-Distance Variation in the Community Structure of Coral Reefs. *Bulletin of Marine Science*, *69*, 331–332.

Asad, T. (1986). El concepto de traducción cultural en la antropología social británica. In J. Clifford & G. E. Marcus (Eds.), *Writing culture: The poetics and politics of ethnography.* Prensa de la Universidad de California. doi:10.1525/9780520946286-009

Asamblea General de las Naciones Unidas. (2017). Resolución 71/312 Nuestro océano, nuestro futuro: Llamamiento a la acción. In *Resolución adoptada por la Asamblea General el 6 de julio de 2017.* Naciones Unidas.

Aschemann, R. (2004). Lessons Learned from Austrian SEAs. *European Environment*, *14*(3), 165–174. doi:10.1002/eet.347

Asheim, B., Grillitsch, M., & Trippl, M. (2017). Introduction: Combinatorial knowledge bases, regional innovation, and development dynamics. *Economic Geography*, *93*(5), 429–435. doi:10.1080/00130095.2017.1380775

Ashe, J. (2002). The Australian Regional Forest Agreement process: A Case Study in Strategic Natural Resource Assessment, PP 156-181. In S. Marsden & S. Dovers (Eds.), *Strategic Environmental Assessment in Australia.* The Federation Press.

Ashok, K. (2013). Butterfly (Lepidoptera: Insecta) Diversity from Different Sites of Jhagadia, Ankleshwar, District-Bharuch, Gujarat, *Oct. Jour. Env. Res. 1*(1), 09-18

Ashok, K. (2014). Environmental Management Plan for Chemical Industries Especially Resin Manufacturing Unit, *Oct. Jour. Env. Res*, *2*(3), 262–273.

Ashok, K., & Goyal, A. S. (2013a). Ecology and Biodiversity status of Sachin gidc and its surroundings with Special reference to Conservation measures for Indian Peafowl (Pavo cristatus) schedule –I Bird species. *Oct. Jour. Env. Res.*, *2*(1), 82–100.

Ashok, K., & Goyal, A. S. (2013b). Study of Common Property Resources (Cpr) With Special Reference To Water And Biological Resources At Projected Area Near Village Ninat, Bardoli, District-Surat. *Oct. Jour. Env. Res.*, *1*(4), 319–331.

Ashok, K., Meena, S., & Savita, G. (2013). The Biodiversity At Sandi Bird Sanctuary, Hardoi With Special Reference to Migratory Birds. *Oct. Jour. Env. Res*, *1*(3), 173–181.

Ashraf, S. N., & Singh, A. K. (2022). Implications of appropriate technology and farm inputs in the agricultural sector of Gujarat: Empirical analysis based on primary data. *Agricultural Economics Research Review*, *35*(2), 59–77. doi:10.5958/0974-0279.2022.00031.3

Aşıcı, A. A., & Acar, S. (2016). Does income growth relocate ecological footprint? *Ecological Indicators*, *61*(2), 707–7014. doi:10.1016/j.ecolind.2015.10.022

Aslam, R. W., Shu, H., Naz, I., Quddoos, A., Yaseen, A., Gulshad, K., & Alarifi, S. S. (2024). Machine Learning-Based Wetland Vulnerability Assessment in the Sindh Province Ramsar Site Using Remote Sensing Data. *Remote Sensing (Basel)*, *16*(5), 928. doi:10.3390/rs16050928

Asner, G. P., Knapp, D. E., Kennedy-Bowdoin, T., Jones, M. O., Martin, R. E., Boardman, J., & Hughes, R. F. (2009). Carnegie Airborne Observatory: In-flight fusion of hyperspectral imaging and waveform light detection and ranging for three-dimensional studies of ecosystems. *Journal of Applied Remote Sensing*, *3*(1), 033543. doi:10.1117/1.3223675

Ayeche, M. B., Barhoumi, M., & Hammas, M. A. (2016). Causal linkage between economic growth, financial development, trade openness and CO2 emissions in European countries. *American Journal of Environmental Engineering*, *6*(4), 110–122. http://article.sapub.org/10.5923.j.ajee.20160604.02.html

Aytar, E. C., & Özdener Kömpe, Y. (2023). Effect of different substrates on in vitro symbiotic seed germination for soilless production of Anacamptis laxiflora orchid. *Acta Botanica Croatica, 82*(2), 0-0.

Aytar, E. C., Harzli, I., & Özdener Kömpe, Y. (2023). Phytochemical Analysis of Anacamptis coriophora Plant Cultivated Using Ex Vitro Symbiotic Propagation. *Chemistry & Biodiversity*, *20*(12), e202301218. doi:10.1002/cbdv.202301218 PMID:37870554

Ayvazyan, S. A., Afanase, M. Y., & Rudenko, V. A. (2012). Some questions of specification of three-factor models of the company's production potential, taking into account intellectual capital. *Applied Economics*, *3*(27), 56–66.

Baabou, W., Grunewald, N., Ouellet-Plamondon, C., Gressot, M., & Gallib, A. (2017). The ecological footprint of mediterranean cities: Awareness creation and policy implications. *Environmental Science & Policy*, *69*(1), 94–104. doi:10.1016/j.envsci.2016.12.013

Babenko, V.; Perevozova, I.; Mandych, O.; Kvyatko, T.; Maliy, O. & Mykolenko, I. (2019). World informatization in conditions of international globalization: factors of influence. *Global. J. Environ. Sci. Manage.*, *5*(SI), 172-179. doi:10.22034/gjesm.2019.SI.19

Bailey, J., & Rention, S. (1997). Redseigning EIA to Fit the Future: SEA and the Policy Process. *Impact Assessment*, *15*(4), 319–334. doi:10.1080/07349165.1997.9726145

Bakun, A. (1990). Global climate change and intensification of coastal ocean upwelling. *Science*, *247*(4939), 198–201. doi:10.1126/science.247.4939.198 PMID:17813287

Balmford, A., Bradbury, R. B., Bauer, J. M., Broad, S., Burgess, G., Burgman, M., Byerly, H., Clayton, S., Espelosin, D., Ferraro, P. J., Fisher, B., Garnett, E. E., Jones, J. P. G., Marteau, T. M., Otieno, M., Polasky, S., Ricketts, T. H., Sandbrook, C., Sullivan-Wiley, K., & Nielsen, K. S. (2021). Making more effective use of human behavioural science in conservation interventions. *Biological Conservation*, *261*, 109256. doi:10.1016/j.biocon.2021.109256

Barca, F., McCann, P., & Rodríguez-Pose, A. (2012). The case for regional development intervention: Place-based versus place-neutral approaches. *Journal of Regional Science*, *52*(1), 134–152. doi:10.1111/j.1467-9787.2011.00756.x

Barkin, D. (1998). Sostenibilidad: La economía política del desarrollo autónomo. *Organización y Medio Ambiente*, *11*(1), 5–32.

Bartiaux, F. (2008). ¿La información ambiental supera la compartimentación de las prácticas y cambia los comportamientos de los consumidores? *Revista de producción más limpia*, 16(11), 1170–1180. doi:10.1016/j.jclepro.2007.08.013

Basarir, A., & Arman, H. (2014). The effects of economic growth on environment: An application of environmental Kuznets curve in United Arab Emirates. *The Online Journal of Science and Technology*, *4*(1), 53–59. https://tojsat.net/journals/tojsat/articles/v04i01/v04i01-07.pdf

Bass, D. K., & Miller, I. R. (1998). Crown-of-Thorns Starfish and Coral Surveys Using the Manta Tow and Scuba Search Techniques. Long-*term Monitoring of the Great Barrier Reef. Standard Operations Procedure Number 1*. www.aims.gov.au/pages/research/reef-monitoring/methods.html

Batáry, P., Dicks, L. V., Kleijn, D., & Sutherland, W. J. (2015). The role of agri-environment schemes in conservation and environmental management. *Conservation Biology*, *29*(4), 1006–1016. doi:10.1111/cobi.12536 PMID:25997591

Bateson, G. (1972). Metalogue: ¿Por qué las cosas se meten en un lío? En Pasos hacia una ecología de la mente; Bateson, G., Ed.; Chandler: Nueva York, NJ, EE.UU., 1972; págs. 13–18

Batten, S. D., Clarke, R., & Flinkman, J. (2003). CPR sampling: The technical background, material and methods, consistency and comparability. *Progress in Oceanography*, *58*(2-4), 193–215. doi:10.1016/j.pocean.2003.08.004

Bauman, Z. (2000). *Liquid Modernity*. Polity.

Bayraktarov, E., Saunders, M. I., Abdullah, S., Mills, M., Beher, J., Possingham, H. P., Mumby, P. J., & Lovelock, C. E. (2016). The cost and feasibility of marine coastal restoration. *Ecological Applications*, *26*(4), 1055–1074. doi:10.1890/15-1077 PMID:27509748

Beck, U., & Beck-Gernsheim, E. (2002). *Individualization: Institutionalized Individualism and Its Social and Political Consequences* (1st ed.). SAGE Publications Ltd.

Beechie, T. J., Sear, D. A., Olden, J. D., Pess, G. R., Buffington, J. M., Moir, H., Roni, P., & Pollock, M. M. (2010). Process-based principles for restoring river ecosystems. *Bioscience*, *60*(3), 209–222. doi:10.1525/bio.2010.60.3.7

Begon, M., Harper, J. L., & Townsend, C. R. (1996). *Ecology: Individuals, Populations, and Communities* (3rd ed.). Blackwell Science Ltd. doi:10.1002/9781444313765

Beierle, T. C. (1999). Using social goals to evaluate public participation in environmental decisions. *The Review of Policy Research*, *16*(3-4), 75–103. doi:10.1111/j.1541-1338.1999.tb00879.x

Bellard, C., Bertelsmeier, C., Leadley, P., Thuiller, W., & Courchamp, F. (2012). Impacts of climate change on the future of biodiversity. *Ecology Letters*, *15*(4), 365–377. doi:10.1111/j.1461-0248.2011.01736.x PMID:22257223

Belz, F.-M., & Peattie, K. (2010). Marketing de sostenibilidad: Una concepción innovadora del marketing. *Marketing Review St. Gallen*, *27*(5), 8–15. doi:10.1007/s11621-010-0085-7

Berg, A., Ehnstrom, B., Gustafsson, L., Hallingback, T., Jonsell, M., & Weslien, J. (1994). Threatened Plant, Animal, and Fungus Species in Swedish Forests: Distribution and Habitat Associations. In Conservation Biology, 8(3). doi:10.1046/j.1523-1739.1994.08030718.x

Berry, J. W. (2004). Una perspectiva ecocultural sobre el desarrollo de la competencia. en Cultura y Competencia. Contextos de Éxito en la Vida; Sternberg, R., Grigorenko, E., Eds.; Asociación Americana de Psicología: Washington, DC, EE.UU.

Bhuian, S. N., Amyx, D. A., & Shamma, H. M. (2014). Una extensión de la investigación del comportamiento ambiental del consumidor entre los expatriados. *Revista Internacional de Comercio y Gestión*, *24*(1), 63–84. doi:10.1108/IJCoMA-01-2012-0005

Biasetti, P., & De Mori, B. (2021). The ethical matrix as a tool for decision-making process in conservation. *Frontiers in Environmental Science*, *9*, 584636. doi:10.3389/fenvs.2021.584636

Biehler, D. D. (2013). *Plagas en la ciudad: moscas, chinches, cucarachas y ratas*. University of Washington Press.

Bilsborrow, R. E. (1992). Population growth, internal migration, and environmental degradation in rural areas of developing countries. *European Journal of Population*, *8*(1), 125–148. doi:10.1007/BF01797549 PMID:12158965

Bina, O. (2003). *Re-Conceptualising Strategic Environmental Assessment: Theoretical Overview and Case Study from Chile*. [PhD thesis, University of Cambridge].

BirdLife International. (2000). *Threatened Birds of the World*. Lynx Edicions and BirdLife International.

BirdLife International. (2004a). *Threatened Birds of the World 2004. CD-ROM*. BirdLife International.

BirdLife International. (2004b). *State of the World's Birds 2004-Indicators for our changing world*. BirdLife International.

BirdLife International. (2010). *The BirdLife checklist of the birds of the world, with conservation status and taxonomic sources*. BirdLife International. http://www.birdlife.info/docs/SpcChecklist/Checklist_v3_June10.zip

Bohil, C. J., Alicea, B., & Biocca, F. A. (2011). Virtual reality in neuroscience research and therapy. *Nature Reviews. Neuroscience*, *12*(December), 752–762. doi:10.1038/nrn3122 PMID:22048061

Bohnsack, J. A., & Bannerot, S. P. (1986). A Stationary Visual Census Technique for Quantitatively Assessing Community Structure of Coral Reef Fishes. *NOAA Technical Report NMFS*, *41*, 1–15.

Bojórquez-Tapia, L. A., & Ezcurra, E. (2009). Assessing environmental impacts of large construction projects: An integrative approach. *Environmental Impact Assessment Review*, *29*(4), 219–227.

Bondi, E., Fang, F., Hamilton, M., Kar, D., Dmello, D., Choi, J., Hannaford, R., Iyer, A., Joppa, L., Tambe, M., & Nevatia, R. (2018). Spot poachers in action: Augmenting conservation drones with automatic detection in near real time. *32nd AAAI Conference on Artificial Intelligence, AAAI 2018*, (pp. 7741–7746). IEEE. 10.1609/aaai.v32i1.11414

Bonita, J. (2011). Progreso interdisciplinario en enfoques para abordar los sistemas socio-ecológicos y ecoculturales. *Entorno. Conservar*, *38*, 127–139.

Borel-Saladin, J. M., & Turok, I. N. (2013). La economía verde: ¿cambio incremental o transformación? *Environmental Policy and Governance*, *23*(julio-agosto), 209–220. doi:10.1002/eet.1614

Bowden, A. R., Lane, M. R., & Martin, J. H. (2010). *Triple Bottom Line Risk Management: Enhancing Profit. Environmental Performance, and Community Benefit*. John Wiley.

Bradshaw, A., & Zwick, D. (2016). El campo de la sostenibilidad empresarial y el impulso de muerte: Una intervención radical. *Journal of Business Ethics*, *136*(junio), 267–279. doi:10.1007/s10551-014-2443-x

Briassoulis, H. (2002). El turismo sostenible y la cuestión de los bienes comunes. *Anales de Investigación Turística*, *29*(4), 1065–1085. doi:10.1016/S0160-7383(02)00021-X

Brightman, M., & Lewis, J. (Eds.). (2017). *La antropología de la sostenibilidad. Más allá del desarrollo y el progreso; Palgrave*. Reino Unido.

Brittain, S., Ibbett, H., de Lange, E., Dorward, L., Hoyte, S., Marino, A., Milner-Gulland, E. J., Newth, J., Rakotonarivo, S., Veríssimo, D., & Lewis, J. (2020). Ethical considerations when conservation research involves people. *Conservation Biology*, *34*(4), 925–933. doi:10.1111/cobi.13464 PMID:31953971

Brock, R. E. (1982). A Critique of the Visual Census Method for Assessing Coral Reef Fish Populations. *Bulletin of Marine Science*, *18*, 276–297.

Bromley, D. W. (1995). *Natural Resource Issues in Environmental Policy in South Africa*. Land and Agriculture Policy Centre.

BrondizioE.DíazS. M.SetteleJ.NgoH.GuezeM.Aumeeruddy-ThomasY.BaiX.GeschkeA.MolnárZ.NiamirA. (2019). Assessing a planet in transformation: rationale and approach of the IPBES Global Assessment on Biodiversity and Ecosystem Services. doi:10.5281/zenodo.3831852

Brown, P., & Daigneault, A. (2014). Cost-benefit analysis of managing the Papuana uninodis (Coleopetra: Scarbaeidae) taro beetle in Fiji. *Journal of Economic Entomology*, *107*(5), 1866–1877. doi:10.1603/EC14212 PMID:26309277

Bruckner, A. W. (2002). Priorities for Effective Management of Coral Diseases, U.S. Department of Commerce. *NOAA Technical Memorandum NMFS-OPR-22*.

Buffel, T., & Phillipson, Ch. (2016). ¿Pueden las ciudades globales ser ciudades amigables con las personas mayores? *Desarrollo urbano y envejecimiento de la población. Ciudades*, *55*, 94–100.

Bull, A. T., & Stach, J. E. M. (2014). An Overview of Biodiversity-Estimating the Scale. In A. T. Bull (Ed.), *Microbial Diversity and Bioprospecting* (pp. 15–28). ASM Press. doi:10.1128/9781555817770.ch2

Bunce, L., Townsley, P., Pomeroy, R., & Pollnac, R. (2000). *Socioeconomic Manual for Coral Reef Management*. Global Coral Reef Monitoring Network and Australian Institute of Marine Science Townsville Australia.

Busch, J., & Cullen, R. (2009). Effectiveness and cost-effectiveness of yellow-eyed penguin recovery. *Ecological Economics*, *68*(3), 762–776. doi:10.1016/j.ecolecon.2008.06.007

Buselich, K. (2002). *An Outline of Current Thinking on Sustainability Assessment. Background paper prepared for the Western Australian State Sustainability Strategy*. Institute for Sustainability and Technology Policy, Murdoch University.

Cahen-Fourot, L. (2020). Los capitalismos contemporáneos y su relación social con el medio ambiente. *Economía ecológica*, *172* (junio), 106634. doi:10.1016/j.ecolecon.2020.106634

Calcagni, F., Amorim Maia, A. T., Connolly, J. J. T., & Langemeyer, J. (2019). Digital co-construction of relational values: Understanding the role of social media for sustainability. *Sustainability Science*, *14*(5), 1309–1321. doi:10.1007/s11625-019-00672-1

Callison, C. (2014). *Cómo importa el cambio climático: la vida comunitaria de los hechos*. Prensa de la Universidad de Duke.

Campbell, S. K., & Butler, V. L. (2010). Evidencia arqueológica de la resiliencia de las poblaciones de salmón del noroeste del Pacífico y el sistema socioecológico en los últimos ~ 7,500 años. *Ecology and Society*, *15*, 1–17.

Canter, L. (1996). *Environmental impact assessment* (2nd ed.). McGraw-Hill.

Canter, L. W. (1996). *Environmental Impact Assessment* (2nd ed.). McGraw-Hill.

Carayannis, E. G., & Campbell, D. F. J. (2010). Triple Helix, Quadruple Helix and Quintuple Helix and how do knowledge, innovation and the environment relate to each other? A proposed framework for a trans-disciplinary analysis of sustainable development and social ecology. *International Journal of Social Ecology and Sustainable Development*, *1*(1), 41–69. https://www.igi-global.com/bookstore/article.aspx?titleid=41959. doi:10.4018/jsesd.2010010105

Carayannis, E. G., & Campbell, D. F. J. (2011). Open Innovation Diplomacy and a 21st Century Fractal Research, Education, and Innovation (FREIE) Ecosystem: Building on the Quadruple and Quintuple Helix Innovation Concepts and the "Mode 3" Knowledge Production System. *Journal of the Knowledge Economy*, *2*(3), 327–372. doi:10.1007/s13132-011-0058-3

Carleton, J. H., & Done, T. J. (1995). Quantitative Video Sampling of Coal Reef Benthos: Large Scale Application. *Coral Reefs*, *14*(1), 35–46. doi:10.1007/BF00304070

Carpenter, S. R., Mooney, H. A., Agard, J., Capistrano, D., DeFries, R. S., Díaz, S., & Lubchenco, J. (2009). Science for managing ecosystem services: Beyond the Millennium Ecosystem Assessment. *Proceedings of the National Academy of Sciences of the United States of America*, *106*(5), 1305–1312. doi:10.1073/pnas.0808772106 PMID:19179280

Cash, D. W., Clark, W. C., Alcock, F., Dickson, N. M., Eckley, N., Guston, D. H., Jäger, J., & Mitchell, R. B. (2003). Knowledge systems for sustainable development. *Proceedings of the National Academy of Sciences of the United States of America*, *100*(14), 8086–8091. doi:10.1073/pnas.1231332100 PMID:12777623

Catney, P., MacGregor, S., Dobson, A., Hall, S. M., Royston, S., Robinson, Z., Ormerod, M., & Ross, S. (2014). Big society, little justice? Community renewable energy and the politics of localism. *Local Environment*, *19*(7), 715–730. doi:10.1080/13549839.2013.792044

CBD (Convention on Biological Diversity). (1992). *Convention on Biological Diversity*. CBD. https://www.cbd.int/doc/legal/cbd-en.pdf

CBD. (2003). *Interlinkages Between Biological Diversity and Climate Change, CBD Technical Series 10*. Secretariat of the Convention of Biological Diversity.

Ceballos, G., Ehrlich, P. R., & Dirzo, R. (2017). Biological annihilation via the ongoing sixth mass extinction signaled by vertebrate population losses and declines. *Proceedings of the National Academy of Sciences of the United States of America*, *114*(30), E6089–E6096. doi:10.1073/pnas.1704949114 PMID:28696295

Cerri, F., Louis, Y. D., Fallati, L., Siena, F., Mazumdar, A., Nicolai, R., Zitouni, M. S., Adam, A. S., Mohamed, S., Lavorano, S., & Galli, P. (2024). Mangroves of the Maldives: A review of their distribution, diversity, ecological importance and biodiversity of associated flora and fauna. *Aquatic Sciences*, *86*(2), 44. doi:10.1007/s00027-024-01061-2

ChalmersC.FergusP.WichS.MontanezA. C. (2019). *Conservation AI: Live Stream Analysis for the Detection of Endangered Species Using Convolutional Neural Networks and Drone Technology*. http://arxiv.org/abs/1910.07360

Change, P. C. (2018). *Global warming of 1.5° C*. World Meteorological Organization.

Chape, S., Harrison, J., Spalding, M., & Lysenko, I. (2005). Measuring the extent and effectiveness of protected areas as an indicator for meeting global biodiversity targets. *Philosophical Transactions of the Royal Society of London. Series B, Biological Sciences*, *360*(1454), 443–455. doi:10.1098/rstb.2004.1592 PMID:15814356

Chen, J., Wang, C., Zhang, Y., & Li, D. (2022). Measuring spatial accessibility and supply-demand deviation of urban green space: A mobile phone signaling data perspective. *Frontiers in Public Health*, *10*, 1029551. doi:10.3389/fpubh.2022.1029551 PMID:36339177

Chen, Y.-W., Milstein, T., Anguiano, C., Sandoval, J., & Knudsen, L. (2012). Desafíos y beneficios de la investigación participativa basada en la comunidad para la justicia ambiental: Un caso de examen colaborativo de las luchas ecoculturales. *Entorno. Comuna*, *6*, 403–421.

Cherp, A., & Golubeva, S. (2004). Environmental Assessment in the Russian Federation. *Impact Assessment and Project Appraisal*, *22*(2), 121–130. doi:10.3152/147154604781766030

Cherrier, H., Szuba, M., & Özçağlar-Toulouse, N. (2012). Barreras a la reducción de las emisiones de carbono: Explorando el consumo sostenible frente al piso de vidrio. *Revista de Gestión de Marketing*, *28*(3–4), 397–419.

Christensen, N. L., Bartuska, A. M., Brown, J. H., Carpenter, S., D'Antonio, C., Francis, R., Franklin, J. F., MacMahon, J. A., Noss, R. F., Parsons, D. J., Peterson, C. H., Turner, M. G., & Woodmansee, R. G. (1996). The report of the ecological society of America committee on the scientific basis for ecosystem management. *Ecological Applications*, *6*(3), 665–691. doi:10.2307/2269460

Coenen, L., Hansen, T., & Rekers, J. V. (2015). Innovation policy for grand challenges: An economic geography perspective. *Geography Compass*, *9*(9), 483–496. doi:10.1111/gec3.12231

Coenen, L., & Morgan, K. (2019). Evolving geographies of innovation: Existing paradigms, critiques, and possible alternatives. *Norsk Geografisk Tidsskrift*.

Colebrook, J. M. (1960). Continuous Plankton records: Methods of analysis, 1950-59. *Bulletins of Marine Ecology*, *5*, 51–64.

Colin, P. L., Sadovy, Y. J., & Domeier, M. L. (2003). *Manual for the Study and Conservation of Reef Fish Spawning Aggregations*. Society for the Conservation of Reef Fish Aggregations Special Publication.

Comisión Europea. (2017). Informe sobre la estrategia de crecimiento azul hacia un crecimiento más sostenible y empleo en la economía azul. *Bruselas: Comisión Europea*.

Contos, P., Wood, J. L., Murphy, N. P., & Gibb, H. (2021). Rewilding with invertebrates and microbes to restore ecosystems: Present trends and future directions. *Ecology and Evolution*, *11*(12), 7187–7200. doi:10.1002/ece3.7597 PMID:34188805

Convention on Biological Diversity (CBD). (2004). *Handbook of the convention on biological diversity*. CBD. https://www.cbd.int/doc/handbook/

Coolen, H., & Meesters, J. (2012). Private and public green spaces: Meaningful but different settings. *Journal of Housing and the Built Environment*, *27*(1), 49–67. doi:10.1007/s10901-011-9246-5

Co-operation on Health and Biodiversity (2010). *Biodiversity and Human Health Policy Brief*. Co-operation on Health and Biodiversity.

Costanza, R., d'Arge, R., De Groot, R., Farber, S., Grasso, M., Hannon, B., & Paruelo, J. (1997). The value of the world's ecosystem services and natural capital. *Nature*, *387*(6630), 253–260. doi:10.1038/387253a0

Costanza, R., de Groot, R., Braat, L., Kubiszewski, I., Fioramonti, L., Sutton, P., Farber, S., & Grasso, M. (2017). Twenty years of ecosystem services: How far have we come and how far do we still need to go? *Ecosystem Services*, *28*, 1–16. doi:10.1016/j.ecoser.2017.09.008

Couchman, O., Wulffeld, E., Muurmans, M., Steer, M., & Fanning, E. (Eds.). (2009) Society for Environmental Exploration UK. Sea Turtle Monitoring Methods Manual. LIDER Foundation.

Creager, A. N. H. (2013). Paisajes de tiempo de trazadores radiactivos en bioquímica y ecología. *Historia y filosofía de las ciencias de la vida, 35*(1). https://www.jstor.org/stable/43862145

Crosby, M. P., & Reese, E. S. (1996). *A Manual for Monitoring Coral Reefs with Indicator Species: Butterflyfishes as Indicators of Change on Indo-Pacific Reefs*. Office of Ocean and Coastal Resource Management, National Oceanic and Atmospheric Administration.

Crosti, R., Arcangeli, A., Campagnuolo, S., Castriota, L., Falautano, M., Maggio, T., & Andaloro, F. (2020). Assessing worth of marine protected areas for the protection of threatened biodiversity using IUCN Red List and Red List Index. A pilot study in six mediterranean areas. *Ecological Indicators*, *119*, 106765. doi:10.1016/j.ecolind.2020.106765

Crutzen, P. J., & Eugene, F. S. (2000). El Antropoceno. *IGBP Newsl*, *41*, 17–18.

Cruz-Cruz, C. A., González-Arnao, M. T., & Engelmann, F. (2013). Biotechnology and conservation of plant biodiversity. *Resources*, *2*(2), 73–95. doi:10.3390/resources2020073

CSIR. (2002), Strategic Environmental Assessment (SEA) Resource Document Introduction to the Process, Principle and Application of SEA version1. *CSIR Report ENV-S-C- 2002-073, August 2002.*

Czech, B. (2003). Technological Progress and Biodiversity Conservation: A Dollar Spent, a Dollar Burned. *Conservation Biology*, *17*(5), 1455–1457. doi:10.1046/j.1523-1739.2003.02370.x

D'avila Garcez, C. M. (2000). Sistemas Locais de Inovação na Economia do Aprendizado: Uma abordagem conceitual. *Revista do BNDES*, *14*(7), 351–366.

Dahake, S. (2020). La creación de identidades ecoculturales fluidas en la India urbana. En Routledge Handbook of Eco-cultural Identity; Milstein, T., Castro-Sotomayor, J., Eds.; Routledge: Abingdon, Reino Unido; Nueva York, NY, EE.UU.

Dallimer, M., Tang, Z., Bibby, P. R., Brindley, P., Gaston, K. J., & Davies, Z. G. (2011). Temporal changes in greenspace in a highly urbanized region. *Biology Letters*, *7*(5), 763–766. doi:10.1098/rsbl.2011.0025 PMID:21429910

Daly, H. E., & Farley, J. (2004). *Ecological Economics: Principles and Applications*. Island Press.

Daniels, S. E., & Walker, G. B. (2001). *Working through environmental conflict: The collaborative learning approach*. Praeger Publishers.

Danish, H., Hassan, S. T., Baloch, M. A., Mahmood, N., & Zhang, J. W. (2019). Linking economic growth and ecological footprint through human capital and biocapacity. *Sustainable Cities and Society*, *47*(1), 13–24. doi:10.1016/j.scs.2019.101516

Danish, Z. W., & wang, Z. (2019). Investigation of the ecological footprint's driving factors: What we learn from the experience of emerging economies. *Sustainable Cities and Society*, *49*(1), 1–12. doi:10.1016/j.scs.2019.101626

Daryl Slack, J. (2008). Resistiendo los estudios ecoculturales. *Cultural Studies*, *2008*(22), 477–497. doi:10.1080/09502380802012575

Davison, C. W., Rahbek, C., & Morueta-Holme, N. (2021). Land-use change and biodiversity: Challenges for assembling evidence on the greatest threat to nature. *Global Change Biology*, *27*(21), 5414–5429. doi:10.1111/gcb.15846 PMID:34392585

de Melo, M. A. C. (1997). *Processo de Planejamento e as Inovações Tecnológicas e Sociais: uma perspectiva sócio-ecológica*. Anais do 5o. Seminário de Modernização Tecnológica.

de Melo, M.A.C. (1977b). *Articulated incrementalism: a strategy for planning (with special reference to the design of an information system as an articulative task).* [Tese de Doutorado, University of Pennsylvania, Filadélfia].

de Melo, M.A.C. (2002b). Inovação e Modernização Tecnológica e Organizacional nas MPMEs: o domínio interorganizacional. *Seminário Internacional: Políticas para Sistemas Produtivos Locais de MPME*: 2002

de Melo, A. (2001). *The Innovation Systems of Latin America and the Caribbean*. Inter-American Development Bank. Agosto.

de Melo, M. A. C. (2002a). *Enriquecendo a Atuação de Incubadora de Emrpesas. Tecnologia e Inovação: experiências de gestão nas micro e pequenas empresas*. PGT/USP.

De Palma, A., Sanchez-Ortiz, K., Martin, P. A., Chadwick, A., Gilbert, G., Bates, A. E., Börger, L., Contu, S., Hill, S. L., & Purvis, A. (2018). Challenges with inferring how land-use affects terrestrial biodiversity: study design, time, space and synthesis. In *Advances in ecological research* (Vol. 58, pp. 163–199). Elsevier. doi:10.1016/bs.aecr.2017.12.004

Dedovic, K., Renwick, R., Mahani, N. K., Engert, V., Lupien, S. J., & Pruessner, J. C. (2005). The Montreal Imaging Stress Task: Using functional imaging to investigate the effects of perceiving and processing psychosocial stress in the human brain. *Journal of Psychiatry & Neuroscience*, *30*(5), 319–325. PMID:16151536

Deguignet, M., Juffe-Bignoli, D., Harrison, J., MacSharry, B., Burgess, N., & Kingston, N. (2014). *United Nations list of protected areas*. UNEP-WCMC.

Del Campo, J. I., Lorenzo, C. M., & Castaño, A. M. (2019). La transparencia sobre Responsabilidad Social Corporativa y su implicación con el comportamiento económico-financiero: efectos en las empresas de Economía Social. *REVESCO: revista de estudios cooperativos,* (132), 122-143.

Deng, X., Huang, J., Rozelle, S., Zhang, J., & Li, Z. (2015). Impact of urbanization on cultivated land changes in China. *Land Use Policy*, *45*, 1–7. doi:10.1016/j.landusepol.2015.01.007

Deniz, İ. G., Kömpe, Y. Ö., Harzli, I., Aytar, E. C., Mutlu, V. A., & Uysal, D. İ. (2022). From seed to flowering tuberous orchid using ex vitro symbiotic seed germination: A breakthrough study with Anacamptis sancta. *Rhizosphere*, *24*, 100597. doi:10.1016/j.rhisph.2022.100597

Department of Prime Minister and Cabinet. (2003). *Sustainable Development for New Zealand: Programme of Action*. Department of Prime Minister and Cabinet.

Devictor, V., van Swaay, C., Brereton, T., Brotons, L., Chamberlain, D., Heliölä, J., Herrando, S., Julliard, R., Kuussaari, M., Lindström, Å., Reif, J., Roy, D., Schweiger, O., Settele, J., Stefanescu, C., van Strien, A., van Turnhout, C., Vermouzek, Z., WallisDeVries, M., & Jiguet, F. (2012). Differences in the climatic debts of birds and butterflies at a continental scale. *Nature Climate Change*, *2*(2), 121–124. doi:10.1038/nclimate1347

Dezembro.

Di Giminiani, P. y Haines, S. (2020). Introducción: Traducción de entornos. *Ethnos*, *85*(1), 1-16. doi:10.1080/00141844.2018.1542409

Dias, A. B., Rosentha, L. D., & de Melo, M. A. C. (2000) Enriquecendo a Atuação de Incubadoras de Empresas. XXI Simpósio de Gestão da Inovação Tecnológica. Co-autores: Adriano Batista Dias e David Rosenthal. São Paulo: 2000.

Díaz, S., & Malhi, Y. (2022). Biodiversity: Concepts, patterns, trends, and perspectives. *Annual Review of Environment and Resources*, *47*(1), 31–63. doi:10.1146/annurev-environ-120120-054300

Dietz, T., Rosa, E. A., & York, R. (2007). Driving the human ecological footprint. *Frontiers in Ecology and the Environment*, *5*(1), 13–18. https://www.jstor.org/stable/20440554. doi:10.1890/1540-9295(2007)5[13:DTHEF]2.0.CO;2

Dimitrova, A. (2013). *Literature review on fundamental concepts and definitions, objectives, and policy goals as well as instruments relevant for socio-ecological transition*, (WWWforEurope Working Paper, 40.).

Dixit, R. D. (1984). A census of the Indian Pteridophytes. Flora of India Series 4. Botanical Survey of India, Howrah (Calcutta).

Dobhal, R., Kumar, A., & Rawat, S. (2011). Conservation and Management of Bioresources in Uttarakhand, India. *Community based bio-diversity conservation in the Himalay, The Energy and Resource Institute (TERI).*

Dociu, M. (2012). The Socio-Economic Impact of Urbanization. *International Journal of Academic Research in Accounting, Finance and Management Sciences, 2*(1).

DoE. (1994). *Environmental Appraisal in Government Departments.* The Stationary Office.

Dogan, E., Majeed, M. T., & Luni, T. (2022). Revisiting the nexus of ecological footprint, unemployment, and renewable and non-renewable energy for South Asian economies: Evidence from novel research methods. *Renewable Energy, 194*(1), 1060–1070. doi:10.1016/j.renene.2022.05.165

Dong, Q., Cai, J., Chen, S., He, P., & Chen, X. (2022). Spatiotemporal Analysis of Urban Green Spatial Vitality and the Corresponding Influencing Factors: A Case Study of Chengdu, China. *Land (Basel), 11*(10), 1820. doi:10.3390/land11101820

Dooren, V., Aiking, H., & Vellinga, P. (2018). In search of indicators to assess the environmental impact of diets. *The International Journal of Life Cycle Assessment, 23*(1), 1297–1314. doi:10.1007/s11367-017-1371-2

Doran, N. E., & Richardson, A. M. (2010). History of biodiversity conservation, protected areas and the conservation movement. *History and Philosophy of Science and Technology, 2*, 55–88.

Dornelas, M., Gotelli, N. J., McGill, B., Shimadzu, H., Moyes, F., Sievers, C., & Magurran, A. E. (2014). Assemblage time series reveal biodiversity change but not systematic loss. *Science, 344*(6181), 296–299. doi:10.1126/science.1248484 PMID:24744374

Doughty, C. A. (2016). Building climate change resilience through local cooperation: A Peruvian Andes case study. *Regional Environmental Change, 16*(8), 2187–2197. doi:10.1007/s10113-015-0882-2

Driver, A., Sink, K. J., Nel, J. L., Holness, S., Van Niekerk, L., Daniels, F., Jonas, Z., Majiedt, P. A., Harris, L., & Maze, K. (2012). *National Biodiversity Assessment 2011: An assessment of South Africa's biodiversity and ecosystems.*

Druilhe, C., & Garnsey, E. (2000). The incubation of Academic Spin-off Companies. *Proceedings of the High Technology Small Firm Conference 2000*, University of Twente, Enschede, The Netherlands.

Dulsat-Masvidal, M., Ciudad, C., Infante, O., Mateo, R., & Lacorte, S. (2023). Water pollution threats in important bird and biodiversity areas from Spain. *Journal of Hazardous Materials, 448*, 130938. doi:10.1016/j.jhazmat.2023.130938 PMID:36860036

Dumont, E. (2012). Estimated impact of global population growth on future wilderness extent. *Earth System Dynamics Discussions, 3*(1), 433–452. doi:10.5194/esdd-3-433-2012

Dunkley, R. A., & Franklin, A. (2017). Failing better: The stochastic art of evaluating community-led environmental action programs. *Evaluation and Program Planning, 60*, 112–122. doi:10.1016/j.evalprogplan.2016.11.005 PMID:27842275

Dunn, D. C., Ardron, J., Bax, N., Bernal, P., Cleary, J., Cresswell, I., Donnelly, B., Dunstan, P., Gjerde, K., Johnson, D., Kaschner, K., Lascelles, B., Rice, J., von Nordheim, H., Wood, L., & Halpin, P. N. (2014). The convention on biological diversity's ecologically or biologically significant areas: Origins, development, and current status. *Marine Policy, 49*, 137–145. doi:10.1016/j.marpol.2013.12.002

Duro, J. A., & Teixidó-Figueras, J. (2013). Ecological footprint inequality across countries: The role of environment intensity, income and interaction effects. *Ecological Economics, 93*(1), 34–41. doi:10.1016/j.ecolecon.2013.04.011

EC. (2002). *Trans-European Transport Network – TEN-T priority projects.* European Commission.

Eckert, S., & Kohler, S. (2014). Urbanization and health in developing countries: a systematic review. In World health & population, 15(1). doi:10.12927/whp.2014.23722

Edelman, M. (2013). Messy hectares: Questions about the epistemology of land grabbing data. *The Journal of Peasant Studies*, *40*(3), 485–501. doi:10.1080/03066150.2013.801340

Edgar, G. J., Stuart-Smith, R. D., Willis, T. J., Kininmonth, S., Baker, S. C., Banks, S., Barrett, N. S., Becerro, M. A., Bernard, A. T. F., Berkhout, J., Buxton, C. D., Campbell, S. J., Cooper, A. T., Davey, M., Edgar, S. C., Försterra, G., Galván, D. E., Irigoyen, A. J., Kushner, D. J., & Thomson, R. J. (2014). Global conservation outcomes depend on marine protected areas with five key features. *Nature*, *506*(7487), 216–220. doi:10.1038/nature13022 PMID:24499817

Edmondson, W. T. (1974). A simplified method for counting phytoplankton. In R. E. Vollenmeider (Ed.), *A manual on methods for measuring primary production in Aquatic environments* (pp. 14–16). Balckwell Sci. Pub.

EEA. (1999). Environment in European Union at Turn of the Century. *Environmental Assessment Report No.2*. European Environment Agency, Copenhagen

Egan, D., & Howell, E. A. (Eds.). (2001). The Historical Ecology Handbook: A Restorationist's Guide to Reference Ecosystems. Island Press. doi:10.5751/ES-00468-070111

Ehler, C., & Douvere, F. (2009). Marine Spatial Planning: A Step-by-Step Approach toward Ecosystem-Based Management. Intergovernmental Oceanographic Commission and Man and the Biosphere Programme, IOC Manual and Guides No. 53, ICAM Dossier No. 6. UNESCO.

Elith, J., Mark, A. B., & Helen, M. R. (2002). Mapping epistemic uncertainties and vague concepts in predictions of species distributions. *Ecological Modelling*, *157*(2–3), 313–329. doi:10.1016/S0304-3800(02)00202-8

English, S., Wilkinson, C., & Baker, V. (1997). *Survey Manual for Tropical Marine Resources. Townsville, Australia.* Australian Institute of Marine Science.

Epstein, M. J. (2010). El desafío de mejorar simultáneamente el rendimiento social y financiero: nuevos resultados de investigación, en Epstein, M.J., Manzoni, J.-F. y Davila, A. (Eds), Política pluriversal: lo real y lo possible. Duke University Press: Durham

Epstein, M. J. (2008). *Making Sustainability Work: Best Practices in Managing and Measuring Corporate Social, Environmental and Economic Impacts*. Greenleaf Publishing y Berrett-Koehler Publishers.

Essl, F., Latombe, G., Lenzner, B., Wilson, J. R. U., Genovesi, P., Pagad, S., Seebens, H., & Smith, K. (2020). The Convention on Biological Diversity (CBD)'s Post-2020 target on invasive alien species – what should it include and how should it be monitored? *NeoBiota*, *62*, 99–121. doi:10.3897/neobiota.62.53972

European Commission. (2012). *Guide to Research and Innovation Strategies for Smart Specialisations (RIS 3)*. Publications Office of the European Union.

Fagerberg, J. (2018). Mobilizing innovation for sustainability transitions: A comment on transformative innovation policy. *Research Policy*, *47*(9), 1568–1576. doi:10.1016/j.respol.2018.08.012

Fahrig, L. (2003). Effects of Habitat Fragmentation on Biodiversity. *Annual Review of Ecology, Evolution, and Systematics*, *34*(1), 487–515. doi:10.1146/annurev.ecolsys.34.011802.132419

FAO. (2010). *The State of World Fisheries and Aquaculture: 2010 Accessed March 2013*. FAO.

Feltynowski, M. (2023). Urban green spaces in land-use policy – types of data, sources of data and staff – the case of Poland. *Land Use Policy*, *127*, 106570. doi:10.1016/j.landusepol.2023.106570

Fernandes, J. P. (2000). Landscape ecology and conservation management – evaluation of alternatives in a highway EIA process. *Environmental Impact Assessment Review*, *20*(6), 665–680. doi:10.1016/S0195-9255(00)00060-3

Ferrier, S., & Guisan, A. (2006). Spatial modelling of biodiversity at the community level. *Journal of Applied Ecology*, *43*(3), 393–404. doi:10.1111/j.1365-2664.2006.01149.x

Fiala, N. (2008). Measuring sustainability: Why the ecological footprint is bad economics and bad environmental science. *Ecological Economics*, *67*(4), 519–525. doi:10.1016/j.ecolecon.2008.07.023

Figge, L., Oebels, K., & Offermans, A. (2016). The effects of globalization on ecological footprints: An empirical analysis. *Environment, Development and Sustainability*, *19*(1), 863–876. doi:10.1007/s10668-016-9769-8

Fischer, F. (2000). Citizen participation and the democratization of policy expertise: From theoretical inquiry to practical cases. [DOI: 10.1023/A:1004819900491]. *Policy Sciences*, *33*(3-4), 283–320.

Fischer, J., Lindenmayer, D. B., & Manning, A. D. (2006). Biodiversity, ecosystem function, and resilience: Ten guiding principles for commodity production landscapes. *Frontiers in Ecology and the Environment*, *4*(2), 80–86. doi:10.1890/1540-9295(2006)004[0080:BEFART]2.0.CO;2

Fiske, S. J., Crate, S. A., Crumley, C., Galvin, K., Lazrus, H., Lucero, L., Oliver-Smith, A., Orlove, B., Strauss, S., & Wilk, R. R. (2014). *Cambiando la atmósfera. Antropología y cambio climático. Informe final del grupo de trabajo sobre el cambio climático global de la AAA*. Asociación Americana de Antropología.

Fox, N. J. y Alldred, P. (2020). La sostenibilidad, el humanismo feminista y las capacidades inusuales de los humanos (post). *Sociología ambiental*, *6*(2), 121–131. doi:10.1080/23251042.2019.1704480

Fox, S. (2012). Urbanization as a global historical process: Theory and evidence from sub-saharan africa. *Population and Development Review*, *38*(2), 285–310. doi:10.1111/j.1728-4457.2012.00493.x

Françoso, R. D., Brandão, R., Nogueira, C. C., Salmona, Y. B., Machado, R. B., & Colli, G. R. (2015). Habitat loss and the effectiveness of protected areas in the Cerrado Biodiversity Hotspot. *Natureza & Conservação*, *13*(1), 35–40. doi:10.1016/j.ncon.2015.04.001

Freitas, C. T., Lopes, P. F., Campos-Silva, J. V., Noble, M. M., Dyball, R., & Peres, C. A. (2020). Co-management of culturally important species: A tool to promote biodiversity conservation and human well-being. *People and Nature*, *2*(1), 61–81. doi:10.1002/pan3.10064

Friedmann, J. (1966). Two Concepts of urbanization: A Comment. *Urban Affairs Quarterly*, *1*(4), 78–84. doi:10.1177/107808746600100406

Fuller, R. A., Irvine, K. N., Devine-Wright, P., Warren, P. H., & Gaston, K. J. (2007). Psychological benefits of greenspace increase with biodiversity. *Biology Letters*, *3*(4), 390–394. doi:10.1098/rsbl.2007.0149 PMID:17504734

Galbally, I., Meyer, C. P., Wang, Y.-P., & Kirstine, W. (2010). Soil–atmosphere exchange of CH4, CO, N2O and NOx and the effects of land-use change in the semiarid Mallee system in Southeastern Australia. *Global Change Biology*, *16*(9), 2407–2419. doi:10.1111/j.1365-2486.2010.02161.x

Galea, S., & Vlahov, D. (2005). URBAN HEALTH: Evidence, Challenges, and Directions. *Annual Review of Public Health*, *26*(1), 341–365. doi:10.1146/annurev.publhealth.26.021304.144708 PMID:15760293

Galli, A., Iha, K., Pires, S. M., Mancini, M. S., Alves, A. A., Zakai, G., Lin, D., Murthy, A., & Wackernagel, M. (2020). Assessing the ecological footprint and biocapacity of portuguese cities: Critical results for environmental awareness and local management. *Cities (London, England)*, *96*(1), 1–13. doi:10.1016/j.cities.2019.102442

Gamble, J. S. (1924). *The Flora of Presidency of Madras 2*. Botanical Survey of India.

Gaston, K. J., & Spicer, J. I. (2013). *Biodiversity: An introduction*. John Wiley & Sons. https://books.google.com/books?hl=en&lr=&id=0Bjp2o5CVnQC&oi=fnd&pg=PT6&dq=Biodiversity:+An+overview&ots=Y2X6YSmcEx&sig=KHEre9jRGafSgjzx7GfRWb1tK-w

Gavin, M. C., McCarter, J., Berkes, F., Mead, A. T. P., Sterling, E. J., Tang, R., & Turner, N. J. (2018). Effective biodiversity conservation requires dynamic, pluralistic, partnership-based approaches. *Sustainability (Basel)*, *10*(6), 1846. doi:10.3390/su10061846

GCB. (2020). *Global carbon budget report*. GCB. https://essd.copernicus.org/articles/12/3269/2020/essd-12-3269-2020.pdf

Geels, F. (2002). Technological transitions as evolutionary reconfiguration processes: A multilevel perspective and a case-study. *Research Policy*, *31*(8-9), 1257–1274. doi:10.1016/S0048-7333(02)00062-8

Geneletti, D., Beinat, E., Chung, C. F., Fabbri, A. G., & Scholten, H. J. (2003). Accounting for uncertainty factors in biodiversity impact assessment: Lessons from a case study. *Environmental Impact Assessment Review*, *20*(6), 665–680. doi:10.1016/S0195-9255(03)00045-3

Gerwing, T. G., Hawkes, V. C., Gann, G. D., & Murphy, S. D. (2022). Restoration, reclamation, and rehabilitation: on the need for, and positing a definition of, ecological reclamation. *Restoration Ecology*, *30*(7), 0–2. doi:10.1111/rec.13461

Ghosh, S. R., Ghosh, B., Biswas, A., & Ghosh, R. K. (2004). The Pteridophytic Flora of Eastern India. Flora of India Series 4. Botanical Survey of India, Kolkata.

Gibson, C., Head, L., & Carr, C. (2015). From incremental change to radical disjuncture: Rethinking everyday household sustainability practices as survival skills. *Annals of the Association of American Geographers*, *105*(2), 416–424. doi:10.1080/00045608.2014.973008

Girma, Y., Terefe, H., Pauleit, S., & Kindu, M. (2019). Urban green spaces supply in rapidly urbanizing countries: The case of Sebeta Town, Ethiopia. *Remote Sensing Applications: Society and Environment*, *13*, 138–149. doi:10.1016/j.rsase.2018.10.019

Gizachew, G. T. (2021). Spatial-temporal and factors influencing the distribution of biodiversity: A Review. *Scientific Reports in Life Sciences*, *2*(4), 1–19.

Glasson, J., Therivel, R., & Chadwick, A. (2012). *Introduction to environmental impact assessment*. Routledge.

Glasson, J., Therivel, R., & Chadwick, A. (2012). *Introduction to Environmental Impact Assessment: Principles and Procedures, Process, Practice and Prospects*. Routledge.

Global Biodiversity. (n.d.). National Geographic. https://education.nationalgeographic.org/resource/global-biodiversity

Godfrey, R., & Julien, M. (2005). Urbanisation and health. In Clinical Medicine, Journal of the Royal College of Physicians of London, 5(2). doi:10.7861/clinmedicine.5-2-137

Goldfarb, K. E. (2016). Coming to Look Alike: Materializing Affinity in Japanese Foster and Adoptive Care. *Social Analysis*, *60*(2), 47–64. doi:10.3167/sa.2016.600204

Gómez Hinojosa, J. F. (1990). De la ecología a la ecofilía. *Apuntes para una ecología liberadora. Pasos (El Sauzal)*, *30*(July–August).

González Álvarez, M. D. (1997). *Processos de Planejamento nos Pólos Tecnológicos: um enfoque adaptativo. Tese de Doutorado, Departamento de Engenharia Industrial*. PUC-Rio.

Gonzalez, A., Cardinale, B. J., Allington, G. R., Byrnes, J., Arthur Endsley, K., Brown, D. G., Hooper, D. U., Isbell, F., O'Connor, M. I., & Loreau, M. (2016). Estimating local biodiversity change: A critique of papers claiming no net loss of local diversity. *Ecology*, *97*(8), 1949–1960. doi:10.1890/15-1759.1 PMID:27859190

Goodchild, M. F. (2007). Citizens as sensors: The world of volunteered geography. In GeoJournal. doi:10.1007/s10708-007-9111-y

Google Académico Gray, R. (2006). ¿Informes sociales, ambientales y de sostenibilidad y creación de valor organizacional? ¿Valor de quién? ¿De quién es la creación? *Accounting, Auditing & Accountability Journal*, *19*(6), 793–819.

Gorenflo, L. J., & Romaine, S. (2021). Linguistic diversity and conservation opportunities at UNESCO World Heritage Sites in Africa. *Conservation Biology*, *35*(5), 1426–1436. doi:10.1111/cobi.13693 PMID:33448452

Gouverneur, J., & Netzer, N. (2014). Take the Wheel and Steer! Trade Unions and the Just Transition. In *State of the World 2014. State of the World.* Island Press., doi:10.5822/978-1-61091-542-7_21

Gow, L. (1998). *New Zealandís approach to strategic environmental assessment, paper to international seminar on strategic environmental appraisal.*

Gray, R. (2010). ¿Es la contabilidad de la sostenibilidad realmente responsable de la sostenibilidad? ¿Y cómo lo sabremos? Una exploración de las narrativas de las organizaciones y el planeta. *Contabilidad, organizaciones y sociedad*, *35*(1), 47–62

Green, E. P., Mumby, P. J., Edwards, A. E., & Clark, C. D. (2000). *Remote Sensing: Handbook for Tropical Coastal Management.* UNESCO Publishing.

Griffiths, H. I., Kryštufek, B., & Reed, J. M. (Eds.). (2004). *Balkan Biodiversity.* Springer Netherlands., doi:10.1007/978-1-4020-2854-0

Grinberger, A. Y., Minghini, M., Juhász, L., Yeboah, G., & Mooney, P. (2022). OSM Science—The Academic Study of the OpenStreetMap Project, Data, Contributors, Community, and Applications. In ISPRS International Journal of Geo-Information, 11(4). doi:10.3390/ijgi11040230

Groom, M. J., Meffe, G. K., & Carroll, C. R. (2006). *Principles of Conservation Biology* (3rd ed.). Sinauer Associates.

Grumbine, R. E. (1994). What is ecosystem management? *Conservation Biology*, *8*(1), 27–38. doi:10.1046/j.1523-1739.1994.08010027.x

Guisan, A., & Zimmermann, N. E. (2000). Predictive habitat distribution models in ecology. *Ecological Modelling*, *135*(2–3), 147–186. doi:10.1016/S0304-3800(00)00354-9

Guxens, M., Lubczyńska, M. J., Muetzel, R. L., & Dalmau-Bueno, A., Jaddoe, Vincent W., Hoek, G., et al. (2018). Exposición a la contaminación del aire durante la vida fetal, morfología cerebral y función cognitiva en niños en edad escolar. *Psiquiatría Biológica : Publicación Oficial de la Sociedad Española de Psiquiatría Biológica*, *84*(4).

Haberl, H., Winiwarter, V., Andersson, K., Ayres, R. U., Boone, C., Castillo, A., Cunfer, G., Fischer-Kowalski, M., Freudenburg, W. R., Furman, E., Kaufmann, R., Krausmann, F., Langthaler, E., Lotze-Campen, H., Mirtl, M., Redman, C. L., Reenberg, A., Wardell, A., Warr, B., & Zechmeister, H. (2006). From LTER to LTSER: Conceptualizing the socio-economic dimension of long-term socio-ecological research. *Ecology and Society*, *11*(2), 13. www.ecologyandsociety.org/vol11/iss2/art13/. doi:10.5751/ES-01786-110213

Habibullah, M. S., Din, B. H., Tan, S. H., & Zahid, H. (2022). Impact of climate change on biodiversity loss: Global evidence. *Environmental Science and Pollution Research International*, *29*(1), 1073–1086. doi:10.1007/s11356-021-15702-8 PMID:34341937

Haklay, M. (2013). Citizen science and volunteered geographic information: Overview and typology of participation. In Crowdsourcing Geographic Knowledge: Volunteered Geographic Information (VGI) in Theory and Practice. Springer. doi:10.1007/978-94-007-4587-2_7

Hald-Mortensen, C. (2023). The Main Drivers of Biodiversity Loss: A Brief Overview. *Journal of Ecology and Natural Resources*, *7*(3), 000346. doi:10.23880/jenr-16000346

Halpern, B. S., Walbridge, S., Selkoe, K. A., Kappel, C. V., Micheli, F., D'Agrosa, C., & Watson, R. (2008). A global map of human impact on marine ecosystems. *Science*, *319*(5865), 948–952. doi:10.1126/science.1149345 PMID:18276889

Hamraie, A. (2017). *Acceso a edificios: diseño universal y la política de la discapacidad*. University of Minnesota Press.

Han, J., & Dalaibaatar, E. (2023). A study on the influencing factors of China's ecological footprint based on EEMD-GaoDetector. *Sustainability (Basel)*, *15*(8), 1–15. doi:10.3390/su15086680

Hansen, M. C., & Loveland, T. R. (2012). A review of large area monitoring of land cover change using Landsat data. *Remote Sensing of Environment*, *122*, 66–74. doi:10.1016/j.rse.2011.08.024

Haraway, D. J. (2016). *Staying with the trouble: making kin en el chthulucene*. Duke University Press.

Haraway, D., Ishikawa, N., Gilbert, S. F., Olwig, K., & Tsing, A. L. (2016). Los antropólogos están hablando – sobre el Antropoceno. *Ethnos*, *81*(3), 535–564. doi:10.1080/00141844.2015.1105838

Harekrushna, S., & Kumar, D.C. (2012). A review on bioremediation. *International Journal of chemistry and environment, 2*(1), 13-21.

Harel, F., Bourcier, A., & Skriabine, P. (2003). A French Methodological and Exploratory Study of SEA at the Regional Scale [newsletter published by Transport Research Laboratory, UK]. *SEA and Transport Planning, Issue*, (5), 34–35.

Harfoot, M. B., Johnston, A., Balmford, A., Burgess, N. D., Butchart, S. H., Dias, M. P., Hazin, C., Hilton-Taylor, C., Hoffmann, M., & Isaac, N. J. (2021). Using the IUCN Red List to map threats to terrestrial vertebrates at global scale. *Nature Ecology & Evolution, 5*(11), 1510-1519. doi:10.1038/s41559-021-01542-9

Haward, M. (2021). Biodiversity in Areas Beyond National Jurisdiction (BBNJ): The Commission for the Conservation of Antarctic Marine Living Resources (CCAMLR) and the United Nations BBNJ agreement. *The Polar Journal*, *11*(2), 303–316. doi:10.1080/2154896X.2021.1984658

Heather-Clark, S. (2002). *Strategic Integrated Port Planning: Moving from EIA to SEA*. Kuwait Conference, Kuwait.

Hedblom, M., Gunnarsson, B., Iravani, B., Knez, I., Schaefer, M., Thorsson, P., & Lundström, J. N. (2019). Reduction of physiological stress by urban green space in a multisensory virtual experiment. *Scientific Reports*, *9*(1), 1–11. doi:10.1038/s41598-019-46099-7 PMID:31300656

Heets, V. M., & Seminozhenko, V. P. (2006). Innovative Prospects of Ukraine, Kh.: Constant.

Heikinheimo, V., Tenkanen, H., Bergroth, C., Järv, O., Hiippala, T., & Toivonen, T. (2020). Understanding the use of urban green spaces from user-generated geographic information. *Landscape and Urban Planning*, *201*, 103845. doi:10.1016/j.landurbplan.2020.103845

Heino, J. (2009). Biodiversity of aquatic insects: Spatial gradients and environmental correlates of assemblage-level measures at large scales. *Freshwater Reviews*, *2*(1), 1–29. doi:10.1608/FRJ-2.1.1

Henderson, R. (1994). Managing Innovation in the Information Age. *Harvard Business Review*, *72*(01), 100–107.

Hermoso, V., Carvalho, S., Giakoumi, S., Goldsborough, D., Katsanevakis, S., Leontiou, S., Markantonatou, V., Rumes, B., Vogiatzakis, I., & Yates, K. (2022). The EU Biodiversity Strategy for 2030: Opportunities and challenges on the path towards biodiversity recovery. *Environmental Science & Policy*, *127*, 263–271. doi:10.1016/j.envsci.2021.10.028

Hiby, A. R. and Hammond, P. S. (1989). *Survey techniques for estimating abundance of ceta- ceans*. Rep. Int. Whal. Commn.

Hilbert, D. W., Hill, R., Moran, C., Turton, S. M., Bohnet, I., Marshall, N. A., Pert, P. L., Stoeckl, N., Murphy, H. T., & Reside, A. E. (2014). *Climate change issues and impacts in the Wet Tropics NRM cluster region*. James Cook University.

Hill, J., & Wilkinson, C. (2004). *Methods for Ecological Monitoring of Coral Reefs (Version 1): A Resource for Managers*. Australian Institute of Marine Science.

Hilty, J., Merenlender, A. M., & Daily, G. C. (Eds.). (2018). *The Ecology of Climate Change: Responses and Consequences of Natural and Managed Ecosystems*. Academic Press.

Hobbs, R. J., Higgs, E., Hall, C. M., & Carol, M. (2014). Managing the whole landscape: Historical, hybrid, and novel ecosystems. *Frontiers in Ecology and the Environment*, *12*(10), 557–564. doi:10.1890/130300

Hoegh-Guldberg, O., Mumby, P. J., Hooten, A. J., Steneck, R. S., Greenfield, P., Gomez, E., Harvell, C. D., Sale, P. F., Edwards, A. J., Caldeira, K., Knowlton, N., Eakin, C. M., Iglesias-Prieto, R., Muthiga, N., Bradbury, R. H., Dubi, A., & Hatziolos, M. E. (2007). Coral reefs under rapid climate change and ocean acidification. *Science*, *318*(5857), 1737–1742. doi:10.1126/science.1152509 PMID:18079392

Hoffmann, J. A. (2020). Memoria, vías fluviales e identidad ecocultural. En The Routledge Handbook of Ecocultural Identity; Milstein, T., Castro-Sotomayor, J., Eds.; Routledge: Abingdon, Reino Unido.

Hoke, M. K., & Schell, L. M. (2020). Haciendo antropología biocultural: *Continuidad y cambio. American Journal of Human Biology*, *32*, E23471. doi:10.1002/ajhb.23471 PMID:32681558

Hollands, R. G. (2008). Will the real smart city please stand up? Intelligent, progressive, or entrepreneurial? *City (London, England)*, *12*(3), 303–320. doi:10.1080/13604810802479126

Holt, D. B. (2012). Construyendo un consumo sostenible: De los valores éticos a la transformación cultural de mercados insostenibles. *Los ANNALS de la Academia Americana de Ciencias Políticas y Sociales*, *644*(1), 236–255. doi:10.1177/0002716212453260

Hominick, W. M., Reid, A. P., Bohan, D. A., & Briscoe, B. R. (1996). Entomopathogenic Nematodes: Biodiversity, Geographical Distribution and the Convention on Biological Diversity. *Biocontrol Science and Technology*, *6*(3), 317–332. doi:10.1080/09583159631307

Honarparvar, S., Forouzandeh Jonaghani, R., Alesheikh, A. A., & Atazadeh, B. (2019). Improvement of a location-aware recommender system using volunteered geographic information. *Geocarto International*, *34*(13), 1496–1513. doi:10.1080/10106049.2018.1493155

Hooper, L. V., Midtvedt, T., & Gordon, J. I. (2002). How host-microbial interactions shape the nutrient environment of the mammalian intestine. *Annual Review of Nutrition*, *22*(1), 283–307. doi:10.1146/annurev.nutr.22.011602.092259 PMID:12055347

Hopwood, A. G. (2009). Contabilidad y medio ambiente Contabilidad. Organizaciones y Sociedad, 2009, vol. 34, números 3-4, 433-439

Hopwood, A. G. (2010). In memoriam, Anthony G. Hopwood, 1944-2010. *Accounting, Organizations and Society*, *35*(4), 496–497. doi:10.1016/j.aos.2010.05.002

Hornborg, A. (2019). *Naturaleza, sociedad y justicia en el Antropoceno: desentrañando el complejo dinero-energía-tecnología; Cambridge University Press*. Reino Unido., doi:10.1017/9781108554985

Hoyle, H., Jorgensen, A., & Hitchmough, J. D. (2019). What determines how we see nature? Perceptions of naturalness in designed urban green spaces. *People and Nature*, *1*(2), 167–180. doi:10.1002/pan3.19

Huang, C., Yang, J., Clinton, N., Yu, L., Huang, H., Dronova, I., & Jin, J. (2021). Mapping the maximum extents of urban green spaces in 1039 cities using dense satellite images. *Environmental Research Letters*, *16*(6), 064072. doi:10.1088/1748-9326/ac03dc

Huang, H., Yao, X. A., Krisp, J. M., & Jiang, B. (2021). Analytics of location-based big data for smart cities: Opportunities, challenges, and future directions. *Computers, Environment and Urban Systems*, *90*, 101712. doi:10.1016/j.compenvurbsys.2021.101712

Huang, Y., Lin, T., Zhang, G., Zhu, Y., Zeng, Z., & Ye, H. (2021). Spatial patterns of urban green space and its actual utilization status in China based on big data analysis. *Big Earth Data*, *5*(3), 391–409. doi:10.1080/20964471.2021.1939990

Hubbard, G. (2008). Beyond Accounting: Assessing the Impact of Sustainability Reporting on Tomorrow's Business. *University of Adelaide Business School, mimeo*.

Hughes, T. P., Baird, A. H., Bellwood, D. R., Card, M., Connolly, S. R., Folke, C., & Nystrom, M. (2003). Climate change, human impacts, and the resilience of coral reefs. *Science*, *301*(5635), 929–933. doi:10.1126/science.1085046 PMID:12920289

Hult, G., & Tomás, M. (2011). Sostenibilidad centrada en el mercado: ¡Orientación al mercado plus! *Revista de la Academia de Ciencias de Marketing*, *29*(febrero), 1–6. doi:10.1007/s11747-010-0223-4

Hunter, W. W. (1879). *Statlstlcal Account of Assani* (Vol. II). Trubner and Co.

Hutto, R. L., Pletsechel, S. M., & Hendrick, P. (1986). A fixed radius point count method for non-breeding season use. *The Auk*, *103*(3), 593–602. doi:10.1093/auk/103.3.593

IEA. (2018). *Energy information administration. International Energy Outlook*. Paris, France: US Department of Energy. http://energyatlas.iea. org/#!/profile/WORLD/IND

Índice de sostenibilidad (2010), índices de sostenibilidad Dow Jones en colaboración con Robeco SAM, disponible en: *J. Peasant Stud.*, *44*(6). doi:10.1080/03066150.2017.1308353

Intergovernmental Panel on Climate Change (IPCC). (2019). *Special report on climate change and land*. IPCC. https://www.ipcc.ch/srccl/

IPBES. (2019). *Global assessment report on biodiversity and ecosystem services of the Intergovernmental Science-Policy Platform on Biodiversity and Ecosystem Services* (E. S. Brondizio, J. Settele, S. Díaz, & H. T. Ngo, Eds.). IPBES Secretariat.

IUCN. (1994). IUCN Red List Categories. IUCN Species Survival Commission.

IUCN. (2001). IUCN Red List Categories and Criteria: Version 3.1. IUCN Species Survival Commission.

IUCN. (2003). Guidelines for Application of IUCN Red List Criteria at Regional Levels: Version 3.0. IUCN Species Survival Commission.

IUCN. (2008). *Red List of Threatened Species*. IUCN. (www.iucnredlist.org)

IUCN. (2010). *Guidelines for Using the IUCN Red List Categories and Criteria, version 8.1 (August 2010), prepared by the Standards and Petitions Subcommittee of the IUCN Species Survival Commission*. IUCN. http://intranet.iucn.org/webfiles/doc/SSC/RedList/ RedListGuidelines.pdf

IUCN. (2019). *Guidelines for Using the IUCN Red List Categories and Criteria. Version 14*. Standards and Petitions Subcommittee. https://www.iucnredlist.org/resources/redlistguidelines

Jabbar, M., Yusoff, M. M., & Shafie, A. (2022). Assessing the role of urban green spaces for human well-being: a systematic review. In GeoJournal, 87(5). doi:10.1007/s10708-021-10474-7

Jain, S. K. (1992). The Problem of Endangered Species: Concepts, Problems and Solutions. In Tropical Ecosystems: Ecolosv and Management (Eds. K.P.Singh and J.S.Singh.), Iiley Eastern limited.

Jain, S. K., & Sastry, A. R. K. (1984). Safeguarding Plant diversity in threatened Natural Habitats. In A. Hall (ed.) Conservation of Threatened Natural Habitats. African nat. Sci. Prog. Report. 92.

Jain, S. K. (1968). *Medicinal Plants Nation Book Trust, New Delhi. Jain, S.K. 1983. Rare and Endangered Specles: Observation on rare, imperfectly known endemic plants. In the sacred groves of Western Maharashtra*. Bot. Sur of India.

Jain, S. K. (1991). *Dictionary of Indian folk medicine and ethnobotany*. Deep publications.

Jain, S. K., & Rao, R. K. (1983). *An assessment of threatened plants of India*. Bot. Surv. of India.

Jain, S. K., & Sastry, A. R. K. (1980). *Threatened plants of India - A State of the Alf Report Bot*. Surv. of India.

Jambeck, J. R., Geyer, R., Wilcox, C., Siegler, T. R., Perryman, M., Andrady, A., Narayan, R., & Law, K. L. (2015). Plastic waste inputs from land into the ocean. *Science*, *347*(6223), 768–771. doi:10.1126/science.1260352 PMID:25678662

Jaureguiberry, P., Titeux, N., Wiemers, M., Bowler, D. E., Coscieme, L., Golden, A. S., Guerra, C. A., Jacob, U., Takahashi, Y., Settele, J., Díaz, S., Molnár, Z., & Purvis, A. (2022). The direct drivers of recent global anthropogenic biodiversity loss. *Science Advances*, *8*(45). doi:10.1126/sciadv.abm9982 PMID:36351024

Jensen, M. B., Johnson, B., Lorenz, E., & Lundvall, B. Å. (2007). Forms of knowledge and modes of innovation. *Research Policy*, *36*(5), 680–693. doi:10.1016/j.respol.2007.01.006

Jetz, W., McPherson, J. M., & Guralnick, R. P. (2012). Integrating biodiversity distribution knowledge: Toward a global map of life. *Trends in Ecology & Evolution*, *27*(3), 151–159. doi:10.1016/j.tree.2011.09.007 PMID:22019413

Jiao, Z. (2024). The Application of Remote Sensing Techniques in Ecological Environment Monitoring. *Highlights in Science. Engineering and Technology*, *81*, 449–455.

Jo, H., Kim, D.-K., Park, K., & Kwak, I.-S. (2019). Discrimination of spatial distribution of aquatic organisms in a coastal ecosystem using eDNA. *Applied Sciences (Basel, Switzerland)*, *9*(17), 3450. doi:10.3390/app9173450

Johnson, C. N., Balmford, A., Brook, B. W., Buettel, J. C., Galetti, M., Guangchun, L., & Wilmshurst, J. M. (2017). Biodiversity losses and conservation responses in the Anthropocene. *Science*, *356*(6335), 270–275. doi:10.1126/science.aam9317 PMID:28428393

Jonas, H. D., Lee, E., Jonas, H. C., Matallana-Tobon, C., Wright, K. S., Nelson, F., & Enns, E. (2017). Will "other effective area-based conservation measures" increase recognition and support for ICCAs. *Parks*, *23*(2), 63–78. doi:10.2305/IUCN.CH.2017.PARKS-23-2HDJ.en

Joppa, L. N., Visconti, P., Jenkins, C. N., & Pimm, S. L. (2013). Achieving the Convention on Biological Diversity's Goals for Plant Conservation. *Science*, *341*(6150), 1100–1103. doi:10.1126/science.1241706 PMID:24009391

Jungalwalla, N. (1968). Urbanization and health. *Israel Journal of Medical Sciences*, *4*(3). doi:10.4038/jccpsl.v14i1.2941 PMID:5698240

Juujärvi, S., & Pesso, K. (2013). Actor Roles in an Urban Living Lab: What Can We Learn from Suurpelto, Finland? *Technology Innovation Management Review*, *11*(3), 22–27. http://timreview.ca/article/742. doi:10.22215/timreview/742

Jyoti, B., & Singh, A. K. (2023). Does digitalization have a causal relationship with economic development? An experience from a country-wise panel data statistical analysis. In R. S. Nazlı & G. Sari (Eds.), *Perspectives on Society and Technology Addiction* (pp. 34–61). IGI Global., doi:10.4018/978-1-6684-8397-8.ch003

Jyoti, B., Singh, A. K., & Ashraf, S. N. (2023). Appropriate technologies and their implications in the agricultural sector. In P. H. Alex Khang (Ed.), *Advanced Technologies and AI-Equipped IoT Applications in High-Tech Agriculture* (pp. 65–87). IGI Global., doi:10.4018/978-1-6684-9231-4.ch004

Kabir, S. M. Z., Rabbi, F., & Chowdhury, M. B. (2015). Mine Closure Planning and Practice in Canada and Australia : A Comparative Review. *World Review of Business Research*, *5*(3), 1–22.

Kaliaperumal, N., & Kalimuthu, S. (1997). *Seaweed potential and its exploitation in India. Sea- weed Res. Utiln. 19(1&2):33-40. Koskimies, P. and Väisänen, R.A. 1991. Monitoring bird populations*. Zoological Museum, Finnish.

Kareiva, P., Groves, C., & Marvier, M. (2014). The evolving linkage between conservation science and practice at T he N ature C onservancy. *Journal of Applied Ecology*, *51*(5), 1137–1147. doi:10.1111/1365-2664.12259 PMID:25641980

Karlsson, M., & Edvardsson Björnberg, K. (2021). Ethics and biodiversity offsetting. *Conservation Biology*, *35*(2), 578–586. doi:10.1111/cobi.13603 PMID:32757221

Karvonen, A., Cugurullo, F., & Caprotti, F. (Eds.). (2019). *Inside Smart Cities: Place, Politics and Urban Innovation*. Routledge.

Kasperowicz, R. (2015). Economic growth and CO2 emissions: The ECM analysis. *Journal of International Students*, *8*(3), 91–98. https://www.jois.eu/files/07_Kasperowicz.pdf

Kazemzadeh, E., Fuinhas, J. A., Salehnia, N., Koengkan, M., & Silva, N. (2023). Assessing influential factors for ecological footprints: A complex solution approach. *Journal of Cleaner Production*, *414*(15), 32–43. doi:10.1016/j.jclepro.2023.137574

Kehoe, R., Frago, E., & Sanders, D. (2021). Cascading extinctions as a hidden driver of insect decline. In Ecological Entomology, 46(4). doi:10.1111/een.12985

Kempera, J. A., & Ballantine, P. W. (2019). ¿Qué entendemos por marketing de sostenibilidad? *Journal of Marketing Management*, *35*(3–4), 277–309. doi:10.1080/0267257X.2019.1573845

Kemp, R., Loorbach, D., & Rotmans, J. (2007). Transition management as a model for managing processes of co-evolution towards sustainable development. *International Journal of Sustainable Development and World Ecology*, *14*(1), 78–91. doi:10.1080/13504500709469709

Kenney, M. (Ed.). (2000). *Understanding Silicon Valley: the anatomy of an entrepreneurial region*. Stanford University Press. doi:10.1515/9781503618381

Kevin, J. Gaston & John I. Spicer. (n.d.). *Biodiversity: An Introduction, 2nd Edition | Wiley* (2nd ed.). Wiley. https://www.wiley.com/en-in/Biodiversity:+An+Introduction,+2nd+Edition-p-9781405118576

Kholia, B. S., & Bhakuni, K. (2009). Western Himalaya a new range of distribution for a critically endangered fern, Dryopsis manipurensis (Bedd.) Holttum et P. J. Edwards. Nelumbo. *Bulletin of the Botanical Survey of India*, *51*, 245–248.

Kiehn, M., Fischer, F., & Smith, P. P. (2021). The Nagoya protocol and access and benefit sharing regulations of the Convention on Biological Diversity (CBD) and its impacts on botanic gardens' collections and research. *Perspectives in Agriculture, Veterinary Science, Nutrition and Natural Resources, 16*(34), PAVSNNR202116034. doi:10.1079/PAVSNNR202116034

Kirikkaleli, D., Adebayo, T. S., Khan, Z., & Ali, S. (2021). Does globalization matter for ecological footprint in Turkey? Evidence from dual adjustment approach. *Environmental Science and Pollution Research International, 28*(1), 14009–14017. doi:10.1007/s11356-020-11654-7 PMID:33201501

Kitchin, R., Lauriault, T. P., & Wilson, M. W. (2018). Understanding Spatial Media. In Understanding Spatial Media. Sage. doi:10.4135/9781526425850.n1

Kitchin, R. (2014). The real-time city? Big data and smart urbanism. *GeoJournal, 79*(1), 1–14. doi:10.1007/s10708-013-9516-8

Kitzes, J., Wackernagel, M., Loh, J., Peller, A., Goldfinger, S., Cheng, D., & Tea, K. (2008). Shrink and share: Humanity's present and future ecological footprint. *Philosophical Transactions of the Royal Society of London. Series B, Biological Sciences, 363*(1491), 467–475. doi:10.1098/rstb.2007.2164 PMID:17652075

Kızılgöl, Ö., & Öndes, H. (2022). Factors affecting the ecological footprint: A study on the OECD countries. *The Science of the Total Environment, 849*(25), 34–45. doi:10.1016/j.scitotenv.2022.157757 PMID:35926619

Klabbers, J. H. G., Swart, R. J., Janssen, R., Van Vellinga, P., & van Ulden, A. P. (1996). Climate science and climate policy: Improving the science/policy interface. *Mitigation and Adaptation Strategies for Global Change, 1*(1), 73–93. doi:10.1007/BF00625616

Klebanova, T.S. & Kizima, N.A. (2010). *Models of estimation, analysis and forecasting of social and economic systems.* Monogr. Kh.: FOP Liborkina L.M., VD "INZHEK".

Kleiner, G. B. (2013). System Economics and System-Oriented Modeling. *Economics and Mathematical Methods, 3*, 71–93.

Koehler, T., Weber, O., Fenchel, M., & Scholtz, R. (2005). Principles for sustainability rating of investment of funds. *Business Strategy and the Environment, 14*(1), 54–70. doi:10.1002/bse.423

Kolawole, A. S., & Iyiola, A. O. (2023). *Environmental Pollution: Threats, Impact on Biodiversity, and Protection Strategies.* doi:10.1007/978-981-19-6974-4_14

Kosnik, M. B., Schuwirth, N., & Rico, A. (2024). Harnessing Computational Methods to Characterize Chemical Impacts on Biodiversity. *Environmental Science & Technology Letters, 11*(3), 185–194. doi:10.1021/acs.estlett.3c00865

Kothencz, G., Kolcsár, R., Cabrera-Barona, P., & Szilassi, P. (2017). Urban green space perception and its contribution to well-being. *International Journal of Environmental Research and Public Health, 14*(7), 766. doi:10.3390/ijerph14070766 PMID:28704969

Krabbendam, L., & Van Os, J. (2005). Schizophrenia and Urbanicity : A Major Environmental Influence — Conditional on Genetic Risk. *Schizophrenia Bulletin, 31*(4), 795–799. doi:10.1093/schbul/sbi060 PMID:16150958

Kramer, J. M. (2021). De depredador a parásito: En la propiedad privada y nuestro desastre ecológico. *Journal of Economic Issues, 55*(junio), 416–422. doi:10.1080/00213624.2021.1908804

Krausmann, F., Fischer-Kowalski, M., Grunbuhel, C., & Krausmann, F. (2008). Socio-ecological regime transitions in Austria and the United Kingdom. *Ecological Economics, 65*(1), 187–201. doi:10.1016/j.ecolecon.2007.06.009

Kühl, H. S., Bowler, D. E., Bösch, L., Bruelheide, H., Dauber, J., Eichenberg, D., Eisenhauer, N., Fernández, N., Guerra, C. A., Henle, K., Herbinger, I., Isaac, N. J. B., Jansen, F., König-Ries, B., Kühn, I., Nilsen, E. B., Pe'er, G., Richter, A., Schulte, R., & Bonn, A. (2020). Effective Biodiversity Monitoring Needs a Culture of Integration. *One Earth*, *3*(4), 462–474. doi:10.1016/j.oneear.2020.09.010

Kuhn, T. S. (1962). *La estructura de las revoluciones científicas*. Prensa de la Universidad de Chicago.

Kujala, H., Burgman, M. A., & Moilanen, A. (2013). Treatment of uncertainty in conservation under climate change. *Conservation Letters*, *6*(2), 73–85. doi:10.1111/j.1755-263X.2012.00299.x

Kulick, D. (2019). Una muerte en la selva: cómo un idioma y una forma de vida llegaron a su fin en Papúa Nueva Guinea. Algonquin Books: Chapel Hill, NC, USA.

Kullberg, P., & Moilanen, A. (2014). How do recent spatial biodiversity analyses support the convention on biological diversity in the expansion of the global conservation area network? *Natureza & Conservação*, *12*(1), 3–10. doi:10.4322/natcon.2014.002

Kumar Ashok and Srivastava Meena. (2012). Diversity of medicinal Plants in Uttarakhand and their conservation Strategy with special reference to Orchids, *In*: Proceeding of National Conference on Environementla Health: Challaneges and Management. PG College Rajajipuram, Lucknow.

Kumar, A. (2022). 4. Global Environment Facility (GEF). *Yearbook of International Environmental Law*, *31*(1), 296–301. doi:10.1093/yiel/yvab055

Kumar, A., Ahmad, M. M., & Sharma, P. (2017). Influence of climatic and non-climatic factors on sustainable food security in India: A statistical investigation. *International Journal of Sustainable Agricultural Management and Informatics*, *3*(1), 1–30. doi:10.1504/IJSAMI.2017.082917

Kumar, A., Sharma, P., & Ambrammal, S. K. (2015). Climatic effects on sugarcane productivity in India: A stochastic production function application. *International Journal of Economics and Business Research*, *10*(2), 179–203. doi:10.1504/IJEBR.2015.070984

Kusumoto, B., Costello, M. J., Kubota, Y., Shiono, T., Wei, C., Yasuhara, M., & Chao, A. (2020). Global distribution of coral diversity: Biodiversity knowledge gradients related to spatial resolution. *Ecological Research*, *35*(2), 315–326. doi:10.1111/1440-1703.12096

Lachman, D. (2013). A survey and review of approaches to study transitions. *Energy Policy*, *58*, 269–276. doi:10.1016/j.enpol.2013.03.013

Lachowycz, K., & Jones, A. P. (2011). Greenspace and obesity: A systematic review of the evidence. *Obesity Reviews*, *12*(5). doi:10.1111/j.1467-789X.2010.00827.x PMID:21348919

Lackey, J. B. (1938). The manipulation and counting of river plankton and changes in some organisms due to formalin preservation. *Public Health Reports*, *53*(47), 2081–2093. doi:10.2307/4582717

Lalkaka, R., & Abetti, P. (1999). Business Incubation and Entreprise Support Systems in Restructuring Countries. *Creativity and Innovation Management*, *8*(3), 197–209. doi:10.1111/1467-8691.00137

Lal, R. (2004). Soil carbon sequestration impacts on global climate change and food security. *Science*, *304*(5677), 1623–1627. doi:10.1126/science.1097396 PMID:15192216

Lambert, A. (2019). ¿Psicótico, acrítico y precario? Una exploración lacaniana del sujeto del consumidor neoliberal. *Teoría del marketing*, *19*(3), 329–346. doi:10.1177/1470593118796704

Lamberton, G. (2005). Contabilidad de sostenibilidad: Una breve historia y marco conceptual. *Accounting Forum*, *29*(1), 7–26. doi:10.1016/j.accfor.2004.11.001

Landres, P. B., Morgan, P., & Swanson, F. J. (1999). Overview of the use of natural variability concepts in managing ecological systems. *Ecological Applications*, *9*, 1179–1188.

Langton, M., & Rhea, Z. M. (2005). Traditional indigenous biodiversity-related knowledge. *Australian Academic and Research Libraries*, *36*(2), 45–69. doi:10.1080/00048623.2005.10721248

Larson, M. A., Thompson, F. R. III, Millspaugh, J. J., Dijak, W. D., & Shifley, S. R. (2004). Linking population viability, habitat suitability, and landscape simulation models for conservation planning. *Ecological Modelling*, *180*(1), 103–118. doi:10.1016/j.ecolmodel.2003.12.054

Łaszkiewicz, E., Czembrowski, P., & Kronenberg, J. (2020). Creating a map of the social functions of urban green spaces in a city with poor availability of spatial data: A sociotope for lodz. *Land (Basel)*, *9*(6), 183. Advance online publication. doi:10.3390/land9060183

Latour, B. (2018). *Con los pies en la tierra: la política en el nuevo régimen climático. Política*. Reino Unido.

Lausch, A., Selsam, P., Pause, M., & Bumberger, J. (2024). Monitoring vegetation-and geodiversity with remote sensing and traits. *Philosophical Transactions. Series A, Mathematical, Physical, and Engineering Sciences*, *382*(2269), 20230058. doi:10.1098/rsta.2023.0058 PMID:38342219

Le Saout, S., Hoffmann, M., Shi, Y., Hughes, A., Bernard, C., Brooks, T. M., Bertzky, B., Butchart, S. H., Stuart, S. N., Badman, T., & Rodrigues, A. S. L. (2013). Protected areas and effective biodiversity conservation. *Science*, *342*(6160), 803–805. doi:10.1126/science.1239268 PMID:24233709

Le Texier, M., Schiel, K., & Caruso, G. (2018). The provision of urban green space and its accessibility: Spatial data effects in Brussels. *PLoS One*, *13*(10), e0204684. doi:10.1371/journal.pone.0204684 PMID:30332449

Leach, W. D., & Pelkey, N. W. (2001). Making watershed partnerships work: A review of the empirical literature. [DOI: 10.1061/(ASCE)0733-9496(2001)127:6(378)]. *Journal of Water Resources Planning and Management*, *127*(6), 378–385. doi:10.1061/(ASCE)0733-9496(2001)127:6(378)

Leberger, R., Rosa, I. M. D., Guerra, C. A., Wolf, F., & Pereira, H. M. (2020). Global patterns of forest loss across IUCN categories of protected areas. *Biological Conservation*, *241*, 108299. doi:10.1016/j.biocon.2019.108299

Leddin, D. (2024). The impact of climate change, pollution and biodiversity loss on digestive health and disease. *Gastro Hep Advances*. doi:10.1016/j.gastha.2024.01.018

Lee, A. C. K., & Maheswaran, R. (2011). The health benefits of urban green spaces: A review of the evidence. In Journal of Public Health (Vol. 33, Issue 2). doi:10.1093/pubmed/fdq068

Leeuwis, C., & van den Ban, A. (2004). Communication for Innovation in Agricultural and Rural Resource Management. Building on the Tradition of Agricultural Extension. Oxford: Blackwell Science (in cooperation with CTA).

Lehmann, A., Overton, J. M., & Leathwick, J. R. (2003). GRASP: Generalized regression analysis and spatial prediction. *Ecological Modelling*, *160*(1–2), 165–183. doi:10.1016/S0304-3800(02)00354-X

Lester, R. K. (2005). *Universities, Innovation, and the Competitiveness of Local Economies A Summary Report from the Local Innovation Systems Project.* Industrial Performance Center Massachusetts Institute of Technology.

Lewthwaite, G.R. (1966). Ecologismo y determinismo: Una búsqueda de aclaración. *Ann. Assoc. Am. Geogr. 56*, 1–23.

Li, T.M. (2017). Desplazamiento intergeneracional en la zona de plantaciones de palma aceitera de Indonesia.

Li, F., Altermatt, F., Yang, J., An, S., Li, A., & Zhang, X. (2020). Human activities' fingerprint on multitrophic biodiversity and ecosystem functions across a major river catchment in China. *Global Change Biology*, *26*(12), 6867–6879. doi:10.1111/gcb.15357 PMID:32936984

Li, F., Li, M., & Feng, X. (2022). Ecological Benefit Analysis of Urban Green Space Based on Three-Dimensional Green Quantity. *Photonirvachak (Dehra Dun)*, *50*(11), 2267–2276. doi:10.1007/s12524-022-01603-x

Lim, D. Y., Starnes, T., & Plumptre, A. J. (2023). Global priorities for biodiversity conservation in the United Kingdom. *Biological Conservation*, *277*, 109798. doi:10.1016/j.biocon.2022.109798

Lindbladh, M., Bradshaw, R., & Holmqvist, B. H. (2000). Pattern and process in south Swedish forests during the last 3000 years, sensed at stand and regional scales. *Journal of Ecology*, *88*(1), 113–128. doi:10.1046/j.1365-2745.2000.00429.x

Linhares, Y., Kaganski, A., Agyare, C., Kurnaz, I. A., Neergheen, V., Kolodziejczyk, B., Kędra, M., Wahajuddin, M., El-Youssf, L., dela Cruz, T. E., Baran, Y., Pešić, M., Shrestha, U., Bakiu, R., Allard, P. M., Rybtsov, S., Pieri, M., Siciliano, V., & Flores Bueso, Y. (2023). Biodiversity: The overlooked source of human health. *Trends in Molecular Medicine*, *29*(3), 173–187. doi:10.1016/j.molmed.2022.12.002 PMID:36585352

Lira, A. (2018). Introducción al concepto de sostenibilidad. *Unidades de Apoyo para el Aprendizaje. CUAED/Facultad de Arquitectura-UNAM.* https://uapa.cuaieed.unam.mx/sites/default/files/minisite/static/693ee8e8-f02c-43c2-8222-498e1e8b8814/ConceptoSostenibilidad/index.html

Livi Bacci, M. (2017). *Nuestro planeta menguante (Il Pianeta Stretto); Política.* Reino Unido.

Locke, H., Ellis, E. C., Venter, O., Schuster, R., Ma, K., Shen, X., Woodley, S., Kingston, N., Bhola, N., Strassburg, B. B., Paulsch, A., Williams, B., & Watson, J. E. M. (2019). Three global conditions for biodiversity conservation and sustainable use: An implementation framework. *National Science Review*, *6*(6), 1080–1082. doi:10.1093/nsr/nwz136 PMID:34691979

Löfsten, H., & Lindelof, P. (2001). Science parks in Sweden — Industrial renewal and development? *R & D Management*, *5*(3), 309–322. doi:10.1111/1467-9310.00219

Lomolino, M., Riddle, B., & Whittaker, R. (2017). *Biogeography: Biological diversity across time and space.* Sinauer Associates.

Loreau, M. (2014). Reconciling utilitarian and non-utilitarian approaches to biodiversity conservation. *Ethics in Science and Environmental Politics*, *14*(1), 27–32. doi:10.3354/esep00149

Lotze, H. K., Guest, H., O'Leary, J., Tuda, A., & Wallace, D. (2018). Public perceptions of marine threats and protection from around the world. *Ocean and Coastal Management*, *152*, 14–22. doi:10.1016/j.ocecoaman.2017.11.004

Ludwig, C., Hecht, R., Lautenbach, S., Schorcht, M., & Zipf, A. (2021). Mapping Public Urban Green Spaces Based on OpenStreetMap and Sentinel-2 Imagery Using Belief Functions. *ISPRS International Journal of Geo-Information*, *10*(4), 251. Advance online publication. doi:10.3390/ijgi10040251

Luo, Y., Salman, M., & Lu, Z. (2021). Heterogeneous impacts of environmental regulations and foreign direct investment on green innovation across different regions in China. *The Science of the Total Environment*, *759*, 143744. doi:10.1016/j.scitotenv.2020.143744 PMID:33341514

Lushington, A. W. (1915). *Vernacular list of trees, shrubs and woody cl~rnbers of the Madras Presidency.* Govt. Press.

Macdonald, D. W. (2023). *Biodiversity Conservation: A Very Short Introduction.* Oxford University Press., doi:10.1093/actrade/9780199592272.001.0001

Macdonald, E. K., & Uncles, M. D. (2007). Conocimiento del consumidor: Conceptualización y medición. *Revista de Gestión de Marketing*, *23*(5–6), 497–517. doi:10.1362/026725707X212793

Madureira, H., Nunes, F., Oliveira, J. V., Cormier, L., & Madureira, T. (2015). Urban residents' beliefs concerning green space benefits in four cities in France and Portugal. *Urban Forestry & Urban Greening*, *14*(1), 56–64. Advance online publication. doi:10.1016/j.ufug.2014.11.008

Maestre, F. T., Benito, B. M., Berdugo, M., Concostrina-Zubiri, L., Delgado-Baquerizo, M., Eldridge, D. J., Guirado, E., Gross, N., Kéfi, S., Le Bagousse-Pinguet, Y., Ochoa-Hueso, R., & Soliveres, S. (2021). Biogeography of global drylands. *The New Phytologist*, *231*(2), 540–558. doi:10.1111/nph.17395 PMID:33864276

Maffi, L. (2005). LINGUISTIC, CULTURAL, AND BIOLOGICAL DIVERSITY. *Annual Review of Anthropology*, *34*, 599–617. doi:10.1146/annurev.anthro.34.081804.120437

Magazzino, C. (2024). Ecological footprint, electricity consumption, and economic growth in China: Geopolitical risk and natural resources governance. *Empirical Economics*, *66*(1), 1–25. doi:10.1007/s00181-023-02460-4

Magurran, A. E. (1988). *Ecological Diversity and its Measurement*. Princeton University Press. doi:10.1007/978-94-015-7358-0

Magurran, A. E. (2004). *Measuring Biological Diversity*. Blackwell Publishing.

Majeed, M. T., & Mazhar, M. (2019). Financial development and ecological footprint: A global panel data analysis. *Pakistan Journal of Commerce and Social Sciences*, *13*(2), 487–514. http://hdl.handle.net/10419/201002

Malik, C., Singhal, N., & Tiwari, S. (2017). Antecedentes de la actitud ambiental del consumidor y la intención de comprar productos verdes: Papel moderador de la necesidad percibida del producto. *Revista Internacional de Tecnología y Gestión Ambiental*, *20*(5/6), 259–279. doi:10.1504/IJETM.2017.091290

Mancini, M. S., Galli, A., Niccolucci, V., Lin, D., Bastianoni, S., Wackernagel, M., & Marchettini, N. (2016). Ecological footprint: Refining the carbon footprint calculation. *Ecological Indicators*, *61*(2), 390–403. doi:10.1016/j.ecolind.2015.09.040

Mann, C. C. (2011). 1493: Descubriendo el Nuevo Mundo que Colón creó. Vintage: Nueva York.

Marcuse, H. (1972). *Contrarrevolución y revuelta*. Boston Beacon Press.

Marshall, J., Adcroft, A., Hill, C., Perelman, L., & Heisey, C. (1997). A finite-volume, incompressible Navier Stokes model for studies of the ocean on parallel computers. *Journal of Geophysical Research*, *102*(C3), 5753–5766. doi:10.1029/96JC02775

Marxt, C., & Brunner, C. (2012). *Analyzing and improving the national innovation system of highly developed countries: the case of Switzerland Technol.* Forecast. Soc. Chang., doi:10.1016/j.techfore.2012.07.008

Masco, J. (2006). *The Nuclear Borderlands: The Manhattan Project in Post-Cold War New Mexico*. Princeton University Press. doi:10.1515/9781400849680

Mathews, D., & Turner, N. J. (2017). *Culturas oceánicas: ecosistemas de la costa noroeste y sistemas de gestión indígena. Levin PS, Poe MR editores. Conservación para el Océano Antropoceno: Ciencia Interdisciplinaria en Apoyo de la Naturaleza y las Personas*. Academic Press.

Mathur, N. (2017). La tarea del traductor climático. *Economic and Political Weekly*, *LII*(31), 77–84.

Maxwell, S. L., Cazalis, V., Dudley, N., Hoffmann, M., Rodrigues, A. S. L., Stolton, S., Visconti, P., Woodley, S., Kingston, N., Lewis, E., Maron, M., Strassburg, B. B. N., Wenger, A., Jonas, H. D., Venter, O., & Watson, J. E. M. (2020). Area-based conservation in the twenty-first century. *Nature 2020 586:7828, 586*(7828), 217–227. doi:10.1038/s41586-020-2773-z

Maxwell, S. L., Fuller, R. A., Brooks, T. M., & Watson, J. E. (2016). Biodiversity: The ravages of guns, nets and bulldozers. *Nature*, *536*(7615), 143–145. doi:10.1038/536143a PMID:27510207

Mazel, F., Guilhaumon, F., Mouquet, N., Devictor, V., Gravel, D., Renaud, J., Cianciaruso, M. V., Loyola, R., Diniz-Filho, J. A. F., Mouillot, D., & Thuiller, W. (2014). Multifaceted diversity–area relationships reveal global hotspots of mammalian species, trait and lineage diversity. *Global Ecology and Biogeography*, *23*(8), 836–847. doi:10.1111/geb.12158 PMID:25071413

Mcallister, E., Bhullar, N., & Schutte, N. S. (2017). Into the Woods or a Stroll in the Park : How Virtual Contact with Nature Impacts Positive and Negative Affect. *International Journal of Environmental Research and Public Health*, *14*(786), 1–11. doi:10.3390/ijerph14070786 PMID:28708074

McCann, P., & Ortega-Argilés, R. (2013). Modern regional innovation policy. *Cambridge Journal of Regions, Economy and Society*, *6*(2), 187–216. doi:10.1093/cjres/rst007

McCarthy, D. P., Donald, P. F., Scharlemann, J. P., Buchanan, G. M., Balmford, A., Green, J. M., Bennun, L. A., Burgess, N. D., Fishpool, L. D., Garnett, S. T., Leonard, D. L., Maloney, R. F., Morling, P., Schaefer, H. M., Symes, A., Wiedenfeld, D. A., & Butchart, S. H. M. (2012). Financial costs of meeting global biodiversity conservation targets: Current spending and unmet needs. *Science*, *338*(6109), 946–949. doi:10.1126/science.1229803 PMID:23065904

McDonagh, P., & Prothero, A. (2014). Investigación de mercados de sostenibilidad: Pasado, presente y futuro. *Revista de Gestión de Marketing*, *30*(11–12), 1186–1219. doi:10.1080/0267257X.2014.943263

McElroy, A. (1990). Modelos bioculturales en estudios de salud humana y adaptación. *Anthropological Quarterly*, *4*, 243–265. doi:10.1525/maq.1990.4.3.02a00010

McIntosh, E. J., Pressey, R. L., Lloyd, S., Smith, R. J., & Grenyer, R. (2017). The impact of systematic conservation planning. *Annual Review of Environment and Resources*, *42*(1), 677–697. doi:10.1146/annurev-environ-102016-060902

McKinney, M. L. (2008). Effects of urbanization on species richness: A review of plants and animals. *Urban Ecosystems*, *11*(2), 161–176. doi:10.1007/s11252-007-0045-4

McNeill, J., & Peter, E. (2016). La Gran Aceleración: Una Historia Ambiental del Antropoceno desde 1945. Harvard University Press.

Mehmood, U., Muhammad, U. A., & Muhammad, A. J. (2023). Associating economic growth and ecological footprints through human capital and biocapacity in south Asia. *WORLD (Oakland, Calif.)*, *4*(3), 598–611. doi:10.3390/world4030037

Mci, R., & Liu, W. T. (2019). Quantifying thc contribution of microbial immigration in engineered water systems. *Microbiome*, *7*(1), 144–156. doi:10.1186/s40168-019-0760-0 PMID:31694700

Millennium Ecosystem Assessment. (2005). Ecosystems and Human Well-being: Biodiversity Synthesis. World Resources Institute.

Millennium Ecosystem Assessment. (2005). *Ecosystems and human well-being: Synthesis*. Island Press.

Millennium Ecosystem Assessment. (2005). *Ecosystems and Human Well-being: Synthesis*. Island Press.

Milstein, T., & Castro-Sotomayor, J. (Eds.). (2020). *Routledge Handbook of Ecocultural Identity. Routledge.* Reino Unido. doi:10.4324/9781351068840

Minteer, B. A., & Collins, J. P. (2005). Ecological ethics: Building a new tool kit for ecologists and biodiversity managers. *Conservation Biology*, *19*(6), 1803–1812. doi:10.1111/j.1523-1739.2005.00281.x

Minton, C. D. T. (1968). Pairing and breeding in mute swans. [Morgan, D.G., Estimating Vertebrate Population Densities by Line Transect Methods, Melbourne College of Advanced Education, Australia.]. *Wildfowl*, *19*, 41–60.

Mishra, A. K., & Dash, A. K. (2022). Connecting the carbon ecological footprint, economic globalization, population density, financial sector development, and economic growth of five South Asian countries. *Energy Research Letters*, *3*(2), 1–6. doi:10.46557/001c.32627

Mishra, S., Anuradha, J., Tripathi, S., & Kumar, S. (2016). In vitro antioxidant and antimicrobial efficacy of Triphala constituents: Emblica officinalis, Terminalia belerica and Terminalia chebula. *Journal of Pharmacognosy and Phytochemistry*, *5*(6), 273–277.

Misra, R. (2013). Ecology Workbook. Scientific Publishers. (Page 31 to 45). Moran, D; Laycock, H; White, P C.L. The role of cost-effectiveness analysis in conservation decision-making. *Biological Conservation*, *143*(4), 826–827.

Mitani, J. C., Abwe, E., Campbell, G., Giles-Vernick, T., Goldberg, T., McLennan, M. R., Preuschoft, S., Supriatna, J., & Marshall, A. J. (2024). Future coexistence with great apes will require major changes to policy and practice. *Nature Human Behaviour*, *2024*(4), 1–12. doi:10.1038/s41562-024-01830-x PMID:38374442

Mittelstaedt, J. D., Shultz, C. J., Kilbourne, W. E. y Peterson, M. (2014). La sostenibilidad como megatendencia: Dos escuelas de pensamiento de macromarketing. *Revista de macromarketing*, *34*(3), 253–264. doi:10.1177/0276146713520551

Mi, X., Feng, G., Hu, Y., Zhang, J., Chen, L., Corlett, R. T., Hughes, A. C., Pimm, S., Schmid, B., Shi, S., Svenning, J.-C., & Ma, K. (2021). The global significance of biodiversity science in China: An overview. *National Science Review*, *8*(7), nwab032. doi:10.1093/nsr/nwab032 PMID:34694304

Mladenoff, D. J. (2004). LANDIS and forest landscape models. *Ecological Modelling*, *13*(2–3), 147–159.

Mog, J. M. (2004). Luchando con la sostenibilidad. Un marco comparativo para evaluar los programas de desarrollo sostenible. *World Development*, *32*, 2139–2160. doi:10.1016/j.worlddev.2004.07.002

Monck, C. S. P., Porter, R. B., Quintas, P. R., & Storey, D. J. (1988). *Science Parks and the Growth of High Technology Firms*. Routledge.

Mondière, A., Corson, M. S., Auberger, J., Durant, D., Foray, S., Glinec, J.-F., Green, P., Novak, S., Signoret, F., & van der Werf, H. M. (2024). Trade-offs between higher productivity and lower environmental impacts for biodiversity-friendly and conventional cattle-oriented systems. *Agricultural Systems*, *213*, 103798. doi:10.1016/j.agsy.2023.103798

Montesino Pouzols, F., Toivonen, T., Di Minin, E., Kukkala, A. S., Kullberg, P., Kuusterä, J., Lehtomäki, J., Tenkanen, H., Verburg, P. H., & Moilanen, A. (2014). Global protected area expansion is compromised by projected land-use and parochialism. *Nature*, *516*(7531), 383–386. doi:10.1038/nature14032 PMID:25494203

Moore, J. (Ed.). (2016). *Antropoceno o Capitaloceno? Naturaleza, historia y crisis del capitalism*. Kairos Press.

Morgan, R. K., & Henrion, M. (1990). *Uncertainty: A guide to dealing with uncertainty in quantitative risk and policy analysis*. Cambridge University Press. doi:10.1017/CBO9780511840609

Morimoto, R. (2022) Una persecución de jabalíes: Ecología del daño y la política de HalfLife en la costa de Fukushima. *Antropología cultural*, *37*(1), pp. 69-98.

Morrison-Saunders, A., & Fischer, T. B. (2006). What is wrong with EIA and SEA – a diagnostic approach. *Environmental Impact Assessment Review*, *26*(6), 552–566.

Mörtberg, U., & Anders, K. (2005). Predicting forest grouse distribution taking account of spatial autocorrelation. *Journal for Nature Conservation*, *13*(2–3), 147–159. doi:10.1016/j.jnc.2005.02.008

Moss, M. L. (2011). *Costa Noroeste: Arqueología como Historia Profunda*. Society for American Archaeology Press.

Mossman, H. L., Panter, C. J., & Dolman, P. M. (2015). Modelling biodiversity distribution in agricultural landscapes to support ecological network planning. *Landscape and Urban Planning*, *141*, 59–67. doi:10.1016/j.landurbplan.2015.04.010

Mulongoy, K., & Chape, S. (2004). *Protected areas and biodiversity: An overview of key issues*. Policy Commons. https://policycommons.net/artifacts/1376714/protected-areas-and-biodiversity/1990978/

Murguía, D. I., Bringezu, S., & Schaldach, R. (2016a). Global direct pressures on biodiversity by large-scale metal mining: Spatial distribution and implications for conservation. *Journal of Environmental Management*, *180*, 409–420. doi:10.1016/j.jenvman.2016.05.040 PMID:27262340

Museum of Natural History & Helsinki, P. (1959). Zur Methodik der quantitativen Vogelforschung in den Binnengewässern. *Ornis Fennica*, *36*(3–4), 66–78.

Nan, M. A., Lun, Y. A. N. G., Qingwen, M. I. N., Keyu, B., & Wenhua, L. (2021). The significance of traditional culture for agricultural biodiversity—Experiences from GIAHS. *Journal of Resources and Ecology*, *12*(4), 453–461. doi:10.5814/j.issn.1674-764x.2021.04.003

Nautiyal, D. C., Sharmaand, S. K., & Pandit, M. K. (2009). Notes on the taxonomic history of two rare species of Begonia (Begoniaceae) from Sikkim Himalaya and their conservation. *Journal of the Botanical Research Institute of Texas*, *3*(2), 823–830.

Nautiyal, D. C., Sharma, S. K., & Pandit, M. K. (2009). Notes on the taxonomic history, rediscovery and conservation status of two endangered species of Ceropegia (Asclepiadaceae) from Sikkim Himalaya. *Journal of the Botanical Research Institute of Texas*, *3*(2), 815–822.

Navarro, J., Coll, M., Cardador, L., Fernández, Á. M., & Bellido, J. M. (2015). The relative roles of the environment, human activities and spatial factors in the spatial distribution of marine biodiversity in the Western Mediterranean Sea. *Progress in Oceanography*, *131*, 126–137. doi:10.1016/j.pocean.2014.12.004

Nayar, M. P. (1996). Hotspots of Endemic Plants of India, Nepal and Bhutan. Thiruvanathapuram: Tropical Botanical Garden and Research Institute. 204 pp.

Nayar, M. P. (1980). Endemism and patterns of distribution of endemic genera (Angiosperms) in India. *Journal of Economic and Taxonomic Botany*, *I*, 99–110.

Nayar, M. P., & Sastry, A. R. K. (1987). *Red Data Book of Indian Plants* (Vol. I). Botanical Survey of India.

Neagu, O. (2020). Economic complexity and ecological footprint: Evidence from the most complex economies in the world. *Sustainability (Basel)*, *12*(21), 1–18. doi:10.3390/su12219031

Nguyen, H. M., & Nguyen, L. D. (2018). The relationship between urbanization and economic growth an empirical study on ASEAN countries. *International Journal of Social Economics*, *45*(2), 316–339. Advance online publication. doi:10.1108/IJSE-12-2016-0358

Nguyen, M. H., & Vuong, Q. H. (2021). Evaluation of the Aichi Biodiversity Targets: The international collaboration trilemma in interdisciplinary research. *Pacific Conservation Biology*, *28*(6), 517–531. doi:10.1071/PC21026

Nicholson, E., Andrade, A., Brooks, T. M., Driver, A., Ferrer-Paris, J. R., Grantham, H., Gudka, M., Keith, D. A., Kontula, T., Lindgaard, A., Londono-Murcia, M. C., Murray, N., Raunio, A., Rowland, J. A., Sievers, M., Skowno, A. L., Stevenson, S. L., Valderrabano, M., Vernon, C. M., & Obura, D. (2024). Roles of the Red List of Ecosystems in the Kunming-Montreal Global Biodiversity Framework. *Nature Ecology & Evolution*, *8*(4), 1–8. doi:10.1038/s41559-023-02320-5 PMID:38332025

Nielsen, K. S., Marteau, T. M., Bauer, J. M., Bradbury, R. B., Broad, S., Burgess, G., Burgman, M., Byerly, H., Clayton, S., Espelosin, D., Ferraro, P. J., Fisher, B., Garnett, E. E., Jones, J. P. G., Otieno, M., Polasky, S., Ricketts, T. H., Trevelyan, R., van der Linden, S., & Balmford, A. (2021). Biodiversity conservation as a promising frontier for behavioural science. *Nature Human Behaviour, 5*(5), 550–556. doi:10.1038/s41562-021-01109-5

Nongbri, T. (2006). *Culture and biodiversity: Myths, legends and the conservation of nature in the hills of North-East India.* Indian Anthropologist, 1-21.

Nor, A. N. M., Aziz, H. A., Nawawi, S. A., Jamil, R. M., Abas, M. A., Hambali, K. A., Yusoff, A. H., Ibrahim, N., Rafaai, N. H., Corstanje, R., Harris, J., Grafius, D., & Perotto-Baldivieso, H. L. (2021). Evolution of green space under rapid urban expansion in southeast Asian cities. *Sustainability (Basel), 13*(21), 12024. doi:10.3390/su132112024

Nor, A. N. M., Corstanje, R., Harris, J. A., Grafius, D. R., & Siriwardena, G. M. (2017). Ecological connectivity networks in rapidly expanding cities. *Heliyon*, *3*(6), e00325. doi:10.1016/j.heliyon.2017.e00325 PMID:28706999

Novack, T., Wang, Z., & Zipf, A. (2018). A system for generating customized pleasant pedestrian routes based on openstreetmap data. *Sensors (Basel), 18*(11), 3794. doi:10.3390/s18113794 PMID:30404175

Nozawa, S. (2015). Phatic Traces: Sociality in Contemporary Japan. *Anthropological Quarterly*, *88*(2), 373–400. https://muse.jhu.edu/article/582266. doi:10.1353/anq.2015.0014

NRC. (National Research Council). (2008). Public participation in environmental assessment and decision making. Washington, DC: National Academies Press.

Nsamenang, A. B. (2015). Teorías ecoculturales del desarrollo. En la Enciclopedia Internacional de las Ciencias Sociales y del Comportamiento. Elsevier.

Nunn, P. (2017). Dejando de lado a Dios: Por qué los proyectos climáticos seculares en las islas del Pacífico están fallando. La conversación. https://theconversation.com/sidelining-god-why-secular-climate-projects-in-the-pacific-islands-are-failing-77623

O'Connor, R. J. (1985). Long-term monitoring of British bird populations. *Ornis Fennica*, *62*(2), 73–79.

O'Hara, C. C., Frazier, M., & Halpern, B. S. (2021). At-risk marine biodiversity faces extensive, expanding, and intensifying human impacts. *Science*, *372*(6537), 84–87. doi:10.1126/science.abe6731 PMID:33795456

OCDE. (2001). *The DAC Journal: Development Co-operation. 2000 Report.* OCDE.

OECD. (1996). Guidelines for Aid Agencies for Improved Conservation and Sustainable Use of Tropical and Sub-Tropical Wetlands. Guidelines on Aid and Environment, 9. OECD.

OECD. (1999a). *Managing National Innovation Systems.* OECD.

OECD. (2001a). Innovative Clusters: drivers of national innovation systems. Paris: OECD, 2001a

OECD. (2001b). *Drivers of Growth: information technology, innovation, and entrepreneurship. Science and Industry Outlook*. OECD.

OECD. (2001c). *Innovative Networks: co-operation in national innovation systems.* OECD.

OECD. (2011). *Regions and innovation policy, OECD Reviews of regional innovation*. Author.

Ogawa, M., Soga, M., & Yoshida, T. (2021). Participation of diverse actors and usage of traditional and local knowledge in local biodiversity strategies and action plans of Japanese municipalities. *Ecology and Society*, *26*(3), art26. doi:10.5751/ES-12612-260326

Ohansen, K. L., Boertmann, D., Mosbech, A., & Hansen, T. B. (2012). *Manual for Seabird and Marine mammal survey on Seismic Vessels in Greenland* (3rd Revised edition).

Ohasi, H. (1975). Flora of Eastern Himalaya, Third Report. *University Museum of University of Tokyo Bulletin*, *8*, 1–458.

Oladapo, B. I., Ismail, S. O., Afolalu, T. D., Olawade, D. B., & Zahedi, M. (2021). Review on 3D printing: Fight against COVID-19. *Materials Chemistry and Physics*, *2021*(258), 123943. doi:10.1016/j.matchemphys.2020.123943 PMID:33106717

Oldfield, S., Lusty, C., & MacKinven, A. (1998). *The World List of Threatened Trees*. World Conservation Press.

Omann, I., Stocker, A., & Jäger, J. (2009). Climate change as a threat to biodiversity: An application of the DPSIR approach. *Ecological Economics*, *69*(1), 24–31. doi:10.1016/j.ecolecon.2009.01.003

Organización de las Naciones Unidas para la Agricultura y la Alimentación. (2017). *Iniciativa de crecimiento azul de la Organización de las Naciones Unidas para la Alimentación y la Agricultura. Asociación con los países para alcanzar los objetivos de desarrollo sostenible*. Naciones Unidas.

Orlove, B., Lazrus, H., Hovelsrud, G. K. y Giannini, A. (2014). Sobre los orígenes y las consecuencias de la atención desigual al cambio climático en todo el mundo: Sobre los orígenes y las consecuencias de la atención desigual al cambio climático en todo el mundo. *Antropología actual*, *55*(3), 249-275. doi:10.1086/676298

Otake, Y., Yamamichi, M., Hirata, Y., Odagiri, H., & Yoshida, T. (2024). Different photoperiodic responses in diapause induction can promote the maintenance of genetic diversity via the storage effect in *Daphnia pulex. Proceedings of the Royal Society B: Biological Sciences, 291*(2016), 20231860. 10.1098/rspb.2023.1860

Palese, E. (2013). Ethics without Morality, Morality without Ethics—Politics, Identity, Responsibility in Our Contemporary World. *Open Journal of Philosophy*, *03*(03), 366–371. doi:10.4236/ojpp.2013.33055

Palmer, M. A., Zedler, J. B., & Falk, D. A. (2017). Ecological theory and restoration ecology. Foundations of Restoration Ecology: Second Edition, 3–26. doi:10.5822/978-1-61091-698-1_1

Pandey, P. C., & Pandey, M. (2023). Highlighting the role of agriculture and geospatial technology in food security and sustainable development goals. *Sustainable Development (Bradford)*, *31*(5), 3175–3195. doi:10.1002/sd.2600

Papachristos, G., Sofianos, A., & Adamides, E. (2013). System Interactions in Socio-technical Transitions: Extending the Multi-level Perspective. *Environmental Innovation and Societal Transitions*, *7*, 53–69. doi:10.1016/j.eist.2013.03.002

Park, Y., Kim, C. J., Choi, M. B., & Do, Y. (2020). A comparative study of carabid beetles in green spaces and former natural habitats. *Diversity*, *12*(12), 479. doi:10.3390/d12120479

Partidário, M. R. (2008). Strategic environmental assessment and the management of Natura 2000 sites in Portugal: The experience of a research project. *Impact Assessment and Project Appraisal*, *26*(3), 213–222.

Parviainen, J. (2015). Cultural heritage and biodiversity in the present forest management of the boreal zone in Scandinavia. *Journal of Forest Research*, *20*(5), 445–452. doi:10.1007/s10310-015-0499-9

Pava, M. (2007). A response to getting to the bottom of triple bottom line. *Business Ethics Quarterly*, *17*(1), 105–110. doi:10.5840/beq200717116

Pawankar, R. (2019). Climate change, air pollution, and biodiversity in Asia Pacific: Impact on allergic diseases. *Asia Pacific Allergy*, *9*(2), e11. doi:10.5415/apallergy.2019.9.e11 PMID:31089453

Peattie, K., & Crane, A. (2005), Green marketing: Legend, myth, farce, or prophesy? *Investigación cualitativa de mercado: una revista internacional*, *8*(4), 357–370. doi:10.1108/13522750510619733

Peattie, K. (2001). Hacia la sostenibilidad: La tercera era del marketing verde. *The Marketing Review*, *2*(2), 129–146. doi:10.1362/1469347012569869

Peet, R., & Watts, M. (1996) Ecología de la liberación: desarrollo, sostenibilidad y medio ambiente en una era de triunfalismo de mercado. Richard Peet & Michael Watts (Eds) Liberation Ecologies: Environment, Development, Social Movements. Routledge.

Peh, K. S. (2011). Crop failure signals biodiversity crisis. *Nature*, *473*(7347), 284. doi:10.1038/473284d PMID:21593847

Pelto, P. J. (1973) The Snowmobile revolution: technology and social change in the arctic. Cummings Publishing Company, Menlo Park, CA (revised Waveland Press, Prospects Heights, IL 1987.

Percival, R. V., Miller, A. S., Schroeder, C. H., & Leape, J. P. (1992). *Environ-mental regulation: law, science, and policy*. Little, Brown, and Company.

Pesce, S., Mamy, L., Sanchez, W., Artigas, J., Bérard, A., Betoulle, S., Chaumot, A., Coutellec, M.-A., Crouzet, O., Faburé, J., Hedde, M., Leboulanger, C., Margoum, C., Martin-Laurent, F., Morin, S., Mougin, C., Munaron, D., Nélieu, S., Pelosi, C., & Leenhardt, S. (2024). The use of copper as plant protection product contributes to environmental contamination and resulting impacts on terrestrial and aquatic biodiversity and ecosystem functions. *Environmental Science and Pollution Research International*. Advance online publication. doi:10.1007/s11356-024-32145-z PMID:38324154

Peterson, G. D., Cumming, G. S., & Carpenter, S. R. (2003). Scenario planning: A tool for conservation in an uncertain world. *Conservation Biology*, *17*(2), 358–366. doi:10.1046/j.1523-1739.2003.01491.x

Petterson, M., Kensek, K., & Noble, D. (2017). Pieles flexibles: Considerando la relevancia, escalabilidad y estrategias de diseño para la resiliencia del sistema de fachadas. *Revista de Educación Arquitectónica*, *71*(1), 34–45.

Pettorelli, N., Safi, K., & Turner, W. (Eds.). (2014). *Satellite Remote Sensing for Conservation Action*. Oxford University Press.

Pfaffenbichler, P., Emberger, G., Shepherd, S. (2010) Un enfoque de dinámica de sistemas para el modelado de interacción de transporte de uso de la tierra: el modelo estratégico MARS y su aplicación. *Revisión de dinámica de sistemas*, 26, 262 -282. DO - doi:10.1002/sdr.451

Pikitch, E. K., Santora, C., & Babcock, E. A. (2009). A global overview of the management effectiveness of marine protected areas. *Marine Biology*, *156*, 273–283.

Pimenta-Bueno, J. A. (2002). Parque de Innovación Tecnológica y Cultural de Gávea. *Documento interno*.

Pimenta-Bueno, J.A. (1999). Parque de Inovação Tecnológica e Cultural da Gávea: A Visão da PUC-Rio. *Documento Interno, 23*.

Pimm, S. I., Russell, G. J., Gittelman, J. L., & Brooks, T. M. (1995). The Future of Biodiversity. *Science*, *269*(5222), 347–350. doi:10.1126/science.269.5222.347 PMID:17841251

Pimm, S. L., Jenkins, C. N., Abell, R., Brooks, T. M., Gittleman, J. L., Joppa, L. N., Raven, P. H., Roberts, C. M., & Sexton, J. O. (2014). The biodiversity of species and their rates of extinction, distribution, and protection. *Science*, *344*(6187), 1246752. doi:10.1126/science.1246752 PMID:24876501

Pizzol, M., & Andersen, M. S. (2022). Green tech for green growth? Insights from Nordic environmental innovation. In *Circular Economy Oriented Business Model Innovations: A European Perspective*. Springer. doi:10.1007/978-3-031-08313-6_8

Plumptre, A. J., Baisero, D., Brooks, T. M., Buchanan, G., Butchart, S. H., Bowser, A., Boyd, C., Carneiro, A. P., Davies, T., Elliot, W., Foster, M., Langhammer, P. F., Marnewick, D., Matiku, P., McCreless, E., Raudsepp-Hearne, C., Tordoff, A. W., Azpiroz, A. B., Trisurat, Y., & Upgren, A. (2024). Targeting site conservation to increase the effectiveness of new global biodiversity targets. *One Earth*, *7*(1), 11–17. doi:10.1016/j.oneear.2023.12.007

Polanyi, K. (2001). La gran transformación: Los orígenes políticos y económicos de nuestro tiempo. *Beacon Books*. (Trabajo original publicado en 1947)

Pollock, L. J., Thuiller, W., & Jetz, W. (2017). Large conservation gains possible for global biodiversity facets. *Nature*, *546*(7656), 141–144. doi:10.1038/nature22368 PMID:28538726

Por Melo, M.A.C. (2002). *Innovación y modernización tecnológica y organizacional en las MIPYMES: el dominio interorganizacional*.

Porter, M. (1980). *Competitive Strategy: techniques for analyzing industries and competitors*. Free Press.

Porter, M. (2001). *Clusters of Innovation: regional foundations of U.S. competitiveness. Council on Competitiveness*. Monitor Group.

Prakash, S., & Verma, A. (2022). *Anthropogenic Activities and Biodiversity Threats*. https://papers.ssrn.com/abstract=4048276

Prashantkumar, B. Sathvara, J. Anuradha, Sandeep Tripathi, & R. Sanjeevi. (2023). Impact of climate change and its importance on human performance. In Insights on Impact of Climate Change and Adaptation of Biodiversity (1st ed., pp. 1–9). KD Publication.

Pressey, R. L., Mills, M., Weeks, R., & Day, J. C. (2013). The plan of the day: Managing the dynamic transition from regional conservation designs to local conservation actions. *Biological Conservation*, *166*, 155–169. doi:10.1016/j.biocon.2013.06.025

Press, M., & Arnould, E. J. (2014). Transparencia narrativa. *Journal of Marketing Management*, *30*(13–14), 1353–1376. doi:10.1080/0267257X.2014.925958

Pretty, J., Adams, B., Berkes, F., De Athayde, S. F., Dudley, N., Hunn, E., & Pilgrim, S. (2009). The intersections of biological diversity and cultural diversity: Towards integration. *Conservation & Society*, *7*(2), 100–112. doi:10.4103/0972-4923.58642

Price, R. M. (1996). Technology and Strategic Advantage. *California Management Review*, *38*(3), 38–56. doi:10.2307/41165842

Programa de las Naciones Unidas para el Medio Ambiente. (2012). *Green economy in a blue world: Synthesis report*. Naciones Unidas.

Pugh, D. T. (2004). *Tides, surges, and mean sea-level: A handbook for engineers and scientists*. John Wiley & Sons.

Purvis, A., Molnár, Z., Obura, D., Ichii, K., Willis, K., Chettri, N., Dulloo, M., Hendry, A., Gabrielyan, B., & Gutt, J. (2019). Status and trends-nature.

Purvis, A., & Hector, A. (2000). Getting the measure of biodiversity. *Nature*, *405*(6783), 212–219. doi:10.1038/35012221 PMID:10821281

Pushpangadan, P., George, V., Ijinu, T. P., & Chithra, M. A. (2018). Biodiversity, bioprospecting, traditional knowledge. Sustainable development and value added products: A review. *Journal of Traditional Medicine & Clinical Naturopathy*, *7*(01), 1–7. doi:10.4172/2573-4555.1000256

Radcliffe-Brown, A. R. (1940). Sobre la estructura social. *J. R. Antropol. Inst. Gran H. Irel.*, *1940*(70), 1–12.

Rahman, A., Radulescu, M., Ma, H., Dagar, V., Hussain, I., & Khan, M. K. (2019). The impact of globalization, energy use, and trade on ecological footprint in Pakistan: Does environmental sustainability exist? *Energies*, *14*(17), 1–16. doi:10.3390/en14175234

Raihan, A. (2023). *Original Research Article Artificial intelligence and machine learning applications in forest management and biodiversity conservation. 6*(2), 1–16. doi:10.24294/nrcr.v6i2.3825

Ramakrishnan, S., & Jayaraman, A. (2019). Global Warming and Pesticides in Water Bodies. In *Handbook of Research on the Adverse Effects of Pesticide Pollution in Aquatic Ecosystems* (pp. 421–436). IGI Global. https://www.igi-global.com/chapter/global-warming-and-pesticides-in-water-bodies/213519 doi:10.4018/978-1-5225-6111-8.ch023

Ramazanov, S., Antoshkina, L., Babenko, V., & Akhmedov, R. (2019). Integrated model of stochastic dynamics for control of a socio-ecological-oriented innovation economy. *Periodicals of Engineering and Natural Sciences*, *7*(2). http://pen.ius.edu.ba

Ramazanov, S. K. (2018). *Prediction and management of innovative economics based on the integral static model in phase space.*

Ramsay, M. M., Jackson, A. D., & Porley, R. D. (2000). A pilot study for the ex situ conservation of UK bryophytes. EuroGard.

Rawat, U., & Agarwal, N. (2015). Biodiversity: Concept, threats and conservation. *Environment Conservation Journal*, *16*(3), 19–28. doi:10.36953/ECJ.2015.16303

Rawes, E. (2019). Los mejores secadores para 2019. *Tendencias digitales*. https://www.digitaltrends.com/home/best-dryers/

Raworth, K. (2017). *Economía del donut. Siete maneras de pensar como un economista del siglo 21; Random House*. Estados Unidos.

RayhanA. (2023). AI and the Environment Toward Sustainable Development and Conservation. *Research Gate, July*. doi:10.13140/RG.2.2.12024.42245

Ray, S., & Ray, I. A. (2011). Impact of population growth on environmental degradation: A case of India. *Journal of Economics and Sustainable Development*, *2*(8), 72–77. https://core.ac.uk/reader/234645493

Reed, B. M., Sarasan, V., Kane, M., Bunn, E., & Pence, V. C. (2011). Biodiversity conservation and conservation biotechnology tools. *In Vitro Cellular & Developmental Biology. Plant*, *47*(1), 1–4. doi:10.1007/s11627-010-9337-0

Reed, M. S., Kenter, J., Bonn, A., & Borad, K. (2009). Participatory tools for assessing the impact of land use scenarios on ecosystem services. [DOI: 10.1016/j.jenvman.2009.01.001]. *Journal of Environmental Management*, *90*(5), 1933–1940. doi:10.1016/j.jenvman.2009.01.001 PMID:19231064

Ren, G. Y. (2015). Urbanization as a major driver of urban climate change. In Advances in Climate Change Research, 6(1). doi:10.1016/j.accre.2015.08.003

Reside, A. E., Butt, N., & Adams, V. M. (2018). Adapting systematic conservation planning for climate change. *Biodiversity and Conservation*, *27*(1), 1–29. doi:10.1007/s10531-017-1442-5

Ribeiro-Souza, P., Haji, J., Oshima, J., Lima, F., Lima-Silva, B., Pires, J., Ribeiro, M., & Graipel, M. (2024). Under pressure: Suitable areas for neotropical cats within an under protected biodiversity hotspot. *Remote Sensing Applications: Society and Environment*, *34*, 101155. doi:10.1016/j.rsase.2024.101155

Rivers, W. H. R. (Ed.). (1922). *Ensayos sobre la despoblación de Melanesia; Cambridge University Press*. Reino Unido.

Robinson, J. M., Breed, A. C., Camargo, A., Redvers, N., & Breed, M. F. (2024). Biodiversity and human health: A scoping review and examples of underrepresented linkages. *Environmental Research*, *246*, 118115. doi:10.1016/j.envres.2024.118115 PMID:38199470

Rockström, J., Schellnhuber, H. J., Hoskins, B., Ramanathan, V., Schlosser, P., Brasseur, G. P., Gaffney, O., Nobre, C., Meinshausen, M., Rogelj, J., & Lucht, W. (2016). The world's biggest gamble. *Earth's Future*, *10*(10), 465–470. doi:10.1002/2016EF000392

Rockström, J., Steffen, W., Noone, K., Persson, Å., Chapin, F. S. III, Lambin, E., ... Foley, J. (2009). Límites planetarios: Explorando el espacio operativo seguro para la humanidad. *Ecology and Society*, *14*(2).

Roel, S., Rajvanshi, A., Mathur, V. B., & Kolhoff, A. (2009). *Biodiversity in Environmental Assessment Enhancing Ecosystem Services for Human Well-Being*. Cambridge University Press.

Rogers, C. S., Garrison, G., Grober, R., Hillis, Z. M., & Franke, M. A. (1994). *Coral Reef Monitoring Manu- al for the Caribbean and Western Atlantic*. USVI, Virgin Islands National Park.

Rogers, S. M. (2003). Tissue culture and wetland establishment of the freshwater monocots Carex, Juncus, Scirpus, and Typha. *In Vitro Cellular & Developmental Biology. Plant*, *39*(1), 1–5. doi:10.1079/IVP2002358

Rosenzweig, M. L. (1995). *Species Diversity in Space and Time*. Cambridge University Press. doi:10.1017/CBO9780511623387

Rosser, A. M., & Mainka, S. A. (2002). Overexploitation and Species Extinctions. *Conservation Biology*, *16*(3), 584–586. doi:10.1046/j.1523-1739.2002.01635.x

Roth, D. S., Perfecto, I., & Rathcke, B. (1994). The effects of management systems on ground-foraging ant diversity in Costa Rica. Ecological Applications 4(3):423-436. Botanical Survey of India.

Rowe, G., & Frewer, L. J. (2000). Public participation methods: A framework for evaluation. [DOI: 10.1177/016224390002500101]. *Science, Technology & Human Values*, *25*(1), 3–29. doi:10.1177/016224390002500101

Roy, P., Roy, A., & Karnataka, H. (2012). Contemporary tools for identification, assessment and monitoring biodiversity. *Tropical Ecology*, *53*(3), 261–272.

Rudiak-Gould, P. (2011). Cambio climático y antropología: La importancia de los estudios de recepción. *Antropología Hoy*, *27*(2), 9–12. doi:10.1111/j.1467-8322.2011.00795.x

Sabir, S., & Gorus, M. S. (2019). The impact of globalization on ecological footprint: Empirical evidence from the South Asian countries. *Environmental Science and Pollution Research International*, *26*(32), 33387–33398. doi:10.1007/s11356-019-06458-3 PMID:31522394

Saboori, B., Sulaiman, J. B., & Mohd, S. (2012). An empirical analysis of the environmental Kuznets curve for CO2 emission in Indonesia: The role of energy consumption and foreign trade. *International Journal of Economics and Finance*, *4*(2), 243–251. doi:10.5539/ijef.v4n2p243

Sadler, B. (1996). International study of the effectiveness of environmental assessment. *Impact Assessment and Project Appraisal*, *14*(1), 3–9.

Safford, H. D., Hayward, G., Heller, N., & Wiens, J. A. (2012). Climate change and historical ecology: can the past still inform the future? J.A. Wiens, G. Hayward, H.D. Safford, C.M. Giffen (Eds.), Historical Environmental Variation in Conservation and Natural Resource Management. John Wiley and Sons, New York, NY.

Sahlins, M. (1976). *Razón práctica*. Aldine.

Sahlins, M. (1996). La tristeza de la dulzura: La antropología nativa de la cosmología occidental. *Current Anthropology*, *37*(junio), 395–428. doi:10.1086/204503

Sala, O. E., Chapin, F. S., Armesto, J. J., Berlow, E., Bloomfield, J., Dirzo, R., & Wall, D. H. (2000). Global biodiversity scenarios for the year 2100. *Science*, *287*(5459), 1770–1774. doi:10.1126/science.287.5459.1770 PMID:10710299

Saleem, N., Rahman, S. R., & Jun, Z. (2019). The impact of human capital and biocapacity on environment, environmental quality measure through ecological footprint and greenhouse gases. *Journal of Pollution Effects & Control*, *7*(4), 1–13. doi:10.35248/2375-4397.19.7.237

Samimi, A. J., Kashefi, A., Salatin, P., & Lashkarizadeh, M. (2011). Environmental performance and HDI: Evidence from countries around the world. *Middle East Journal of Scientific Research*, *10*(3), 294–301. https://idosi.org/mejsr/mejsr10(3)11/3.pdf

Sandemose, J. (2016). The Motley World of "International Values": Modes of Production on the World Market. *Theoretical Economics Letters*, *06*(03), 507–533. doi:10.4236/tel.2016.63058

Sanjeevi, R. (2011). *Studies on the treatment of low-strength wastewaters with upflow anaerobic sludge blanket (UASB) reactor: With emphasis on granulation studies*. Centre for Pollution Control and Environmental Engineering.

Sanjeevi, R., Haruna, M., Tripathi, S., Singh, B., & Jayaraman, A. (2017). Impacts of Global Carbon Foot Print on the Marine Environment. *International Journal of Engineering Research & Technology (Ahmedabad)*, *5*, 51–54.

Sarasan, V. (2010). Importance of in vitro technology to future conservation programmes worldwide. *Kew Bulletin*, *65*(4), 549–554. doi:10.1007/s12225-011-9250-7

Sathvara, P. B., Anuradha, J., Sanjeevi, R., Tripathi, S., & Rathod, A. B. (2023). Spatial Analysis of Carbon Sequestration Mapping Using Remote Sensing and Satellite Image Processing. In *Multimodal Biometric and Machine Learning Technologies* (pp. 71–83). John Wiley & Sons, Ltd. doi:10.1002/9781119785491.ch4

Savitz, W., & Weber, K. (2006). *The Triple Bottom Line*. Jossey-Bass.

Saxenian, A. (1994). *Regional Advantage: culture and competition in Silicon Valley and Route 128*. Harvard Business Press.

Schaltegger, M. B., & Burritt, R. (2006), Sustainability Accounting and Reporting. Development, Linkages and Reflection: An Introduction. Springer. doi:10.1080/13511610304521

Scheffers, B. R., De Meester, L., Bridge, T. C., Hoffmann, A. A., Pandolfi, J. M., Corlett, R. T., Butchart, S. H., Pearce-Kelly, P., Kovacs, K. M., Dudgeon, D., Pacifici, M., Rondinini, C., Foden, W. B., Martin, T. G., Mora, C., Bickford, D., & Watson, J. E. M. (2016). The broad footprint of climate change from genes to biomes to people. *Science*, *354*(6313), aaf7671. Advance online publication. doi:10.1126/science.aaf7671 PMID:27846577

Scherer, L., Curran, M., & Alvarez, M. (2017). Expanding Kenya's protected areas under the Convention on Biological Diversity to maximize coverage of plant diversity. *Conservation Biology*, *31*(2), 302–310. doi:10.1111/cobi.12792 PMID:27346759

Schlapfer, A., & Marinova, D. (2001). *Local Innovation Systems: nature, importance and role.* Conference Proceedings: International Summer Academy on Technological Studies: user involvement in technological innovation, Deutschlandsberg. Austria.

Schot, J., & Rip, A. (1997). The past and future of constructive technology assessment. *Technological Forecasting and Social Change*, *54*(2–3), 251–268. doi:10.1016/S0040-1625(96)00180-1

Schot, J., & Steinmueller, W. E. (2018). Three frames for innovation policy: R&D, systems of innovation and transformative change. *Research Policy*, *47*(9), 1554–1567. doi:10.1016/j.respol.2018.08.011

Schultz, P. W. (2001). Assessing the structure of environmental concern: Concern for self, other people, and the biosphere. [DOI: 10.1006/jevp.2001.0227]. *Journal of Environmental Psychology*, *21*(4), 327–339. doi:10.1006/jevp.2001.0227

Schulze, K., Knights, K., Coad, L., Geldmann, J., Leverington, F., Eassom, A., Marr, M., Butchart, S. H., Hockings, M., & Burgess, N. D. (2018). An assessment of threats to terrestrial protected areas. *Conservation Letters*, *11*(3), e12435. doi:10.1111/conl.12435

Schwarz, A. M. (2004). Contribution of photosynthetic gains during tidal emersion to production of Zostera capricorni in a North Island, New Zealand estuary. *New Zealand Journal of Marine and Freshwater Research*, *38*(5), 809–818. doi:10.1080/00288330.2004.9517280

Secretariat of the Convention on Biological Diversity. (2014). *Global Biodiversity Outlook 4*. Secretariat of the Convention on Biological Diversity.

Shaffer, F., & Ginsberg, J. P. (2017). An Overview of Heart Rate Variability Metrics and Norms. *Frontiers in Public Health*, *5*(September), 1–17. doi:10.3389/fpubh.2017.00258 PMID:29034226

Sharma, R., Sinha, A., & Kautish, P. (2020). Does renewable energy consumption reduce ecological footprint? *Evidence from eight developing countries of Asia. MPRA Paper No. 104277.* MPRA. https://mpra.ub.uni-muenchen.de/104277/

Shearmur, R., Carrincazeaux, C., & Doloreux, D. (Eds.). (2016). *Handbook on the Geographies of Innovation*. Edward Elgar. doi:10.4337/9781784710774

Sheehan, P. M. (2001). History of marine biodiversity. *Geological Journal*, *36*(3–4), 231–249. doi:10.1002/gj.890

Shendage, S. M., & Yadav, S. R. (2010). Revision of the Genus Barleria (Acanthaceae) in India. *Rheedea*, *20*(2), 81–230.

Shoo, L. P., Storlie, C., Vanderwal, J., Little, J., & Williams, S. E. (2011). Targeted protection and restoration to conserve tropical biodiversity in a warming world. *Global Change Biology*, *17*(1), 186–193. doi:10.1111/j.1365-2486.2010.02218.x

Short, F. T., McKenzie, L. J., Coles, R. G., & Vidler, K. P. (2002). Sea grass Net Manual for Scientific Monitoring of Sea grass Habitat. QDPI,QFS, Ciarns.

Shove, E. (2003a). Convenciones convergentes de comodidad, limpieza y conveniencia. *Journal of Consumer Policy*, *26*(diciembre), 395–418. doi:10.1023/A:1026362829781

Shove, E. (2003b). *Usuarios, tecnologías y expectativas de confort, limpieza y conveniencia. Innovación: The European Journal of Social Slater, D., & Tonkiss, F. (2001). Sociedad de mercado: mercados y teoría social moderna*. Prensa Política.

Sigwart, J. D., Pollom, R., & Böhm, M. (2023). The IUCN Species Survival Commission launches a new Red List Authority to assess marine invertebrates. *Oryx*, *57*(4), 418–419. doi:10.1017/S003060532300042X

Simberloff, D. (2009). The Role of Propagule Pressure in Biological Invasions. *Annual Review of Ecology, Evolution, and Systematics*, *40*(1), 81–102. doi:10.1146/annurev.ecolsys.110308.120304

Singh, A. K., & Jyoti, B. (2023b). Empirical evidence from different income group countries on the interdependency between digitalization and economic development. In Patricia Ordóñez de Pablos and Xi Zhang (Eds.), 5G, Artificial Intelligence, and Next Generation Internet of Things: Digital Innovation for Green and Sustainable Economies, (pp. 44-85). IGI Global. DOI: doi:10.4018/978-1-6684-8634-4.ch003

Singh, A. K., Kumar, S., & Jyoti, B. (2022a). Association of sustainable food security with climatic and non-climatic factors in Gujarat state of India: A district-wise panel data investigation. In: Leal Filho, W., Kovaleva, M., Popkova, E. (eds.), Sustainable Agriculture and Food Security. World Sustainability Series (pp. 259-282). Springer Cham. doi:10.1007/978-3-030-98617-9_16

Singh, J. S. (2012). Biodiversity: An Overview. *Proceedings of the National Academy of Sciences, India Section B: Biological Sciences*. 10.1007/s40011-012-0112-3

Singh, A. K., Issac, J., & Narayanan, K. G. S. (2019). Measurement of environmental sustainability index and its association with socio-economic indicators in selected Asian economies: An empirical investigation. *International Journal of Environment and Sustainable Development*, *18*(1), 57–100. doi:10.1504/IJESD.2019.098641

Singh, A. K., & Jyoti, B. (2023a). Appropriate technology and adaptation strategies mitigate the adverse impact of climate change on the agricultural sector: A case study in Gujarat, India. In P. H. Alex Khang (Ed.), *Advanced Technologies and AI-Equipped IoT Applications in High-Tech Agriculture* (pp. 389–416). IGI Global., doi:10.4018/978-1-6684-9231-4.ch023

Singh, A. K., Jyoti, B., Kumar, S., & Lenka, S. K. (2021). Assessment of global sustainable development, environmental sustainability, economic development and social development index in selected economies. *International Journal of Sustainable Development and Planning*, *16*(1), 123–138. doi:10.18280/ijsdp.160113

Singh, A. K., Jyoti, B., & Mishra, V. K. (2023b). Policy Implications from selected countries for a sustainable future of the society through green entrepreneurship and economic development. In A. Mohammed El Amine, S. Ul-Durar, & H. Hattab (Eds.), *Entrepreneurship Ecosystems and Their Opportunities and Challenges* (pp. 129–162). IGI Global. doi:10.4018/978-1-6684-7140-1.ch006

Singh, A. K., & Kumar, S. (2022). Exploring the impact of sustainable development on social-economic, and science and technological development in selected countries: A panel data analysis. *Society & Sustainability*, *4*(1), 55–83. doi:10.38157/ss.v4i1.405

Singh, A. K., Kumar, S., & Jyoti, B. (2022b). Does sustainable livelihood security have a relationship with climatic and geographical factors? evidence from a state-wise panel data investigation in India. *The IUP Journal of Applied Economics*, *21*(1), 7–35.

Singh, A. K., Sharma, A. K., & Jyoti, B. (2023a). Does economic development have a causal relationship with environmental degradation? Experience from different income group countries. In G. Sart (Ed.), *Considerations on Education for Economic, Social, and Environmental Sustainability* (pp. 300–333). IGI Global. doi:10.4018/978-1-6684-8356-5.ch015

Singh, A. K., & Singh, S. (2024). A case study from Kanpur Nagar, Uttar Pradesh for assessing the economic value of water and its determinants. In A. Karmaoui (Ed.), *Water-Soil-Plant-Animal Nexus in the Era of Climate Change* (pp. 212–242). IGI Global. doi:10.4018/978-1-6684-9838-5.ch010

Sisaye, S. (2012), An ecological analysis of four competing approaches to sustainability development. *Integration of industrial ecology and ecological anthropology literature.*

Slater, D. y Tonkiss, F. (2001). Sociedad de mercado: mercados y teoría social moderna. Prensa Política.

Slootweg, R., & Kolhoff, A. (2003). A generic approach to integrate biodiversity considerations in screening and scooping for EIA. *Environmental Impact Assessment Review*, *23*(6), 657–681. doi:10.1016/S0195-9255(03)00114-8

Smart, M. (1997). Sehumed. Boletin de la Sedepara el Estudio de los Humedales Mediterraneos.

Smith, V. H., Tilman, G. D., & Nekola, J. C. (1999). Eutrophication: Impacts of excess nutrient inputs on freshwater, marine, and terrestrial ecosystems. *Environmental Pollution (Barking, Essex: 1987), 100*(1–3), 179–196. doi:10.1016/S0269-7491(99)00091-3

Smith, P., Arneth, A., Barnes, D. K. A., Ichii, K., Marquet, P. A., Popp, A., Pörtner, H. O., Rogers, A. D., Scholes, R. J., Strassburg, B., Wu, J., & Ngo, H. (2022). How do we best synergize climate mitigation actions to co-benefit biodiversity? *Global Change Biology*, *28*(8), 2555–2577. doi:10.1111/gcb.16056 PMID:34951743

Soberón, J., & Peterson, A. T. (2009). Monitoring biodiversity loss with primary species-occurrence data: Toward national-level indicators for the 2010 target of the convention on biological diversity. *Ambio*, *38*(1), 29–34. doi:10.1579/0044-7447-38.1.29 PMID:19260344

Solou, R. (2002). Theory of growth. Panorama of the economic thought of the end of the twentieth century. SPb.: Econ. Shk.

Sörlin, S. (2007). *El Antropoceno. Un ensayo sobre las edades del hombre; Weylers*. Estocolmo, Suecia.

Sponsel, L. E. (2013). Human impact on biodiversity, overview. Encyclopedia of Biodiversity, 4, 137–152.

Standard, W., Zielasko, D., Dawans, B. Von, & Domes, G. (2020). *Open TSST VR : Trier Social Stress Test in Virtual Reality*. September.

Steffen, W., Richardson, K., Rockström, J., Cornell, S. E., Fetzer, I., Bennett, E. M., Biggs, R., Carpenter, S. R., de Vries, W., de Wit, C. A., Folke, C., Gerten, D., Heinke, J., Mace, G. M., Persson, L. M., Ramanathan, V., Reyers, B., & Sörlin, S. (2015). Planetary boundaries: Guiding human development on a changing planet. *Science*, *347*(6223), 1259855. doi:10.1126/science.1259855 PMID:25592418

Stephenson, P. J. (2020). Technological advances in biodiversity monitoring: Applicability, opportunities and challenges. *Current Opinion in Environmental Sustainability*, *45*, 36–41. doi:10.1016/j.cosust.2020.08.005

Stoeglehner, G., & Narodoslawsky, M. (2008). Implementing ecological footprinting in decision-making processes. *Land Use Policy*, *25*(1), 421–431. doi:10.1016/j.landusepol.2007.10.002

Stoner, A. M. (2021). Las cosas están empeorando en nuestro camino hacia la catástrofe: el ecologismo neoliberal, la desublimación represiva y el ecoconsumidor autónomo. *Sociología crítica*, *47* (mayo), 491-506. doi:10.1177/0896920520958099

Stoner, A. M. (2021). Las cosas están empeorando en nuestro camino hacia la catástrofe: El ecologismo neoliberal, la desublimación represiva y el ecoconsumidor autónomo. *Sociología crítica*, *47* (mayo), 491-506.

Streck, C. (2023). Synergies between the Kunming-Montreal Global Biodiversity Framework and the Paris Agreement: The role of policy milestones, monitoring frameworks and safeguards. *Climate Policy*, *23*(6), 800–811. doi:10.1080/14693062.2023.2230940

Strengbom, J., Dahlberg, A., Larsson, A., Lindelöw, Å., Sandström, J., Widenfalk, O., & Gustafsson, L. (2011). Introducing intensively managed spruce plantations in Swedish: Forest landscapes will impair biodiversity decline. *Forests*, *2*(3), 610–630. doi:10.3390/f2030610

Stroud, D. A., Davidson, N. C., Finlayson, C. M., & Gardner, R. C. (2022). Development of the text of the Ramsar Convention: 1965–1971. *Marine and Freshwater Research*, *73*(10), 1107–1126. doi:10.1071/MF21312

Suárez, A., Watson, R. T., & Dokken, D. J. (2002). *Climate change and biodiversity*. Research Gate.

Suding, K. N. (2011). Toward an era of restoration in ecology: Successes, failures, and opportunities ahead. *Annual Review of Ecology, Evolution, and Systematics*, *42*(1), 465–487. doi:10.1146/annurev-ecolsys-102710-145115

Taheri, M., Foshtomi, M. Y., Noranian, M., & Mira, S. S. (2012). Spatial distribution and biodiversity of macrofauna in the southeast of the Caspian Sea, Gorgan Bay in relation to environmental conditions. *Ocean Science Journal*, *47*(2), 113–122. doi:10.1007/s12601-012-0012-8

Takov, P., & Djia, V. (2020). Wittgenstein's Ethical Individualism as a Foundation for Environmental Ethics. *Open Journal of Philosophy*, *10*(04), 427–442. doi:10.4236/ojpp.2020.104030

Tarfaoui, D., & Salah Zkim, S. (2017). Brecha actitud-comportamiento ecológico: Un análisis teórico. [ESMB]. *International Journal of Economics & Strategic Management of Business Process*, *8*(1), 33–38.

Tasoff, J., Mee, M. T., & Wang, H. H. (2015). An economic framework of microbial trade. *PLoS One*, *10*(7), e0132907. doi:10.1371/journal.pone.0132907 PMID:26222307

Taylor, C. (2002). Imaginarios sociales modernos. *Cultura pública, 14*(1), 91–124. doi:10.1215/08992363-14-1-91

Taylor, B., Chapron, G., Kopnina, H., Orlikowska, E., Gray, J., & Piccolo, J. J. (2020). The need for ecocentrism in biodiversity conservation. *Conservation Biology*, *34*(5), 1089–1096. doi:10.1111/cobi.13541 PMID:32424955

Teeuwen, R., Milias, V., Bozzon, A., & Psyllidis, A. (2024). How well do NDVI and OpenStreetMap data capture people's visual perceptions of urban greenspace? *Landscape and Urban Planning*, *245*, 105009. doi:10.1016/j.landurbplan.2024.105009

Terrapass. (2023). *Cost of Carbon Offset*. Terrapass. https://terrapass.com/blog/creating-carbon-credits-is-it-profitable/#:~:text=Carbon%20credits%20in%202023%20cost,can%20fluctuate%20greatly%20with%20demand

Tetior, A. (2015). New Philosophy of Plural World, New Ethics of Empathy. *OAlib*, *02*(01), 1–6. doi:10.4236/oalib.1101193

The European Sustainable Development Strategy (2001). *The potential for sustainable development of 2001*. A sustainable Europe for a better World // enterprise of chemical industry. M.: Publishing house Europe – Sustainable Development. SDS 2005-2010.

Thomas, C. D., & Gillingham, P. K. (2015). The performance of protected areas for biodiversity under climate change. *Biological Journal of the Linnean Society. Linnean Society of London*, *115*(3), 718–730. doi:10.1111/bij.12510

Thomas, N. J. R. (2018). Marketing de sostenibilidad. La necesidad de un enfoque realista de sistemas completos. *Journal of Marketing Management*, *34*(17–18), 1530–1556. doi:10.1080/0267257X.2018.1547782

Thommpson, I. D., Davidson, I. J., O' Donnell, S., & Brazeau, F. (1989). Use of track transects to measure the relative occurrence of some arboreal mammals in uncut forest and regeneration stands. *Canadian Journal of Zoology*, *67*(7), 1816–1823. doi:10.1139/z89-258

Thuiller, W., Maiorano, L., Mazel, F., Guilhaumon, F., Ficetola, G. F., Lavergne, S., Renaud, J., Roquet, C., & Mouillot, D. (2015). Conserving the functional and phylogenetic trees of life of European tetrapods. *Philosophical Transactions of the Royal Society B: Biological Sciences, 370*(1662), 20140005. doi:10.1098/rstb.2014.0005

Tidd, J. (1997). *Managing Innovation: integrating technological, market and organizational change*. John Wiley & Sons.

Tilman, D., Clark, M., Williams, D. R., Kimmel, K., Polasky, S., & Packer, C. (2017). Future threats to biodiversity and pathways to their prevention. *Nature, 546*(7656), 73–81. doi:10.1038/nature22900

Tisdale, H. (1942). The process of urbanization. *Social Forces*, *20*(3), 311. doi:10.2307/3005615

Tiwari, G., & Singh, S. P. (2014). Application of bioremediation on solid waste management: A review. *Journal of Bioremediation & Biodegradation*, *5*(6), 248–1172.

Tödtling, F., & Trippl, M. (2005). One size fits all? Towards a differentiated regional innovation policy. *Research Policy*, *34*(8), 203–1219. doi:10.1016/j.respol.2005.01.018

Toledo, V. M. (1994). *Biodiversity islands in a sea of pasturelands: indigenous resource management in the wet tropics of Mexico Etnoecológica*. Etnoecológica. http://www.etnoecologica.org.mx/Etnoecologica_vol2_n3/art_toledo.htm

Toledo, V. M. (1996). Principios etnoecológicos para el desarrollo sustentable de comunidades campesinas e indígenas. *Documentos Red Latino Americana y Caribeña de Ecología Social.*

Toledo, V. M. (1997) Sustainable development at the village community level: a Third World perspective in: Fraser Smith (Ed.) Environmental Sustainability Boca Raton St. Lucie Press (pp. 233–250). doi:10.1201/9781420048742.ch11

Tong, S., Beggs, P. J., Davies, J. M., Jiang, F., Kinney, P. L., Liu, S., Yin, Y., & Ebi, K. L. (2023). Compound impacts of climate change, urbanization and biodiversity loss on allergic disease. *International Journal of Epidemiology*, *52*(3), 655–663. doi:10.1093/ije/dyac197 PMID:36228124

Tran, T. C., Ban, N. C., & Bhattacharyya, J. (2020). A review of successes, challenges, and lessons from Indigenous protected and conserved areas. *Biological Conservation*, *241*, 108271. doi:10.1016/j.biocon.2019.108271

Trist, E. L. (1976). *A Concept of Organizational Ecology: an invited address to the three Melbourne universities.*

Trosper, R. L. (2009). *Resiliencia, reciprocidad y economía ecológica: sostenibilidad de la costa noroeste*. Routledge. doi:10.4324/9780203881996

Trujillo, M. L. N. (2021). Apuntes para una visión crítica y ecológica del capitalismo patriarcal en la red de la vida. *Capital and Class*, *45*(marzo), 21–32. doi:10.1177/0309816820929115

Turner, N. J., Cuerrier, A., & Joseph, L. (2022). Well grounded: Indigenous Peoples' knowledge, ethnobiology and sustainability. *People and Nature*, *4*(3), 627–651. doi:10.1002/pan3.10321

Turner, W., Spector, S., Gardiner, N., Fladeland, M., Sterling, E., & Steininger, M. (2003). Remote sensing for biodiversity science and conservation. *Trends in Ecology & Evolution*, *18*(6), 306–314. doi:10.1016/S0169-5347(03)00070-3

Twohig-Bennett, C., & Jones, A. (2018). The health benefits of the great outdoors: A systematic review and meta-analysis of greenspace exposure and health outcomes. *Environmental Research*, *166*, 628–637. doi:10.1016/j.envres.2018.06.030 PMID:29982151

Ugolini, F., Massetti, L., Calaza-Martínez, P., Cariñanos, P., Dobbs, C., Krajter Ostoić, S., Marin, A. M., Pearlmutter, D., Saaroni, H., Šaulienė, I., Vuletić, D., & Sanesi, G. (2022). Understanding the benefits of public urban green space: How do perceptions vary between professionals and users? *Landscape and Urban Planning*, *228*, 104575. doi:10.1016/j.landurbplan.2022.104575

Ulucak, R., & Lin, D. (2017). Persistence of policy shocks to ecological footprint of the USA. *Ecological Indicators*, *80*(1), 337–343. doi:10.1016/j.ecolind.2017.05.020

UN. I. (1992). Convention on biological diversity. *Treaty Collection*. https://www.ipcc.ch/site/assets/uploads/2001/09/doc4f.pdf

Unerman, J., O'Dwyer, B., & Bebbington, J. (Eds.). (2007). *Contabilidad de sostenibilidad y rendición de cuentas*. Routledge.

United Nations Environment Program. (2005). Millennium Ecosystem Assessment. Ecosystems and Human Well-being: Current State and Trends. CH 4. *Biodiversity (Nepean).*

United Nations Environment Programme (UNEP). (2002). *Biodiversity and environmental assessment: Enhancing ecosystem services for human well-being*. UN. https://www.unep.org/

Valbuena-Carabaña, M., de Heredia, U. L., Fuentes-Utrilla, P., González-Doncel, I., & Gil, L. (2010). Historical and recent changes in the Spanish forests: A socio-economic process. *Review of Palaeobotany and Palynology*, *162*(3), 492–506. doi:10.1016/j.revpalbo.2009.11.003

Valkó, O., Tóth, K., Kelemen, A., Miglécz, T., Radócz, S., Sonkoly, J., Tóthmérész, B., Török, P., & Deák, B. (2018). Cultural heritage and biodiversity conservation–plant introduction and practical restoration on ancient burial mounds. *Nature Conservation*, *24*, 65–80. doi:10.3897/natureconservation.24.20019

Vallejo, V. R. (2009). In S. Bautista, J. Aronson, & V. R. Vallejo (Eds.), *Problems and perspectives of dryland restoration* (pp. 13–22). Land Restoration to Combat Desertification, Fundacion Centro de Estudios Ambientales del Mediterráneo.

Vann, J.W. & White, G.B. (2004). *Sustainability reporting in the accountingcurriculum*. Research Gate.

Velado-Alonso, E., Kleijn, D., & Bartomeus, I. (2024). Reassessing science communication for effective farmland biodiversity conservation. *Trends in Ecology & Evolution*. https://www.cell.com/trends/ecology-evolution/abstract/S0169-5347(24)00032-6

Venter, O., Magrach, A., Outram, N., Klein, C. J., Possingham, H. P., Di Marco, M., & Watson, J. E. M. (2018). Bias in protected-area location and its effects on long-term aspirations of biodiversity conventions. *Conservation Biology*, *32*(1), 127–134. doi:10.1111/cobi.12970 PMID:28639356

Verma, A. K. (2018). Ecological balance: An indispensable need for human survival. *Journal of Experimental Zoology India*, *21*(1), 407–409.

Vetlesen, A. J. (2020). Cosmologías del Antropoceno. Panpsiquismo, animismo y los límites del posthumanismo. Routledge.

Vimal, R., Navarro, L. M., Jones, Y., Wolf, F., Le Moguédec, G., & Réjou-Méchain, M. (2021). The global distribution of protected areas management strategies and their complementarity for biodiversity conservation. *Biological Conservation*, *256*, 109014. doi:10.1016/j.biocon.2021.109014

Viña, A., & Liu, J. (2017). Hidden roles of protected areas in the conservation of biodiversity and ecosystem services. *Ecosphere*, *8*(6), e01864. doi:10.1002/ecs2.1864

Visconti, P., Bakkenes, M., Baisero, D., Brooks, T., Butchart, S. H., Joppa, L., Alkemade, R., Di Marco, M., Santini, L., Hoffmann, M., Maiorano, L., Pressey, R. L., Arponen, A., Boitani, L., Reside, A. E., van Vuuren, D. P., & Rondinini, C. (2016). Projecting global biodiversity indicators under future development scenarios. *Conservation Letters*, *9*(1), 5–13. doi:10.1111/conl.12159

Vitt, L. J., & Pianka, E. R. (2005). Deep history impacts present-day ecology and biodiversity. *Proceedings of the National Academy of Sciences of the United States of America*, *102*(22), 7877–7881. doi:10.1073/pnas.0501104102 PMID:15867150

von Schomberg, R. (2012). Prospects for technology assessment in a framework of responsible research and innovation. Dusseldorp, M. & Beecroft, R. (eds.) Technikfolgen abschätzen lehren, 39–61. Wiesbaden: VS Verlag für Sozialwissenschaften. doi:10.1007/978-3-531-93468-6_2

Wackernagel, M., & Rees, W. (1996). Our ecological footprint: reducing human impact on the earth. *New Society Publishers.* https://escholarship.org/content/qt7730w81q/qt7730w81q.pdf?t=q9ns62

Walker, B. (1995). Conserving Biological Diversity through Ecosystem Resilience. *Conservation Biology*, *9*(4), 747–752. doi:10.1046/j.1523-1739.1995.09040747.x

Wang, S. (2023). International law-making process of combating plastic pollution: Status Quo, debates and prospects. *Marine Policy*, *147*, 105376. doi:10.1016/j.marpol.2022.105376

Wan, X., Jahanger, A., Usman, M., Radulescu, M., Balsalobre-Lorente, D., & Yu, Y. (2022). Exploring the effects of economic complexity and the transition to a clean energy pattern on ecological footprint from the Indian perspective. *Frontiers in Environmental Science*, *9*(1), 1–17. doi:10.3389/fenvs.2021.816519

Watson, J. E., Grantham, H. S., Wilson, K. A., & Possingham, H. P. (2011). Systematic conservation planning: past, present and future. *Conservation biogeography*, 136-160.

Watson, R., Baste, I., Larigauderie, A., Leadley, P., Pascual, U., Baptiste, B., Demissew, S., Dziba, L., Erpul, G., & Fazel, A. (2019). *Summary for policymakers of the global assessment report on biodiversity and ecosystem services of the Intergovernmental Science-Policy Platform on Biodiversity and Ecosystem Services*. IPBES Secretariat.

Weber, P., & Haklay, M. (2008). OpenStreetMap: User-generated street maps. *IEEE Pervasive Computing*, *7*(4).

Webler, T. (2001). Achieving convergence on local environmental disputes: The hard work of soft alternatives. [DOI: 10.2307/2693982]. *The Western Political Quarterly*, *54*(1), 21–47.

Wei, Z., Jian, Z., Sun, Y., Pan, F., Han, H., Liu, Q., & Mei, Y. (2023). Ecological sustainability and high-quality development of the yellow rive delta in China based on the improved ecological footprint model. *Scientific Reports*, *13*(1), 1–16. doi:10.1038/s41598-023-30896-2 PMID:36593249

Welch, F. S. (1948). *Limnological Methods*. McGraw Hill Book Co Inc.

Welsh B. (1987). *Monitoring herpetofauna in woodlands of north western California and south west Oregon: a comparative strategy*.

Welsh, H.H. Jr. & Lind, A. (1991). *The structure of the herpetofaunal assemblage in the Douglas-fir/hardwood forests of northwestern California and south western Oregon*.

Werner, G. D. A., Strassmann, J. E., Ivens, A. B. F., Engelmoer, D. J. P., Verbruggen, E., Queller, D. C., Noë, R., Johnson, N. C., Hammerstein, P., & Kiers, E. T. (2014). Evolution of microbial markets. *Proceedings of the National Academy of Sciences of the United States of America*, *111*(4), 1237–1244. doi:10.1073/pnas.1315980111 PMID:24474743

Weston, A. (1995). Habilitación de la práctica Ambiental. Christine Pierce & Donald VanDeVeer (Eds) People, Penguins and Plastic Trees Belmond Wadsworth.

White, M. P., Yeo, N. L., & Albin, M. (2018). A prescription for "nature" – the potential of using virtual nature in therapeutics. *Neuropsychiatric Disease and Treatment*, *14*, 3001–3013. doi:10.2147/NDT.S179038 PMID:30510422

Wiederkehr, F., Wilkinson, C. L., Zeng, Y., Yeo, D. C. J., Ewers, R. M., & O'Gorman, E. J. (2020). Urbanisation affects ecosystem functioning more than structure in tropical streams. *Biological Conservation*, *249*, 108634. doi:10.1016/j.biocon.2020.108634

Wiedmann, T., & Barrett, J. (2010). A review of the ecological footprint indicator—Perceptions and methods. *Sustainability (Basel)*, *2*(6), 1645–1693. doi:10.3390/su2061645

Wiig, H., & Wood, M. (1995). *What Comprises a Regional Innovation System? An empirical study*. Studies in Technology, Innovation and Economic Policy Group.

Wilson, D. E., & Reeder, D. M. (Eds.). (1993). *Mammal Species of the World a Taxonomic and Geographic reference* (2nd ed.). Smithsonian Institution Press.

Wilson, D. E., & Reeder, D. M. (Eds.). (2005). *Mammal Species of the World. A Taxonomic and Geographic Reference* (3rd ed.). Johns Hopkins University Press. doi:10.56021/9780801882210

Wilson, R. J., Thomas, C. D., Fox, R., Roy, D. B., & Kunin, W. E. (2004). Spatial patterns in species distributions reveal biodiversity change. *Nature*, *432*(7015), 393–396. doi:10.1038/nature03031 PMID:15549106

Wolch, J. R., Byrne, J., & Newell, J. P. (2014). Urban green space, public health, and environmental justice: The challenge of making cities "just green enough.". *Landscape and Urban Planning*, *125*, 234–244. doi:10.1016/j.landurbplan.2014.01.017

Wolford, W. (2021). El Plantationocene: Una contribución lusotrópica a la teoría. *Annals of the American Association of Geographers*, *111*, 1622–1639.

Wondolleck, J. M., & Yaffee, S. L. (2000). *Making collaboration work: Lessons from innovation in natural resource management*. Island Press.

Wong, J. S. Y., Soh, M. C. K., Low, B. W., & Er, K. B. H. (2023). Tropical bird communities benefit from regular-shaped and naturalised urban green spaces with water bodies. *Landscape and Urban Planning*, *231*, 104644. doi:10.1016/j.landurbplan.2022.104644

Wood, E., Harsant, A., Dallimer, M., de Chavez, A. C., McEachan, R. R. C., & Hassall, C. (2018). Not all green space is created equal: Biodiversity predicts psychological restorative benefits from urban green space. *Frontiers in Psychology*, *9*(NOV), 2320. Advance online publication. doi:10.3389/fpsyg.2018.02320 PMID:30538653

World Conservation Monitoring Centre. (1988). *The Conservation of Biological Diversity. WCMC*. I.U.C.N.

World Conservation Monitoring Centre. (2000). *Global Biodiversity: Earth's living resources in the 21st Century. By: Groombridge B. and Jenkins, M.D.* World Conservation Press.

World Wildlife Fund. (2022). Living [Bending the Curve of Biodiversity Loss. World Wildlife Fund International.]. *The Planetary Report*.

Worm, B., Hilborn, R., Baum, J. K., Branch, T. A., Collie, J. S., Costello, C., Fogarty, M. J., Fulton, E. A., Hutchings, J. A., Jennings, S., Jensen, O. P., Lotze, H. K., Mace, P. M., McClanahan, T. R., Minto, C., Palumbi, S. R., Parma, A. M., Ricard, D., Rosenberg, A. A., & Zeller, D. (2009). Rebuilding Global Fisheries. *Science*, *325*(5940), 578–585. doi:10.1126/science.1173146 PMID:19644114

Wu, S., Wang, J., Jia, Y., & Wang, X. (2023). A comparative study on the ecological footprint of living consumption in northwest ethnic regions: 1980-2018. *Frontiers in Environmental Science*, *11*(1), 1–12. doi:10.3389/fenvs.2023.1046203

Wymer, W., & Polonsky, M. J. (2015). Las limitaciones y potencialidades del marketing verde. *Journal of Nonprofit & Public Sector Marketing*, *27*(3), 239–262. doi:10.1080/10495142.2015.1053341

Xue, P., Minasny, B., Wadoux, A. M. J.-C., Dobarco, M. R., McBratney, A., Bissett, A., & De Caritat, P. (2024). Drivers and human impacts on topsoil bacterial and fungal community biogeography across Australia. *Global Change Biology*, *30*(3), e17216. doi:10.1111/gcb.17216 PMID:38429628

Xu, W., Xiao, Y., Zhang, J., Yang, W., Zhang, L., Hull, V., Wang, Z., Zheng, H., Liu, J., Polasky, S., Jiang, L., Xiao, Y., Shi, X., Rao, E., Lu, F., Wang, X., Daily, G. C., & Ouyang, Z. (2017). Strengthening protected areas for biodiversity and ecosystem services in China. *Proceedings of the National Academy of Sciences of the United States of America*, *114*(7), 1601–1606. doi:10.1073/pnas.1620503114 PMID:28137858

Yang, B. Y., Zhao, T., Hu, L. X., Browning, M. H. E. M., Heinrich, J., Dharmage, S. C., Jalaludin, B., Knibbs, L. D., Liu, X. X., Luo, Y. N., James, P., Li, S., Huang, W. Z., Chen, G., Zeng, X. W., Hu, L. W., Yu, Y., & Dong, G. H. (2021). Greenspace and human health: An umbrella review. In Innovation, 2(4). doi:10.1016/j.xinn.2021.100164

Yang, X., Tan, X., Chen, C., & Wang, Y. (2020). The influence of urban park characteristics on bird diversity in Nanjing, China. *Avian Research, 11*(1), 45. doi:10.1186/s40657-020-00234-5

Yilanci, V., & Pata, U. K. (2020). Convergence of per capita ecological footprint among the ASEAN-5 countries: Evidence from a non-linear panel unit root test. *Ecological Indicators*, *113*(1), 13-26. 106178. doi:10.1016/j.ecolind.2020.106178

Yngfalk, C. (2019). Subvertir la sostenibilidad: Trabajos de mantenimiento del mercado y reproducción de la irresponsabilidad corporativa. *Journal ofMarketing Management, 35*(17–18), 1563–1583. doi:10.1080/0267257X.2019.1682031

York University Ecological Footprint Initiative & Global Footprint Network (YUEFI & GFN). (2023). *Public data package of the national footprint and biocapacity accounts*. Produced for the Footprint Data Foundation and distributed by Global Footprint Network. https://data.footprintnetwork.org

Young, C. R. (2015). *The Royal Forests of Medieval England*. University of Pennsylvania Press.

Young, J., Watt, A., Nowicki, P., Alard, D., Clitherow, J., Henle, K., Johnson, R., Laczko, E., McCracken, D., Matouch, S., Niemela, J., & Richards, C. (2005). Towards sustainable land use: Identifying and managing the conflicts between human activities and biodiversity conservation in Europe. *Biodiversity and Conservation, 14*(7), 16411661. doi:10.1007/s10531-004-0536-z

Zambrano, J., Garzon-Lopez, C. X., Yeager, L., Fortunel, C., Cordeiro, N. J., & Beckman, N. G. (2019). The effects of habitat loss and fragmentation on plant functional traits and functional diversity: what do we know so far? In Oecologia, 191(3). doi:10.1007/s00442-019-04505-x

Zambrano-Monserrate, M., Ruano, M. A., Ormeño-Candelario, V., & Sanchez-Loor, D. A. (2020). Global ecological footprint and spatial dependence between countries. *Journal ofEnvironmental Management, 272*(15), 33–42. doi:10.1016/j.jenvman.2020.111069 PMID:32854881

Zameer H, Yasmeen H (2022) *Green innovation and environmental awareness driven green purchase intentions*. Mark Intell Plan.

ZanL.LusianiM. (2011). Managing Machu Picchu: Institutional Settings, Business Model and Master Plans - A Virtual Appendix (22 de enero de 2011). Disponible en SSRN: https://ssrn.com/abstract=1745322 o doi:10.2139/ssrn.1745322

Zedler, J. B., & Stevens, M. L. (2018). Conocimiento ecológico occidental y tradicional en la restauración ecocultural. *San Francisco Estuary and Watershed Science, 16*, 1–18.

Zeng, X., Radford, S., & Mourali, M. (2017). Antecedentes y consecuencias de las creencias ambientales: Un metaanálisis. En *AMA Marketing & Public Policy Academic Conference Proceedings* (*Vol. 27*, pp. 55–56). Asociación Americana de Marketing.

Zhang, C., & Lin, Y. (2012). Panel estimation for urbanization, energy consumption and CO2 emissions: A regional analysis in China. *Energy Policy, 49*(1), 488–498. doi:10.1016/j.enpol.2012.06.048

Zhang, S., Zhou, Y., Yu, R., Xu, X., Xu, M., Li, G., Wang, W., & Yang, Y. (2022). China's biodiversity conservation in the process of implementing the sustainable development goals (SDGs). *Journal of Cleaner Production, 338*, 130595. doi:10.1016/j.jclepro.2022.130595

Zhang, W. J. (2008). A forecast analysis on world population and urbanization process. *Environment, Development and Sustainability, 10*(6), 717–730. doi:10.1007/s10668-007-9081-8

Zhang, Y., Tariq, A., Hughes, A. C., Hong, D., Wei, F., Sun, H., Sardans, J., Peñuelas, J., Perry, G., Qiao, J., Kurban, A., Jia, X., Raimondo, D., Pan, B., Yang, W., Zhang, D., Li, W., Ahmed, Z., Beierkuhnlein, C., & Ma, K. (2023). Challenges and solutions to biodiversity conservation in arid lands. *The Science of the Total Environment*, *857*, 159695. doi:10.1016/j.scitotenv.2022.159695 PMID:36302433

Zhang, Z. H., Jhaveri, D. J., Marshall, V. M., Bauer, D. C., Edson, J., & Narayanan, R. K. (2014). Un estudio comparativo de técnicas para el análisis de expresión diferencial en datos de RNA-Seq. *PLoS One*, *9*(8), e103207. doi:10.1371/journal.pone.0103207 PMID:25119138

Zhao, J. (2020). The role of international organizations in preventing conflicts between the SPS Agreement and the Cartagena Protocol on Biosafety. *Review of European, Comparative & International Environmental Law*, *29*(2), 271–281. doi:10.1111/reel.12326

Zhao, L. (2023). The Economic and Environmental Impacts of Urbanization in China. *Highlights in Business. Economics and Management*, *5*, 142–146. doi:10.54097/hbem.v5i.5041

Zhou, X., & Rana, M. M. P. (2012). Social benefits of urban green space: A conceptual framework of valuation and accessibility measurements. In Management of Environmental Quality, 23(2). doi:10.1108/14777831211204921

Zoffoli, M. L., Topouzelis, K., Traganos, D., Change, G., Traganos, D., Lee, C. B., Blume, A., Poursanidis, D., Čižmek, H., & Deter, J. (2024). The seagrass Posidonia oceanica is the main habitat-forming species of the coastal Mediterranean, providing millennial-scale ecosystem services including habitat provisioning, biodiversity maintenance, food security, coastal protection, and carbon sequestration. *Blue Carbon: Beyond the Inventory*, 186.

Zoomers, A., Gekker, A., & Schäfer, M. T. (2016). Entre dos exageraciones: ¿Ayudará el big data a desentrañar los puntos ciegos en la comprensión de la fiebre global de la tierra? *Geoforum*, *69*, 147–159. doi:10.1016/j.geoforum.2015.11.017

About the Contributors

Ashok K. Rathoure brings an impressive and diverse background for environment protection. His extensive education, with a doctorate degree from Central University (HNBGU Srinagar UK), coupled with over 18 years of combined industry and teaching experience, establishes him as a seasoned expert in the environment field. Currently he is working as Research Director at Saraca Research Inc., India. Dr. Rathoure's expertise lies in Ecology & Biodiversity, Water Pollution Monitoring, Prevention & Control, and EIA (Environmental Impact Assessment) coordination for various sectors, aligning with NABET (QCI, New Delhi) guidelines. His experience encompasses a wide range of industrial sectors, including synthetic organic chemicals, mining, metallurgical industries, power generation, nuclear sector, coal, cement, pesticides, chlor-alkali, chemical fertilizers, paint, sugar, oil & gas, man-made fibers, distilleries, hazardous waste treatment, and more. In his academic and research pursuits, Dr. Rathoure has contributed significantly, with a remarkable record of 100 full-length research papers published in international and national journals. Additionally, he has authored over 50 books from reputed publishers in India and abroad. His dedication to advancing scientific knowledge is further evident through his role as a reviewer for more than 100 research manuscripts in international journals. Being a member of various international organizations such as APCBEES (Hong Kong), IACSIT (Singapore), EFB (Spain), and the Society for Conservation Biology (Washington), Dr. Rathoure has established a global network and stays abreast of international developments in his field. As a founder member of the Scientific Planet Society (Dehradun), he contributes to the promotion of scientific awareness and collaboration. Furthermore, Dr. Rathoure's commitment to the dissemination of knowledge is reflected in his editorial roles. He serves as the Editor-in-Chief for Octa Journal of Environmental Research, Managing Editor for Octa Journal of Biosciences, and Executive Editor for Scientific India. His significant publication record, including research papers, books, and book chapters, underscores his scholarly contributions to advancing knowledge in environmental biotechnology and related fields. Furthermore, his editorial roles for scientific journals demonstrate his dedication to fostering academic discourse and dissemination of research findings within the scientific community.

***

Munir Ahmad is a seasoned professional in the realm of Spatial Data Infrastructure (SDI), Geo-Information Productions, Information Systems, and Information Governance, boasting over 25 years of dedicated experience in the field. With a PhD in Computer Science, Dr. Ahmad's expertise spans Spatial Data Production, Management, Processing, Analysis, Visualization, and Quality Control. Throughout his career, Dr. Ahmad has been deeply involved in the development and deployment of SDI systems specially in the context of Pakistan, leveraging his proficiency in Spatial Database Design, Web, Mobile & Desktop GIS, and Geo Web Services Architecture. His contributions to Volunteered Geographic Information (VGI) and Open Source Geoportal & Metadata Portal have significantly enriched the geospatial community. As a trainer and researcher, Dr. Ahmad has authored over 50 publications, advancing the industry's knowledge base and fostering innovation in Geo-Tech, Data Governance, and Information Infrastructure, and Emerging Technologies. His commitment to Research and Development (R&D) is evident in his role as a dedicated educator and mentor in the field.

J. Anuradha, Assistant Professor at Nims Institute, Nims University Rajasthan, Jaipur. Holds degrees in B.Sc. Botany (2003), M.Sc. Botany (2005), M.Phil Envi. Sci. & Engg. (2006), and Ph.D. Environmental Technology (2014) from Pondicherry University. Specializes in Environmental Technology and Advanced Nanotechnology, evident through 3 patents, 55 publications, 1 authored book, and 9 book chapters. Successfully guided three Ph.D. scholars. Recognized with eight best paper awards in scientific discussions & 4 student best paper awards. Actively contributes as a 'Resource person and Subject Specialist' and Chairperson in national/international scientific meets. Dr. Anuradha's research demonstrates a commitment to advancing environmental science while her mentorship nurtures future scholars, ensuring a lasting impact on the field.

Betül Aydın is a research scientist serving as an Assistant Professor at the Department of Biology, Faculty of Science, Gazi University. Specializing in Biotechnology, Molecular Biology, and Genetics, she focuses her research on various aspects including antimicrobial activities, antioxidant properties, and DNA interactions. Her work often revolves around the synthesis of biological compounds, their structural characterization, and biological activities. She has contributed to a number of research articles published in scientific journals.

Erdi Can Aytar is currently a Research Assistant in the Department of Horticulture, at Usak University, Turkiye. He received her Master of Science degree in biology from Gazi University in 2016, and her Ph.D. in biology from Ondokuz Mayıs University in 2024.He has research experience in Antioxidant and antimicrobial activity, cancer, isolation of organic compounds, plant physiology and germination. He has worked on national and international projects. He has published several research articles.

Hrish Chandra is working as Assistant Professor (Microbiology) in Department of Botany and Microbiology, Gurukula Kangri (Deemed to be University), Haridwar. He has a teaching and research experience of more than 16 years taking classes of UG and PG courses. He has been ex- faculty of H.N.B. Garhwal University (A Central University) Srinagar, Garhwal, G.B. Pant Engineering College, Pauri, Garhwal, G.C.B.M.S, Dehradun and D.I.B.N.S, Dehradun. He has published more than 62 research articles in International and national Journal. He is member of various scientific societies and recipient of fellow awards from scientific societies.

Vneet Kumar is a renowned expert in the field of Malariology, Immunology, Animal physiology and anatomy. He has developed a fascination for the human body and his inquisitive mind and passion for learning led him to pursue a career in teaching and medical research. Dr. Vineet Kumar, has done M.Sc. (Hons.) M.Phil., Ph.D. form a premium institute Panjab University, Chandigarh, India and have a more than eight year of teaching experience in undergraduate and postgraduate level. He is presently serving as Assistant Professor and Head in department of Zoology at Dr. Yashwant Singh Parmar Govt. P.G. College Nahan, Distt. Sirmour, Himachal Pradesh. He researched on parasite immunology and antimalarial vaccines to get his Ph.D., from Panjab University, Chandigarh. His involvement in both teaching and research is active. His area of research is Protozoology (Malariology), Immunology, Entomology, Ecology Animal Physiology and Anatomy, contributing to advancements in medical understanding and technology. Dr. Vineet participated and presented papers at various national and international conferences. Numerous of his research articles have previously appeared in reputable journals on a national and worldwide level. He has awarded with Best faculty award and Young Scientist award. He is a member of Indian Science Congress Association, Vector and Vectors Borne Diseases, Indian Society of Parasitology, Microbiology Society of India and Editorial Board member of Journal of Infectious Disease Research and Guest Editor of Journal of Traditional Medicine Research, New Zealand. He continues to write and continually updates his knowledge in his profession through additional research and study.

Foilan D. Mobo is a Doctor of Public Administration graduate from the Urdaneta City University Class of 2016 and a graduate of the 2nd Doctorate Degree (Ph.D.) in Development Education program at the Central Luzon State University, Nueva Ecija, Philippines, Class of 2022. After obtaining his doctorate degree, he was promoted to the position of Professor II in the Philippine Merchant Marine Academy (PMMA), and this allowed him to work with different international research institutions, such as the Director and Research Consultant of the IKSAD Research Institute, Turkey. At present, he is in the process of finishing his 3rd Masters's degree, leading to Social Studies Education at Bicol University. Recently, Dr. Mobo was able to pass Batch 3—Certified Research Professional—and made it to the top 5 spots.

Bndurani L.G.P.Ram M.Pharm, Pharmacognosy completed her graduation from Pune University(2002) and post-graduation from Bharti Vidyapeeth, Pune(2004). Presently she is working as Associate Professor in Pharmacognosy at SGPSMPM's Dnyanvilas College Of Pharmacy, Dudulgaon PCMC Pune 412105 M.S. India, affiliated to Dr Babasaheb Ambedkar Technological University Lonere. Dr. Bindurani have 20 years of experience at various level in academics. Her major research interest is Herbal drug research, Standardization of herbal drug preparations and phytochemistry. She has presented several research papers in various symposia and conferences and has published more than 30 of quality research paper in national and international journals. She has published two national and one international patents. Dr. Bindurani published two textbooks and three reference books on different subject and also published book chapter in various reputed publications. Dr Bindurani is a life member of All India Pharmacy Teachers Association of India.

R. Sanjeevi, an accomplished academician, serves as Associate Professor in Environmental Science at Nims Institute, NIMS University, Rajasthan. His qualifications include M.Sc., M.Phil., Ph.D., Post-Doctoral Fellowship (CSIR-RA), and PG Diploma in Industrial Waste & Wastewater Treatment, reflecting his expertise in Environmental Science and Engineering. With 19 years of experience, he has published 12 patents, 39 peer-reviewed articles, authored 1 book, contributed 9 book chapters, and received 6 awards. Dr. Sanjeevi's research spans water analysis, wastewater treatment, waste-to-energy, renewable energy, air pollution control, solid waste management, bioprocess technology, environmental biotechnology, and remote sensing & GIS. His multidisciplinary approach addresses pressing environmental challenges. As a mentor, he guides future environmental scientists, emphasizing excellence in research and education.

Rkha Rani is working as Associate Professor at Indira Gandhi National Tribal University. She earned doctorate degree from Dr Bhimrao Ambedkar University, Agra in 2010. She has more than 15 years of experience in teaching & research. She awarded for Vigyan Ratna Purushkar by Asian Biological Research Foundation Prayagraj, (India) during ICNS at RDVV Jabalpur 2023. She has published 27 research & technical manuscript in journal of international and national repute. Her area of expertise is Environmental Pollution, Aquatic Toxicology, Entomology & biodiversity.

Prashantkumar B. Sathvara, a key member of Nims Institute, Nims University Rajasthan, Jaipur, pursues a Ph.D. in Environmental Science. With M.Sc. in Environmental Science and Technology, P.G. Diploma in Geo-informatics System, and a B.Sc. in Fire and Safety Technology, his research focuses on carbon sequestration mapping, vegetation, water and land use change detection and the application using advanced remote sensing and GIS techniques. He has 3 publications, 2 accepted book chapters, and 9 under review. He did training under SAC at ISRO, Ahmedabad, and proficient in ArcGIS, QGIS, ENVI, and Google Earth Pro, Mr. Sathvara is primed to advance environmental science and technology.

Huma Shahid, a young environmentalist and visiting lecturer at Layyah University, is on a mission to protect our planet. Despite being new to the working world, her love for nature runs deep. Growing up in Layyah, She felt at home in nature's embrace. Now, she shares her passion with her students, teaching them to care for our environment. Huma believes nature should be cherished, not exploited. She dreams of exploring the world's beauty as a nature traveler. In a world facing environmental challenges, Huma Shahid inspires us to protect our planet for future generations.

Ajay K. Singh is working as an Associate Professor (Research) in the Department of Humanities and Social Science, Graphic Era (Deemed to be University) Dehradun. He has worked as Assistant Professor (Economics) in the School of Liberal Arts & Management, DIT University Dehradun for 6 years. He did Post-Doctorate Research with EDI of India, Ahmedabad, Gujarat (India). He received MPhil (Economics) from DAVV Indore (India), and PhD (Economics) from IIT Indore (India. He has published several research papers in the diversified area such as climate change, agricultural productivity, assessment of food security; estimation of GFSI, development of environmental sustainability index (ESI) and its association with socio-economic indicators; measurement and determinants of entrepreneurship ecosystem, and dimension of sustainable development and its interlinkages with economic development.

Ravindra Singh has completed his Ph.D. from Banaras Hindu University (Institute of eminence). Currently he is working as an Assistant Professor Psychology Graphic Era Deemed to be University, Dehradun. Earlier he was Indian Youth Ambassador at BRICS. Beijing China 2017 during Indian deligation. Dr. Singh has also worked as Junior Project Fellow, NCERT funded project, New Delhi in 2018. He worked as the Former Teaching Assistant at IIT, BHU. Recently he was the Youth Member at G20 Youth for India's G-20 Presidency. He served as an Independent Reviewer in Frontiers in Psychology Journal and Member at APA, International Affiliate, Washington DC and also the member of prestigious National Academy of Psychology, (NAOP) India. He has 7 Years of teaching experience and supervising 3 PhD candidate along with many posts' graduate thesis supervision.

Santosh Singh is working as an Assistant Professor (English) in the in the School of Liberal Arts & Management, DIT University, Dehradun, Uttarakhand. He received his PhD in English Literature from Gurukul Kangri Vishwavidyalaya, Haridwar in 2015, and MA in English Literature and BA from HNB Garhwal University. He is having more than 22 years' experience in industry and teaching. His research interest lies in Indian Street theatre, Postcolonial Studies, Folk Studies and Indian Writing in English. He has published many research papers and book chapters in his credit in nationally and internationally reputed journals. He has also experience working as Assistant Professor at Raffles University Neemrana, Rajasthan and Uttaranchal University, Dehradun, Uttarakhand.

Sndeep Tripathi, Professor in Biochemistry & Registrar at Nims University Rajasthan, Jaipur. PhD from King George's Medical University, Lucknow. Recognized with "Best PhD thesis award" in 2010. Over 100 research papers, 15 book chapters, 3 patents, and 500+ citations in esteemed journals. Specializes in Environmental and Occupational Toxicity. Established National Referral Centre for Fluoride Poisoning in India (NRCFPI) in 2010. Received prestigious awards like "Niloufer Chinoy" and "Abida Mahdi" for exceptional contributions. Active member of IBRO, ISNR, SOT, IAN, ACBI, SFRR, SOPI, SPGHTN, and InSLAR, contributing to scientific advancement. Prof. Tripathi's dedication to academic excellence, scientific innovation, and public service underscores his status as a trailblazer in the field, leaving a lasting legacy of inspiration.

Oar Vargas-González is a professor, research assistant and former Head of Computer Systems Department at Tecnologico Nacional de Mexico Campus Ciudad Guzman, professor at Telematic Engineering at Centro Universitario del Sur Universidad de Guadalajara with a master degree in Computer Systems. Has been trained in Innovation and Multidisciplinary Entrepreneurship at Arizona State University (2018) and a Generation of Ecosystems of Innovation, Entrepreneurship and Sustainability for Jalisco course by Harvard University T.H. Chan School of Health. At present conduct research on diverse fields such as Entrepreneurship, Economy, Statistics, Mathematics and Information and Computer Sciences. Has colaborated in the publication of many scientific articles and conducted diverse Innovation and Technological Development projects.

Index

T

U

www.ingramcontent.com/pod-product-compliance
Lightning Source LLC
LaVergne TN
LVHW081251011026
849162LV00006BA/1455